Henry Walter Bates

Contributions to an insect fauna of the Amazon Valley

Coleoptera--Longicornes Part I. Lamiares

Henry Walter Bates

Contributions to an insect fauna of the Amazon Valley
Coleoptera--Longicornes Part I. Lamiares

ISBN/EAN: 9783742801968

Manufactured in Europe, USA, Canada, Australia, Japa

Cover: Foto ©Klaus-Uwe Gerhardt /pixelio.de

Manufactured and distributed by brebook publishing software
(www.brebook.com)

Henry Walter Bates

Contributions to an insect fauna of the Amazon Valley

CONTRIBUTIONS

TO AN

INSECT FAUNA

OF THE

AMAZON VALLEY.

COLEOPTERA—LONGICORNES.

PART I.—LAMIAIRES.

BY

HENRY WALTER BATES,

AUTHOR OF 'THE NATURALIST ON THE RIVER AMAZONS.'

[Originally published in the 'Annals and Magazine of Natural History.'

LONDON:

PRINTED BY TAYLOR AND FRANCIS, RED LION COURT, FLEET STREET.

1861–1866

CONTRIBUTIONS

TO AN

INSECT FAUNA OF THE AMAZON VALLEY.

COLEOPTERA—LONGICORNES.

PART I.—LAMIAIRES.

THE number of species of Longicorn Coleoptera which I collected at different stations on the banks of the Amazons amounts to about 705. The collection appeared to me to contain so large a number of curious and interesting forms new to science, that I was anxious to make them known to the entomological public as soon as possible, first determining the already known species, and fixing upon a classification of the genera and groups. I then hoped to be able to give a complete view of the Amazonian productions in this department, incorporating a few general remarks on their natural history, instead of following the usual and much easier practice of giving merely a bare and unfruitful list of diagnoses of the new species.

It has been a difficult task, however, in the absence of a modern monograph on the family, to characterize the genera, and especially to group them into subtribes or groups subordinate to the four tribes of Latreille, which for a long time constituted the only received classification, but are now manifestly insufficient to give a lucid view of the contents of this greatly aug-

mented family of insects. Within each of the tribes the diversity
of forms is so great that it has become absolutely necessary to
subdivide them, and ascertain at the same time the relations of
the subdivisions to each other. I was therefore unwilling to
publish descriptions of the new forms without first attempting
to class the whole in natural groups, as well as to define better
the already known genera. A mere succession of a multitude of
genera treated in an isolated manner, without indications of the
affinities which link them together (such, in fact, as has been
given hitherto in works on the family), could lead to no useful
scientific results.

No general treatise has appeared on this subject (until within
the last few months) since the imperfect one of Audinet-Serville
in 1832-4. In this work the genera are very insufficiently characterized, often from the examination of a single species. Shortly
afterwards appeared the third edition of the Catalogue of Count
Dejean, in which a great number of new genera were introduced
without characters at all. On the uncertain foundation, however, of these two works, a vast number of new species and genera have been published, many of the former being referred, in
a most loose and unsatisfactory manner, to the uncharacterized
genera of Dejean. The want of a good monograph, such as
exists on many other families of Coleoptera, has long been felt.
Faunists, in treating of the family in their special works, and
authors of the numerous works on the zoology of voyages, public
and private, have been obliged to describe great numbers of new
genera and species without reference to a reliable general classification; besides which, many Coleopterists to whom the family
is attractive on account of the great beauty and variety of its
forms, have continually published isolated descriptions of new
species and genera, and this in every variety of natural-history
periodical, and in almost every European language. In this
way at length about 820 genera and 4500 species have been introduced into the science, a very large portion of them without
proper indications of their place in the system.

The general treatise upon the Longicornes which I have
alluded to above as having appeared very lately is by M. J.
Thomson of Paris, and entitled 'Essai d'une Classification de la
Famille des Cérambycides.' It is founded on a previous special
work on the North American Longicornes published by Dr. Leconte in 1852, called 'An Attempt to classify the Longicorn
Coleoptera of America north of Mexico.' The latter essay was a
great step in advance, as it entirely remodelled the previous
knowledge on the subject, and took into account many parts of
the structure of these insects which were left unheeded by pre-

vious writers. Although a faunistic work, it comprehended here and there the results of the examination of genera found in other parts of the world. The treatise of M. Thomson consists of an application of Leconte's classification to the Longicornes in general. Both these essays, however, leave much to be desired, for reasons to be mentioned presently. The only other works which contain considerable modifications of the system of Latreille are Mulsant's 'Coléoptères de France (Longicornes),' 1839, and Blanchard's 'Histoire des Insectes,' 1845. The former, although containing an excellent analysis of the species and genera found in France, added little that could be applied to the family generally. The latter proposed a number of sub-tribes, but with insufficient and inapplicable characters, and without any review of the genera comprehended under them.

Leconte divided each of the tribes of Latreille into a number of subordinate groups, characterized after a searching examination of the whole external structure of the insects. It is doubtful, however, whether his groups can be all maintained: the classification is open to much objection, and, I think, will require considerable emendation before being applied generally. The important discovery of a very constant character for the tribe Lamiaires, viz. the existence of an oblique groove on the inner side of the fore tibiæ, is due to Zimmerman, who first called attention to it. The existence of a smaller similar groove surmounted by a tubercle on the outer side of the middle tibiæ, in most of the divisions of the same tribe, was not mentioned. The form of the anterior acetabula, or sockets of the fore haunches, is employed too rigorously: it is a constant character in some groups of Lamiaires, being a good guide, for instance, in distinguishing the Colobothea from the true Saperditæ, with which they had been confounded by all previous authors; but it separates *Acanthoderes* and its allies too widely from *Oreodera*, *Dryoctenes*, and similar genera, with which they are in all other characters closely connected. In fact, some of these genera are extremely variable in this character. The form of the anterior acetabula depends upon how far the suture which runs from their external rim to the line which separates the pronotum from the pectus is opened or closed. This suture seems to be that which separates the episternum from the epimera, and, according to the shape or manner of action of the fore haunches, it is either quite closed, more or less gaping near the rim of the socket, partly closed but not gaping at its commencement, or widely opened along its whole length. The shape of the acetabula in the Prionidæ was noticed long before the date of Leconte's treatise, viz. by the Marquis Maximilian Spinola, in a paper published in 1842. In this tribe, where the breast is very broad and the haunches cylindrical, the suture is long and widely gaping. When the suture is opened only a little at its commencement near the rim of the socket, the acetabula are termed by Leconte "angulated;" but it is often very difficult (for instance, in the genus *Acanthoderes*) to say when they should be considered angulated and when round.

This work, however, being almost confined to North American productions, could only be a stepping-stone to the desideratum of a sound general classification of the Longicorn family. M. Thomson, in his Essay, adopts the system of Leconte with some slight modifications, and applies it to the Cerambycides of the whole world, for doing which his very large private collection afforded great facilities. He institutes a great number of subtribes, groups, and divisions, arranged in order under the tribes of Latreille as modified by Leconte. This, therefore, is by far the most considerable work that has yet appeared on the subject, and might be expected to form the groundwork and guide which I have alluded to as being the great desideratum in this family. It is, however, disappointing in many respects, although containing much that is very valuable, and forming, upon the whole, a real advance in the science. The greatest objection that can be made to it is that, although there seems at first sight to be a just and well-digested classification, yet the diagnoses of his groups and genera, when examined into, are found not to apply, in most cases, to the majority of the insects they refer to. The characters very often are too general and random, and do not, in fact, serve to characterize at all. The more detailed characters of the numerous new genera, however, are given in a much more satisfactory manner. Part of this obscurity is owing to the innate difficulties that the study of the group presents, as will be mentioned presently. Very many of his groups are natural, and will doubtless stand their ground, but they will mostly still require to be defined. In his fifth group of Lamiaires, viz. the Onciderita, he gives as diagnosis, "Frons apud ♂ sæpissime armata. Tarsorum articulus ultimus longissimus." These two characters apply equally well to many of his thirteenth group, Hypsiomitæ—to several genera of the Apomecynitæ division of his Saperditæ (*Trestonia, Trachysomus,* &c.)—and partly to his fifteenth group, Hippopsitæ. Some features of his classification, however, are very good. Thus, by means of the system adopted, he has been able to ascertain that the curious South-east Asian group, Tmesisternitæ, are true Lamiaires, notwithstanding the porrect direction of the head—a superficial and erroneous guide, which has misled all previous authors. The Callidiitæ approximated to the Spondylidæ is also a good arrangement; and there are many others of the same nature. He has done great service, also, in characterizing most of the remaining genera and species of Dejean which still, as unmeaning names, encumbered the science. Moreover, the work, as bringing together, in something like order, a vast amount of hitherto scattered material, will be of great service.

A few more general remarks on these important works will perhaps not be out of place here, although they do not all strictly apply to the Amazonian fauna. The position of the Lepturitæ as a group subordinate to the Cerambycidæ seems to me untenable. The true Lepturitæ, by the structure of their fore haunches, the shape of the head, the insertion of the antennæ, and other features, appear to me better placed as an independent tribe, according to the system of Latreille. The Distenitæ, for similar reasons, namely the shape of the head and the insertion of the antennæ, I think should also be considered an independent tribe, instead of being intercalated between Rhopalophoritæ and Cerambycitæ. The Pseudolepturitæ of Thomson, as he justly remarks, require much further examination : they are in some respects the most curious forms of the whole family, and will require probably the institution of one or more distinct tribes. It is a merit of M. Thomson's system to have improved very much the constitution of the tribe Prionidæ, which previously was a most heterogeneous assemblage; but it has escaped him as well as other authors that the genera *Cheloderus* and *Oxypeltis*, singular Chilian forms, have a muzzle differently constructed from that of all other Longicornes. They also differ from all in the shortness of the third antennal joint. In the shape of the muzzle they resemble *Sagra* and allied genera in the family Phytophaga. They are especially ill-placed among the Prionidæ. Two Australian genera, viz. *Brachytria* and *Pytheus*, are closely allied to them; and the four, I believe, must be made to constitute another independent tribe.

In the following review of the Amazonian Coleoptera belonging to this family, I have thought it better, on the whole, to adopt the system of M. Thomson, introducing some modifications, and endeavouring to find more suitable characters for the genera, commencing with the tribe Lamiaires. It must not be urged too severely that the groups are not precisely characterized. It is a matter of great difficulty, perhaps impossibility, to find constant characters for the subordinate divisions. It is one of those groups of insects in which Nature, in striving after strong individuality in the species, seems to have changed or adapted those parts of structure on which we rely for characters of genera and groups of genera. The family, too, is found throughout all parts of the world where woody vegetation exists, and has endured probably, under the same laws of modification, throughout long geological epochs. The diversity of specific forms seems endless, running into infinite varieties of grotesque, ornamented, and extraordinary shapes; and nearly every species has structural peculiarities for its specific characters; so that in no family can genera be made so easily and so numerously as here. Analysis is too easy, and has already been pushed, perhaps, to too great an extent.

The Lamiaires, as far as they are represented in the Amazonian fauna, seem to present six different types of form: but in none are the characters quite constant; they can only be considered as very general, but seldom apply to the whole of the species or genera. I have taken into consideration most of the parts of structure employed by Messrs. Leconte and Thomson, and have brought into prominence others which were neglected or only considered subordinate by them, viz. the shape and relative length of the basal joint of the antennæ, the tubercle and groove of the middle tibiæ, and the claw-joint and claws of the tarsi. The parts of the mouth, which offer sure characters in most other families of Coleoptera, are here of scarcely any systematic value. The palpi only occasionally furnish generic characters. The ligula, otherwise a very important organ, varies greatly in species very closely allied in all other characters. Under each subtribe I have quoted such of M. Thomson's groups and divisions subordinate to it as are represented in the Amazon region.

Subtribe 1. Acanthoderitæ. Basal joint of the antennæ shorter than the third, forming an elongate-pyriform club, very slender at the base. Middle tibia with the tubercle and groove on its outer edge conspicuous. Anterior acetabula generally angulated externally, the suture more or less gaping, but sometimes (Steirastoma) entirely closed. Tarsi simple.

Acanthoderitæ, Acrocinitæ, Oreoderitæ, Dryoctenitæ, Polyrhaphitæ, and Anisoceritæ, Thoms.

Subtribe 2. Acanthocinitæ. Basal joint of the antennæ much elongated, as long as or longer than the third. Middle tibia almost always with the tubercle and groove conspicuous. Anterior acetabula circular, the suture being closed or nearly so. Head narrow. Tarsi simple. p. 61.

Acanthocinitæ, Trypanidiitæ, Colobotheitæ, Thoms.

Subtribe 3. Lamiitæ. Basal joint of the antennæ moderate in size, forming an oblong club thickened from base to tip.

Middle tibia with the tubercle and groove always largely developod. Tarsi simple.
Monohammitæ, *Thoms.*

Subtribe 4. Oncideritæ. Basal joint of the antennæ thickened from base to tip; moderate in size (except in Hippopsitæ). Middle tibia with the tubercle and groove conspicuous. Anterior acetabula angular externally. Tarsi with the claw-joint almost always greatly elongated; claws simple. Body elongated.

Oncideritæ, Apomecynitæ, pt. (*Eudesmus, Trachysomus, Trestonia*), Hypsiomitæ, Onocephalitæ, Hippopsitæ, *Thoms.*

Subtribe 5. Desmiphoritæ. Basal joint of the antennæ very slender at the base, abruptly clavate. Middle tibia with the tubercle and groove frequently wanting. Anterior acetabula angulated externally. Tarsi simple. Antennæ filiform, rather short, pilose; muzzle generally very short, and occiput very large, prominent.

Compsosomitæ, Desmiphoritæ, Apomecynitæ, pt. (*Hebestola*), *Thoms.*

Subtribe 6. Saperditæ. Basal joint of the antennæ slender, generally thickened gradually from the base. Middle tibiæ in most of the genera wanting entirely the tubercle and groove. Anterior acetabula widely gaping externally. Tarsi always short; claws very frequently toothed or bifid. Body elongated; thorax very generally cylindric, simple.

Saperditæ, Amphionychitæ, Tapeinitæ, *Thoms.*

It is possible that this classification might be improved by withdrawing the Hippopsitæ from the Oncideritæ, and the Tapeinitæ from the Saperditæ, and instituting with them two additional tribes. I think it would be difficult, however, to form an arrangement which would meet all requirements. Each of the subtribes (except the third) will contain several natural groups, the definition of which I think it better to leave until the whole of the Lamiariæ have been passed under review. The geographical distribution of the six subtribes is interesting, in so far that the first (Acanthoderitæ) is almost peculiar to the New World, a few species of one genus only having yet been recorded from the eastern hemisphere. On the other hand, the third (Lamiitæ), which exist in great number and variety of genera and species in the Old World, is represented in South America, at least in the Amazon region, by one genus only, viz. *Tæniotes.*

Fam. LONGICORNES, Latr.

Tribe LAMIAIRES, Latr.

Subtribe ACANTHODERITÆ, Thoms. (pt.).

Group *Acanthoderinæ.*

Genus ACROCINUS, Illiger.

(Thoms. Class. des Cérambyc. p. 28.)

This genus, as revised by Thomson, is distinguished from *Oreodera* and all the allied genera by the simple femora. To this may be added that in *Oreodera* the basal joint of the antennæ is almost always relatively shorter and more abruptly clubbed than in *Acrocinus*; and the ♂ fore tarsi are naked in the latter, whilst they are always fringed with hairs in the former genus. The anterior acetabular sutures are widely gaping. The face in all the species is short, being nearly twice as broad as long (measuring the length from the top of the antenniferous tubercles); the muzzle is widened from the eyes downwards, and the lower angles are prominent. The eyes above nearly meet on the vertex, being separated only by the central line; below they reach the central line of the forehead only in one species (*A. longimanus*), in the others being widely separated. The fore and middle tibial grooves, with their accompanying tubercles, are removed to very near the apex of the tibiæ in *A. longimanus*; in *A. trochlearis* and *A. accentifer* they are largely developed, especially in the ♂. The fore legs are covered with granulations and elongated in the ♂ of the two species just named, and the tibiæ have a row of tooth-shaped projections along their under-surface. In *A. longimanus* the same legs are tuberculated in both sexes, the denticulations of the tibiæ are very large (extremely so and recurved in the ♀), whilst the fore legs of the ♂ reach an excessive length, the femora having also a strong tooth-shaped projection on the upper surface near the base, which does not exist in the ♂ of the other species. The thoracic lateral spines are long, acute, and retrocurved in *A. longimanus*; in the other species they exist only as points at the apices of the lateral tubercles. There are several other points of difference between *O. longimanus* and its congeners; but I think they are not of a nature to warrant the institution of a separate genus; the species must be viewed rather as a highly developed and exaggerated form of the generic type.

An erroneous statement has been made and repeated by authors with regard to the thoracic tubercles of *A. longimanus*, to the effect that they are moveable. Such a structure would be curious in the highest degree, but it does not seem to have excited attention sufficient to lead to further examination. It is, however, an error, the credit of pointing out which is due to M. Thomson in his recent work on the Cérambycides. A deep depression around the base of the tubercle seems to have given rise to the mistake; but in fact the depression, which is found also in great numbers of Lamiaires, is not continuous, as a slight examination will show.

1. *A. longimanus*, Linn. and authors.

The Amazonian examples of this insect are smaller than those found in other parts of South America. It is not a very common insect, and is not found, as its great size would lead one to suppose, on the larger trees of the forest; I have found it almost always on slender boughs, or on tree trunks of moderate dimensions. I have sometimes cut the insect out of the rather hard wood of such trees, near the centre of which it passes the larva and pupa states. The stridulation of the species is very loud, and can be heard at many yards' distance in the forest. It appears not to be confined to one kind of tree; I have found it on the Inga, a genus of Leguminosae, and on the Jabutí-puhé, a wild fruit-tree of the order Anonaceae, as well as other trees. On the Inga it is sometimes seen in company with *Oreodera glauca,*—the *Oreodera* being coloured in close imitation of the bark, and clinging very closely and flatly to it, thus eluding observation, whilst *A. longimanus* in its bright colours forms a very conspicuous object. It is very slow in motion, but has the habit of bending its long legs rigidly in self-defence on being disturbed. Thus, of two allied species, one has the means of defence and maintenance of existence in one way, and one in another.

2. *A. trochlearis*, Linn.

Cerambyx trochlearis, Linn. Syst. Nat. ii. 622.
Prionus trochlearis, Oliv. Col. iv. 7. 13. 49.

This elegant species seems to be peculiar to Guiana and the Amazon region. Its habits are similar to those of *A. longimanus*, in so far as it is found on the moderate-sized branches of trees blown down in the forest.

The allied *A. accentifer* I did not meet with; it is found in S.E. Brazil and in Venezuela, but not in the intermediate country of Amazonia.

Genus OREODERA.

Serville, Ann. Soc. Ent. Fr. iv. 19.

The body in this, as in the preceding genus, is elongated and flattened; the species, however, are of much smaller size. The prothorax has on its disk three prominent tubercles, arranged in

c

a triangle; but the posterior one is sometimes wanting, and in some of the smaller species the whole are obsolete. The elytra are narrowed from the shoulders to the apex. The muzzle is very short, being prolonged very little beyond the lower margin of the eyes; but it is very broad, and the lower angles are prominent. The antennæ are much longer than the body in both sexes, fringed with hairs beneath; the third joint much the longest, the first being about two-thirds its length, and dilated (chiefly on its inner side), from near the base, into an elongate club. All the femora are strongly clavate; the fore tibiæ of the ♂, in those species which approach nearest the genus *Acrocinus*, are bent near the tip, the tubercle being very prominent, and the first joint of the tarsi much elongated. All the tarsi are remarkably narrow and elongated, especially the claw-joint, more so in some species than in others, a character which distinguishes *Oreodera* from *Acanthoderes* and the allied genera. The ♂ fore tarsi are elongated and fringed with hairs. The sterna are very broad, the anterior acetabula circular, but the sutures are slightly gaping along their whole length. The ligula (in *O. glauca*) is membranous, narrow, deeply and narrowly cleft, and its outer margins are regularly rounded. The lobes of the maxillæ are small and narrow; the mentum extremely short and broad.

The species of this genus are numerous in South America. Their habits are similar to those of *Acrocinus*, with the exception that they are generally found adhering very closely to the twigs or bark of the dead trees on which they are found; and their colours being assimilated to those of the wood or bark, they are with difficulty detected. The smaller species are exclusively confined to the slender branches, the length and slenderness of their tarsal joints and claws being specially adapted for clinging to them. The females deposit their eggs on the bark; and the larvæ, when hatched, penetrate into the wood.

§ Disk of thorax with three or two prominent tubercles: tips of the elytra truncated.

1. *Oreodera undulata*, n. sp.

O. elongata, depressa, tomento tenuissimo holosericeo griseo-olivaceo vestita: elytrorum apicibus obliqua sinuato-truncatis, dimidio basali granulato-punctato, apicali lineis undulatis griseis et fuscis ornato. Long. 7 lin. ♂.

Head sooty-brown, opake: eyes nearly touching the central furrow on the vertex. Antennæ sooty-brown, the base of each joint from the third light grey. Thorax with large lateral tubercles and three discoidal ones—two transverse before, and one longitudinal behind. Elytra tapering slightly from base to apex, the tips rather obliquely and briefly truncated, the external angles of the truncature slightly produced; the basal half studded with acute granulations accompanied by punctures, and in the centre of each, near the base, is a large, obtuse, transverse elevation, dark brown in colour: they are clothed with fine silky changeable olive-grey pile, and are variegated from the middle to the apex with fine grey and fuscous strongly undulating lines, with a grey patch on each side near the middle spotted with black. Under-surface of the body clothed with golden-grey pile. Legs greyish olive, with paler rings.

One individual, taken at Ega. This and the following species resemble much in colour and design *Acrocinus trochlearis* and *accentifer*.

2. *Oreodera fluctuosa*, n. sp.

O. elongata, depressa, tomento tenuissimo holosericeo cinereo vestita: elytris apicibus oblique truncatis et spinosis, plaga laterali pone basin, strigisque undulatis numerosis fuscis. Long. 10 lin. ♀.

Head and thorax grey: eyes nearly touching the central furrow on the vertex. Antennæ grey, apex of each joint from the third dusky. Thorax with the two anterior discoidal tubercles very prominent, the posterior one nearly obsolete. Elytra slightly tapering, the tips rather obliquely truncated, the external angles produced, dentiform; in the middle, near the base, each has a large prominent dark-brown tubercle; the basal half is somewhat sparingly granulate-punctate: ashy grey in colour; across, near the base, is a broad yellowish-grey belt, and on the margins behind the shoulders a long oblique dark-brown patch; there are also two transverse, narrow, strongly undulated belts of the same dark-brown colour,—one behind the middle, the other near the apex. Legs grey, femora varied with dusky; two rings on the tibiæ and claw-joint of the tarsi black. Under-surface of the body densely clothed with a golden-grey pile.

One example, taken at Para. I believe it is also found at Cayenne.

3. *Oreodera glauca*, Linn.

Cerambyx glaucus, Linn. Syst. Nat. ii. 625. 28.
Lamia glauca, Fab. Ent. Syst. ii. 274. 27.
——— *Spengleri*, Fab. Ent. Syst. ii. 291. 93.

This is a very common insect throughout the Amazon region as well as at Cayenne. It is found on the trunks of felled trees of one or more species of *Inga*, the bark of which it resembles in colour. The lateral tubercles of the thorax have indications of the same impressed line around them which is so strongly marked in the *Acrocini*. In the ♂ the fore legs are elongated; the tibiæ bent, rather hooked at the apex on the inner side.

12

The first joint of the tarsi is also much elongated in the same
sex. The elytra are square at the tips, being truncated largely
and transversely.

4. *Oreodera bituberculata*, n. sp.

O. angustata, depressa, tomento tenuissimo holosericeo cinereo-brunneo vestita: elytris maculis tribus lateralibus violaceo-brunneis,
quarum secunda striga undulata transversa emittente, tertiaque
parva notatis. Long. 7–8 lin. ♂ ♀.

Head dusky: eyes nearly touching the central furrow on the
vertex. Antennæ piceous, thinly clothed with grey pile; apices
of the joints dusky. Thorax even, with two prominent discal
tubercles, shining black, the third, posterior, totally wanting;
punctured on the disk as well as on the fore and hind margins;
lateral tubercles large, obtuse. Elytra very long, tapering from
base to apex, the tips obliquely truncated, the external angles
of the truncature much produced and acute; punctured throughout; two short rows of tubercles along the humeral elevations,
two others on the disk near the base, and one of smaller tubercles
along the suture; in some specimens the sutural row and one
of the discal ones are nearly obsolete, in all they consist of a
small number of tubercles: the sinuations between the purple-
brown lateral spots are edged with white scales; the apical spot
is very small, the other two large and semi-oval. The under-
surface of the body and legs ashy; the femora varied with dusky;
two rings round the tibiæ black.

I took this species at Ega and on the banks of the Tapajos.
It is also found at Cayenne. I have received specimens from
Paris labelled *obscurata* and *opaca*; but I cannot find any species
published under those names.

5. *Oreodera rufofasciata*, n. sp.

O. curta, depressa, tomento holosericeo argenteo-griseo vestita: elytris
subtriangularibus, fascia basali rosacea postice late nigro marginata,
prope apicem lineis vermicularibus argenteis et fuscis ornatis.
Long. 6 lin. ♂.

Head dusky: eyes approximating on the vertex. Antennæ wholly
clothed with silvery-grey pile. Thorax short and broad, dusky
grey: lateral tubercles conical; anterior discal ones large, slightly
elevated, transverse, clothed with pile; posterior one slightly
elevated. Elytra wide at the shoulders, tapering to the tips,
which are obliquely truncated, the external angles of the truncature slightly produced and directed outwards; closely granulate-
punctate at the base; silvery-grey, the base with a rose-red
fascia, behind which is a dusky-brown belt shading off posteriorly. Under-surface of body and legs clothed with grey pile.

At Ega, on felled Pamú (a wild fruit) trees in the forest.

6. *Oreodera lacteo-strigata*, n. sp.

O. curta, tomento holosericeo rufo-brunneo vestita: elytris apicem versus attenuatis, pone medium fascia pallidiore strigis lacteis undulatis marginata, prope apicem linea transversa undulata lactea ornatis. Long. 6 lin. ♂.

Head brown : eyes rather distant from the central line on the vertex. Antennæ pitchy-brown, base of joints paler greyish. Thorax punctured on the disk as well as along the fore and hind margins: the anterior pair of tubercles prominent, conical, dusky ; the posterior one slightly elevated ; the lateral ones conical. Elytra rather thickly punctured from the base to three-fourths the length, punctures large, the basal ones accompanied by granulations, each near the base furnished with a longitudinal ridge-shaped tubercle, slightly hooked behind; the basal half is deep red-brown, deepening on the sides to violet-black ; the space between the pale-brown median belt and the subapical transverse undulated line is lighter brown, streaked longitudinally with dark brown ; the subapical milky belt emits short branches, and is edged posteriorly with dark brown ; extreme apex light brown : the apex is obliquely truncated ; the external angles of the truncature acute, but not produced. Legs and under-surface of the body clothed with silky-brown grey pile.

This species was rare on the Upper Amazons. In facies it resembles species of the genus *Alcidion* (group Acanthocinitæ) ; it is readily distinguished, however, by the short clavate basal joint of the antennæ.

7. *Oreodera remota*, Pascoe.

Ægomorphus remotus, Pascoe, Trans. Ent. Soc. n. s. vol. v. pt. 1.

O. elongata, minus depressa, postice valde attenuata, tomento holosericeo violaceo-brunneo vestita : elytris marginibus maculis tribus lateralibus punctisque discalibus nonnullis quorum duobus majoribus pone medium atro-violaceis. Long. 8 lin. ♀.

Head brown: eyes distant on the vertex. Antennæ brown ; basal half of each joint, from the fourth, greyish. Thorax with the disk as well as the fore and hind margins punctured; lateral tubercles prominent, acute ; anterior dorsal ones acute, posterior more obtuse, shining black. Elytra rather elongated, tapering to the apex, which is very obliquely truncated, the external angles of the truncature produced and acute ; the base is densely studded with shining black granulations accompanied by punctures; the small rounded violet spots on the disk, near the apex, cover each a shallow shining puncture ; the lateral spots are merely expansions of the dark violet border, and are placed, one at a third, another at two-thirds the length of the elytra, and the third, much smaller one, near the apex. Under-surface of the body and legs clothed with ashy-brown pile. The tarsi, especially the claw-joints, are remarkably elongated. Taken at St. Paulo on the Upper Amazons, on the slender trunk of a dead standing tree. The pile covering this species is of a much coarser texture than that of most other species of the genus; it resembles in this respect *O. glauca*, but it does not lie so compactly as in that species. The species was referred by Mr. Pascoe to the genus *Ægomorphus* of Dejean, which, however, had not at that time been characterised, and was a loose assemblage of species belonging to four or five different genera.

8. *Oreodera sericata*, n. sp.

O. valde depressa, fulvo-brunnea, tomento tenuissimo holosericeo griseo subtus densiore vestita: elytris plaga magna laterali pone humeros albo-grisea, prope basin punctatis, dimidio apicali lævissimo. Long. 5¼ lin.

Head brown : eyes distant on the vertex : antennæ piceous. Thorax with the lateral tubercles obtuse, the three dorsal ones very slightly elevated and clothed with pile. Elytra obliquely truncated at the apex, sutural angles rounded off, the external ones obtuse; base with a number of large simple punctures, which do not reach beyond one-third the length except along the sides, the rest perfectly smooth and silky; the pile is extremely fine, thin, silky, and changeable. Legs clothed with grey pile. Under-surface densely clothed. The white patches on the elytra reach the suture and occupy nearly one-half the surface. Taken at St. Paulo on the Upper Amazons.

9. *Oreodera cretata*, n. sp.

O. depressa, apicem versus attenuata, tomento tenuissimo holosericeo fulvo vestita: elytris plaga oblonga laterali apud medium cretaceo-alba. Long. 4½ lin.

Head and antennæ fulvous : eyes rather distant on the vertex. Thorax with the lateral tubercles obtuse and the three dorsal ones only slightly indicated; punctured on the hind part of the disk as well as along the fore and hind margins. Elytra truncated obliquely at the apex, sutural angles very obtuse, external ones slightly produced and acute; punctured, partly in lines and sparingly over the basal half. The oblong lateral chalky spot is clear white. Body beneath and legs silky fulvous. On the banks of the Cupari (R. Tapajos), on dried branches.

§ § Disk of thorax with no trace of tubercles. Elytra less distinctly, sometimes scarce perceptibly, truncated at the apex.

a. Elytra depressed.

10. *Oreodera simplex*, n. sp.

O. elongata, angustata, tomento holosericeo varia sordido olivaceo vestita : elytris fasciis tribus abbreviatis indistinctis pallidioribus. Long. 5 lin.

Head and antennæ dark brown, the base of each antennal joint (from the fourth) ringed with grey. Thorax with the lateral tubercles obtuse ; disk uneven, punctured posteriorly and on the lateral tubercles as well as along the fore and hind margins. Elytra narrow and only slightly tapering, the apices slightly truncated, punctured moderately over the basal half and on the disk to the apex. The pile is of a dingy yellowish olivaceous colour, varied with a paler shade, which forms three obscure semi-belts on the elytra. Legs and under-surface ashy-brown.

Ega, on dried branches.

11. *Oreodera griseo-zonata*, n. sp.

O. depressa, apud humeros lata, apicem versus attenuata, tomento holosericeo griseo-brunneo vestita : elytris fascia latissima basali albo-grisea, apices versus lineis flexuosis griseis brunneisque ornatis. Long. 4½ lin.

Head and antennæ brown ; base of each antennal joint (from the fourth) grey : eyes distant on the vertex. Thorax with the disk nearly even ; a few punctures on each side, besides those on the fore and hind margins. The elytra have the shoulders more produced and pointed than in the allied species, they have a few punctures near the base, and the apices are singly rounded : the broad grey belt across the basal half has its fore margin arched posteriorly, so as to leave a space around the scutellum dark brown ; it passes beneath entirely over the mesosternum : the grey and brownish waved lines on the apical half are obscure and silky. Body beneath and legs pitchy-brown, clothed with ashy-brown pile.

Ega and banks of the Tapajos, on dead twigs.

b. Elytra somewhat convex. (Subgenus *Anoreina*.)

12. *Oreodera (Anoreina) nana*, n. sp.

O. curta, convexiuscula, fuliginosa, tomento fusco et flavo-ferrugineo vestita : elytris lateribus rotundatis, apices versus attenuatis, utrinque apud medium macula magna laterali triangulari albo-grisea. Long. 3½ lin.

Head and thorax dark drown : eyes distant on the vertex.

Antennæ brown, the basal part of many of the middle joints pale testaceous. Thorax with the disk smooth, even; lateral tubercles very obtuse. Elytra punctured throughout, towards the base densely, towards the apex sparingly; they are sooty-brown varied with obscure rusty-yellow patches; each has on the side, about the middle, a large greyish-white triangular spot, not generally touching the suture. Under-surface of the body brown; legs pitchy-black, clothed with ashy pile. Tarsi moderately slender.

Santarem and Para; on dried twigs.

Genus ÆGOMORPHUS.

Thomson, Class. des Céramb. p. 336.

Char. emend. Body narrow, thick, and somewhat convex. Head as in *Oreodera*, the muzzle being very slightly prolonged beyond the lower margin of the eyes, its anterior angles obtuse: the eyes distant on the vertex. Antennæ rather shorter; the proportions of the joints the same; but they are not fringed beneath, as in *Oreodera*. Sides of thorax furnished with a large conical tubercle. Prosternum behind and mesosternum in front steeply inclined; clothed with long hairs in the ♂ (at least in *Æ. moniliferus*). Anterior acetabula angulated. Second and third ventral segments contracted in the middle in the ♀; the fifth very large, its apex truncate-emarginate and densely hairy. Tarsi broad, claw-joint long; fore tarsi neither dilated nor fringed in the ♂.

The name of this genus first appears in Dejean's Catalogue, but it was first characterized by M. Thomson in the present year; the characters given, however, although numerous, omit the chief peculiarities of the group. The thickness and convexity of the body, nakedness of fore tarsi in the male, and shape of the sterna are the chief points of distinction. M. Thomson places it in the group Trypanidiitæ,—an arrangement quite unintelligible on his system, as it does not agree at all with the characters of the section to which the Trypanidiitæ belong.

1. *Ægomorphus obesus*, n. sp.

Æ. elongatus, convexus, crassus, nigro-brunneus, tomento griseo tessellato vestitus: thorace nigro bivittato : elytris apicem versus attenuatis, sinuato-truncatis, angulis externis productis. Long. 11 lin. ♀.

Head clothed with a fine grey pile, leaving three narrow longitudinal lines on the vertex brownish black. Antennæ shorter than the body, grey; tips of the joints (from the third) dusky. Thorax with the lateral tubercles conical acute, and with two large slightly-raised dorsal tubercles; the fore and hind margins and sides punctured; clothed with grey pile; the dorsal tubercles and a stripe from each to the hind margin black, or thinly clothed with brownish-black pile. Elytra each with three or four slightly elevated longitudinal ridges, disappearing at about half the length; a row of granulations (accompanied by punctures) on each ridge, besides three or four other rows in the interstices; the sides near the base also densely granulate-punctate: the fine hoary-grey pile is in large patches near the base; elsewhere it forms regular rows of small, distinct, oblong spots. Under-surface of the body clothed with dense silky yellowish-grey pile, longest on the pro- and mesosterna. Legs clothed with grey pile, leaving spots on the femora, the tips of the tibiæ, and of the claw-joints of the tarsi black. Abdomen (in ♀) with the second to the fourth ventral segments contracted in the middle; apical segment very large, tumid near the apex, which latter is truncate-emarginate and densely hairy.

Taken at Para. It resembles very much *Æ. adspersus* (Dej.), Thoms. Class. p. 337. It may be a local form of that species.

2. *Ægomorphus moniliferus*, White.

Ægomorphus moniliferus, White, Cat. Long. Col. in Brit. Mus. ii. p. 374, pl. 9. fig. 7.

This, as will be seen from the excellent figure and description above quoted, is a narrower and more depressed insect than *Æ. obesus*. The ♂ has all the three sterna covered with a dense brush of hairs, more erect and of a brown colour on the metasternum. The apical ventral segment in the ♀ is large, with a longitudinal impressed line ending in a fovea before the apex; the latter is emarginate-truncate and densely hairy.

Found at Para and Santarem, on trunks of felled trees.

Genus MYOXOMORPHA.

White, Cat. Long. Col. in Brit. Mus. ii. p. 355.

Body thick, convex, elongated. Head broad, muzzle short, somewhat narrowed from the eyes; sides rounded, obtuse: eyes distant on the vertex, very large, especially the lower lobes, which advance considerably on the forehead. Antennæ shorter than the body, simple, neither grooved nor fringed. Thorax with large and very acute lateral tubercles. Elytra convex, elongated, their surface without ridges, apices briefly truncated. Legs moderate, tarsi short, femora clavate; the fore tarsi in the males simple, neither dilated nor fringed. Prosternum narrow, simple, with the acetabular sutures angularly gaping; mesosternum broad, quadrate, horizontal. Abdomen with the terminal ventral segment sinuate-truncate in the males.

D

The generic name of *Myoxomorpha* was first applied in French collections to the *Acanthoderes funerarius* of Dejean's Catalogue[*]. Neither the genus nor the species has yet been characterized; but Mr. White, in the Catalogue of the Longicorn Coleoptera of the British Museum, adopted the genus, adding to it the *Acanthoderes funestus* of Erichson. *A. funerarius*, however, is a true *Acanthoderes*, having dilated and fringed fore tarsi in the males: it differs from most of the species only in the rounded tips of the elytra, a character presented by many of its congeners; therefore the generic name can apply only to *A. funestus*. *Myoxomorpha*, as thus defined, is very closely allied to *Acanthoderes*, its chief distinction being the simple fore tarsi in the males. The ungrooved antennal joints, the voluminous eyes, narrow prosternum and horizontal mesosternum also separate it well from the majority of the species.

1. *Myoxomorpha funesta*, Erichs.

Acanthoderes funestus, Erichson in Schomb. Reise, iii. 573.

In facies and colours this species has some resemblance to *A. funerarius*. It is black, clothed beneath and on the legs with a fine silvery hoary tomentum. The forehead, vertex, a broad central vitta on the thorax, the scutellum, and the apical half of the elytra are also clothed with a very fine silky whitish pile,—the apical half of the elytra having a large patch on each side, and a number of small rounded spots of a black colour.

Found throughout the Amazon region, sparingly, under the loose bark of felled trees, chiefly of *Inga* and other Leguminosæ, in newly-made plantations. It is very sluggish in its motions.

Genus ACANTHODERES, Serv.

Serville, Ann. Soc. Ent. Fr. iv. 29.

Char. emend. Body oblong, more or less depressed, narrowed posteriorly. Head rather broad, muzzle transverse-quadrate, much depressed, its anterior angles distinct, front plane; antenniferous tubercles not prominent, consequently there is no concavity between the antennæ: mouth projecting; mandibles long

---

[*] *A. funerarius*, Dej. Cat. *A.* oblongus, crassus, niger, subtilissime punctulatus. Caput nigrum, vertice utrinque macula cana. Antennæ crassæ, nigræ, articulis tertio quartoque maxime elongatis supra sulcatis, reliquis abbreviatis. Thorax niger, tuberibus lateralibus conicis, dorso trituberculato, marginibus cano-maculatis. Elytra simplicia, apice conjunctim rotundata, ubique sparsim granulato-punctata, nigra, basi et pone medium confluenter cano-maculata vel cana nigro-maculata. Long. 5–11 lin. ♂ ♀.

Hab. Mexico.

and flattened: eyes wide apart. Antennæ slightly hairy, never fringed beneath as in *Oreodera*; the basal joint always pyriform clavate, smooth, considerably shorter than the third. Thorax with a simple large conical tubercle on each side, generally ending in a spine. Femora strongly clavate; tarsi moderate, claw-joint short; fore tarsi in the ♂ broadly dilated and ciliated.

The above are the only characters that I find tolerably constant in the thirty-eight species which I have examined. The forms are very variable in most of the parts of structure from which generic characters are derivable, and exemplify well the difficulties which the Longicorn family offers to the classifier. No definition has yet been given founded on a large number of species That of Leconte ("Attempt to classify, &c.," Journ. Ac. N. Sc. Philad. ii. n. s.) is probably the best; but, relating only to the two or three North American species, it is not applicable generally. The rounded outline of the anterior acetabula, which he gives as a character of the section to which *Acanthoderes* belongs, is very variable. In *A. varius*, the European species which may be considered typical of the genus, they are angulated; in other species the acetabular sutures are gaping along their whole length; in a few, however, they are closed. Although they differ in species otherwise closely allied, yet they are more constantly closed in those which approach *Steirastoma*. The head is generally plane in front, the muzzle prolonged considerably below the eyes, the lower lobe of the latter being very small; in some few species, however, the eyes are rather more voluminous below the antennæ, thus reducing the breadth of the forehead and the length of the muzzle. The palpi are always elongated, with the terminal joint obtusely pointed. The ligula has its sides dilated and rounded; the lobes, however, are widely divergent in some species (*A. thoracicus*), and nearly united to their tips in others (*A. bivitta*). The antennæ are very variable in length, thickness, and shape of the joints, being in some species no longer than the body, in others twice the length: the third joint is generally very long, and the fourth considerably longer than any of the following; sometimes the two are as long as the remaining taken together; both are generally filiform, with a longitudinal furrow above, but they are occasionally dilated and produced beneath at their apices, and in a few aberrant species furnished with tufts of hairs: the terminal joints are generally filiform, sometimes short, thickened, and ciliated in the ♂, and sometimes dilated and serriform in both sexes. The thorax has the lateral tubercles, in rare instances, very obtuse; the dorsal surface is uneven, sometimes tuberculated, occasionally furnished with three very prominent tubercles, but generally tricostate. The elytra are generally trigonal, at times oblong, depressed or slightly convex, their surface sometimes even, but generally furnished with a ridge on each at the base, which often projects forwards, and in many species is prolonged posteriorly to the apex : the latter is generally briefly and obliquely truncated, but it is sometimes whole, and at other times largely truncated, with the external angles projecting into a tooth or spine. The pro- and mesosterna are moderately broad, but variable in this respect; the former never very narrow, the latter not contracted between the haunches nor extremely short, but always of a quadrangular shape. Both are plane on their surface in some species, but they are more generally tumid or tuberculated, ridged on the sides, and projecting : in a number of cases the mesosternum is projecting, whilst the prosternum is simple; in many species, however, both project and have their opposing faces steeply inclined. They vary greatly in species otherwise closely allied, although they are similarly constructed in all those species which approximate to *Steirastoma*. The tibiæ are, in one section of the genus, strongly dilated and compressed. The terminal ventral segment is sinuate-truncate in the ♂, and entire in the ♀.

The flattened shape of the muzzle distinguishes this genus from the preceding. There is no character to separate it from *Dryoctenes*, Serv. The shape of the sterna distinguishes it from *Polyrhaphis*. From *Steirastoma* it differs at once in the simple, conical, lateral thoracic tubercles ; and from *Alphus* by the pyriform basal joint of the antennæ. I have incorporated with it the genus *Pteridotelus*, White,—with some hesitation, however, as I think *Pteridotelus* might probably form a natural group if the generic definition were modified so as to include all those species which have the terminal joints of the antennæ shortened and thickened in any degree, or thickened and ciliated in the ♂. The species on which it is founded (*Pteridotelus laticornis*) cannot be generically separated from *A. pupillatus*, Chevrolat, which, again, is closely allied to *A. spectabilis*, n. sp., and *A. pilicornis*, Chevr.*, all four most diversified in ornamentation of the antennæ, but agreeing in the thickening in some way or other of the terminal joints. These species have in common also rounded anterior acetabula, slender fore tibiæ, and steeply inclined sterna. As a genus, however, it would not be sharply limited from *Acanthoderes*: other species have the terminal antennal joints somewhat shortened and ciliated, without

* To these may probably be added *A. extenuatus* of Guérin-Méneville (Ins. Recueillis par Osculati, Verh. des Z. B. Verein in Wien, 1855, p. 599). *A. pupillatus* is from Venezuela, and *A. pilicornis* from Mexico ; both are undescribed. *A. spectabilis* belongs to the Amazonian fauna.

agreeing with *Pteridotelus* in other characters (e. g. *A. maculicollis*) ; others have the joints in question ciliated in the ♂ and at the same time elongated (*A. lateralis*) ; and many species agree in the shape of the sterna, whilst resembling typical *Acanthoderes* in all other characters. I have thought it best on this account to treat *Pteridotelus* as a subgenus or section of *Acanthoderes*.

Acanthoderes and its allies (*Steirastoma, Myoxinus,* &c.) are not, perhaps, so closely allied to the preceding genera as *Polyrhaphis*; it would therefore in some respects be better to place the latter genus after *Ægomorphus*, followed by the Anisocerinæ, with which group it has also an evident connexion; whilst *Acanthoderes* leads through *Alphus* naturally to the Acanthocinitæ. This, however, would be presenting only one suite of affinities amongst several which these insects present : the Acanthocinitæ, for instance, have a certain similarity to *Oreodera* and *Ægomorphus*. It seems almost hopeless to detect the true lines of affinity, and quite so to represent them in a scheme of arrangement when detected.

§ 1. Antennæ with the terminal joints filiform, slender.

a. Fore tibiæ widely dilated and compressed.

1. *Acanthoderes hebes,* n. sp.

A. oblongus, convexiusculus, postice rotundatus, supra tomento fusco, subtus pilis griseis sparsim vestitus : thorace tuberibus lateralibus obtusis, dorsalibus tribus magnis : elytris apicibus parum truncatis, fuscis, fascia abbreviata pone medium nigra velutina, prope apicem ochreo maculatis. Long. 5 lin. ♂ ♀.

Head and thorax sooty-brown, with deep scattered punctures. Antennæ about the length of the body, black ; base of each joint (from the third) and centre of the third with a pale testaceous ring. Thorax with the lateral tubercles obtuse ; three dorsal ones—two anterior very large and prominent, and one posterior smaller and acute. Elytra rounded at the sides, towards the apex very briefly truncated, with a short, tuberculated, longitudinal, slightly elevated ridge in the middle of each near the base; punctured throughout, the punctures accompanied by granulations towards the base : the ochreous spots near the apex are few and irregular. Under surface shining black, with a scanty grey pile. Legs shining black, middle of the tibiæ on the edge, tips of same, and basal joints of the tarsi above greyish ; tarsi beneath yellow, claw-joint pallid. In the ♂ the fore tarsi are black beneath, and densely fringed with black hairs. The fore tibiæ are abruptly dilated from the middle in the ♂, more gradually so in the ♀.

On boughs of dead trees in the forest, Ega. Rare.

2. *Acanthoderes Egaensis,* White.

Scleronotus Egaensis, White, Cat. Long. Col. in Brit. Mus. ii. p. 364, pl. 9. f. 3.

The third to the sixth antennal joints are produced and acute at the apex beneath, the fourth to the eleventh are very slender. The extreme tips of the elytra are distinctly truncated. All the tibiæ are compressed, the anterior pair gradually dilated (wider in the ♂ than in the ♀) from the base to the apex.

This species has a peculiar facies, arising from its short figure, black colour, and the slenderness of its antennæ. . Owing to this, probably, it was placed in a different genus by Mr. White. The genus to which he referred it (*Scleronotus,* Dejean, at that time a mere catalogue name) has since been characterized by M. Thomson (Class. p. 840), and, from the diagnosis, appears to be very closely allied to *Acanthoderes.* M. Thomson places it amongst the Anisoceritæ, regardless of the shape of the anterior acetabula, which he gives as rounded in *Scleronotus* and angulated in the definition of the group to which he refers it.

8. *Acanthoderes fuscicollis,* n. sp.

A. oblongus, fuscus, tomento luteo (capite thoraceque sparsim) vestitus : elytris breviter transverse truncatis, seriatim nigro punctatis, utrinque maculis duabus suturalibus duabusque lateralibus fuscis notatis. Long. 5 lin. ♂.

Head and thorax dusky, with specks of ochreous clay-coloured pile, very scanty on the disk of the latter. Antennæ about the length of the body, the apices of the third to fifth joints produced beneath ; black, the base to the middle of the third joint speckled with ochreous atoms ; the basal half of the fourth and the base of the remaining joints pale testaceous. The lateral tubercles of the thorax are large and slightly pointed ; the disk has a longitudinal smooth line and a large obtuse elevation on each side. The elytra have a few coarse punctures at the base, a few small round black spots arranged in lines, and on each four larger blackish spots, namely two near the suture (one before, one after the middle) and two on the side beyond the middle ; there are also two irregular transverse patches of a paler ochreous colour. Under surface of the body black, shining ; sides of the metasternum and second to fifth ventral segments ochreous. Legs black, shining, speckled with ochrey pile ; middle and hind tibiæ paler, their apices dusky. The fore tibiæ are gradually and widely dilated.

Ega ; on branches of dead trees in the forest. The species has much resemblance to *A. fascialis,* White.

4. *Acanthoderes fascialis*, White.

Acanthoderes fascialis, White, Cat. Long. Col. in Brit. Mus. ii. p. 361.

The external angles of the truncature of the elytra are slightly produced. The fore tibiæ (in the ♀) are moderately and gradually dilated.

Ega. This and the preceding are nearly allied to *A. semigriseus*, a Rio-Janeiro species common in collections*.

5. *Acanthoderes minimus*, n. sp.

A. ovalis, nigricans cinereo irroratus, elytris sinuato-truncatis angulis externis acutis. Long. 3¼ liu. ♀.

Head and thorax blackish, sprinkled with greyish pile. Thorax short, rather narrow, the lateral tubercles not prominent although pointed, punctured on the disk, and with two distinct dorsal tubercles. Elytra with the centro-basal ridge indistinct, granulate-punctate at the base, punctured along the sides to the apex. Body beneath and legs black, sprinkled with grey hairs; claw-joint of the tarsi testaceous. The fore tibiæ (in the ♀) are widely, but not abruptly, dilated from the middle.

Para.

6. *Acanthoderes maculicollis* (Dej.), n. sp.

A. ovalis, tomento variegato vestitus : antennis curtis, articulo tertio maxime elongato apice dilatato, quarto elongato, reliquis abbreviatis : thorace tubere laterali obtuso, fulvo-griseo, maculis magnis lateralibus duabus atro-brunneis velutinis : elytris trigonis apicibus sinuato-truncatis, fulvo-griseis, utrinque pone medium macula elongata transversa discoidali atro-brunnea, fasciisque duabus maculoribus, altera ante medium altera subapicali, flavo-griseis. Long. 4½ lin. ♂ ♀.

Head fulvous mixed with grey, and with scattered brown points. Antennæ about as long as the body, first to third joints pitchy, the third black at the tip, the rest pitchy, with their bases testaceous; the terminal joints are ciliated beneath in the ♂.

* This species is undescribed; therefore the following diagnosis will be useful :—

Acanthoderes semigriseus (Cat. Dej.).

A. oblongo-ovatus, dimidio anteriore tomento fuliginoso, dimidio posteriore cinereo vestitus. Antennæ fuliginosæ, articulis ad basin griseis. Thorax fuliginosus grosse punctatus linea dorsali lævi. Elytra sensim attenuata, apicibus breviter sinuato-truncatis, angulis obtusis, ad basin granulato-punctatis, quarta parte basali fuliginosa coloris margine posteriore retrorsum arcuata, reliquis cinereis utrinque fascia interrupta undulata pone medium maculaque laterali prope apicem fuscis. Subtus niger nitidus, postpectoris lateribus abdomminisque apice luteis; pedibus luteis femorum dimidio basali tibiarum tarsorumque apicibus fuscis. Long. 4½ lin. ♀.

Hab. Rio Janeiro.

The disk of the thorax on each side is very convex, the middle is depressed, with a raised dorsal line; on each side is a large, rounded, velvety, dark-brown spot, which is impunctate, the rest of the surface being punctured. The elytra are grey-fulvous, the suture and some indistinct lines on the disk light grey, covered with small rounded brown spots, confluent in places; the disk behind the middle is crossed by an elongate dark-brown spot, besides which there are two transverse macular lines of a yellowish colour. Body beneath black, scantily clothed with grey; the second to fourth ventral segments have a spot of yellow hairs on each side. Legs pitchy, clothed with grey; the tarsi yellowish, the middle and hind tibiæ ringed with yellowish at the middle and the apex. The fore tibiæ in the ♂ are widely and abruptly dilated from near the base.

Para and Villa Nova. It is found also at Cayenne, and exists in many collections under the name I have adopted.

7. *Acanthoderes alboniger*, n. sp.

A. oblongus, niger; fronte, thoracis vitta mediana, elytrorum plaga magna elongata basali communi fasciaque magna maculari sub-apicali tomentosis ochraceo-albis; ipso apice ochraceo. Long. 7 lin. ♀.

Head black, front ochreous punctured. Antennæ shorter than the body, black, the basal half of the third and the bases of the remaining joints light grey. Thorax with the lateral tubercles prominent, acute; the disk depressed, longitudinally punctured, a broad central stripe and a lateral one below the tubercle ochrey white. Elytra oblong, their sides rounded posteriorly, their apices sinuate-truncate, outer angles produced: the centro-basal ridge is strongly pronounced, and produced forward towards a corresponding sinuation in the hind margin of the thorax; it is smooth and shining, not reaching the middle of the elytra behind, and leaving a broad depression at the base: the elytra are punctured, partly in lines, most thickly so on the sides near the base; black, a broad, basal, common stripe, including the scutellum, notched in the middle externally, and a broad, macular, uneven belt before the apex, not touching the suture, ochrey white; the extreme tips are yellowish brown. The black parts are nearly naked, and have a few grey specks. Body beneath and legs black, thinly covered with grey hairs; tarsi grey. There is a row of ochreous points on each side of the abdomen. The fore tibiæ are widely dilated. The prosternum is produced behind, and the mesosternal tubercles are very prominent.

Santarem. This species bears a superficial resemblance to certain Curculionidæ of the genus *Heilipus*, inhabiting the same district.

b. Fore tibiæ compressed, not dilated.

[To this section belongs the European *A. varius.*]

8. *Acanthoderes maculatissimus*, n. sp.

A. curtus, subdepressus, tomento ochraceo-fulvo vestitus: elytris lituris nonnullis griseis, punctis innumerosis lineaque transversa undata pone medium brunneis. Long. 6 lin. ♂ ♀.

Head punctured, fulvous varied with brown. Antennæ brown, spotted and ringed with grey. Thorax with the lateral tubercles produced and pointed at the apex, and with two obtuse dorsal elevations and a shining central line; the interstices punctured; in colour minutely variegated with fulvous and brown. Elytra subtrigonal, briefly sinuate-truncate at the apex, the external angles produced; punctured throughout, the centro-basal ridge apparent only at the extreme base, ochrey fulvous, silky, studded with small brown spots, which everywhere cover the punctures : there are a few light-grey marks, and behind the middle a transverse dark-brown zigzag line. Body beneath ashy brown. Legs variegated with ashy, dusky brown, and fulvous. The fringe of the male fore tarsi is black. The prosternum is simple, the mesosternum subvertical in front.

At Santarem ; on hanging woody climbers in new plantations.

9. *Acanthoderes thoracicus*, White.

Acanthoderes thoracicus, White, Cat. Long. Col. in Brit. Mus. ii. p. 359.

To the description quoted above I will add that the third to the sixth antennal joints are acutely produced at their apices beneath, as in *A. Egaensis* and other species; the body is depressed; the elytra are subtrigonal, with the apices slightly truncated, and have always an oblique dark-brown streak on the disk; the centro-basal ridges are narrow, disappear about the middle of the elytra, and leave a depressed space between them. The prosternum is simply rounded behind, the mesosternum vertically inclined in front. Long. 6–7½ lin. ♂ ♀.

This is a common species, on branches of felled trees, in the forest throughout the Amazon region. It is also found, I believe, at Cayenne.

c. Fore tibiæ neither dilated nor compressed.

10. *Acanthoderes albolinitus*, n. sp.

A. elongatus, subcylindricus, tenuiter tomentosus, fulvo-brunneus : elytris apice conjunctim rotundatis, pone medium plaga communi antice biramosa griseo-alba et vitta abbreviata nigra utrinque ornatis. Long. 8½ lin. ♀.

Head dingy fulvous, punctured. Antennæ stout, as long as

the body, fulvous brown, each joint from the third ringed with
dusky near the apex. Thorax with the lateral tubercles large,
conical, and pointed ; the two dorsal tubercles connected by
ridges with the hind margin ; the dorsal line strongly elevated,
the interstices coarsely punctured. Elytra elongate, very slightly
narrowed posteriorly ; the apices scarcely perceptibly truncated,
somewhat convex, the centro-basal ridges strongly raised at the
base, subsiding before the middle ; the whole surface punctured,
each puncture having a greyish-white scale : the colour is light
yellowish brown ; behind the middle, over the suture, is an ill-
defined greyish-white patch, prolonged on each side in front
into an oblique streak : on the disk of each elytron, behind the
middle, there is also a short inwardly curved black vitta connected by a zigzag line with the lateral margin. Body beneath
and legs black, shining, clothed with thin ashy pile ; apex of
tibiæ and tarsi fulvous. The prosternum is simply rounded; the
mesosternum bituberculate in front.

One individual only of this aberrant species occurred : I found
it at Ega, on a slender dead branch.

11. *Acanthoderes longispinis*, n. sp.

A. elongatus, subconvexus, tenuiter tomentosus, brunneo-fulvus :
elytris plagis pallidioribus, maculis nonnullis punctisque numerosis
nigro-brunneis ornatis, apice sinuato-truncatis, angulis externis in
spinam longissimam productis. Long. 8 lin. ♀.

Head dingy fulvous, impunctate, front uneven, channeled ;
lower lobe of the eyes very large for this genus, reducing therefore the breadth of the forehead. Antennæ slender, as long as
the body, piceous brown, each joint ringed with testaceous at the
base. Thorax fulvous, varied with dark brown ; lateral tubercles
large, very acute, dorsal ones very large, elongated, and obtuse ;
the disk with a few coarse punctures. Elytra curvilinearly
attenuated posteriorly, the centro-basal ridges slightly raised towards the base only, the surface faintly punctured, the punctures
numerous only near the sides towards the base ; on the disk they
are accompanied by granulations ; the colour is fulvous brown ;
there is an oblique ochreous spot on each, near the base, and a
waved transverse patch of the same colour near the apex, both
edged behind with dark brown ; the punctures are covered by
small dusky spots ; there is a strongly waved transverse spot
behind the middle, and three or four smaller ones on the margins, also dark brown ; the suture is tessellated with black and
grey. Body beneath black, clothed with shining silvery pile.
Femora piceous, clothed with grey pile ; tibiæ dusky, with two
pale rings, their apices and the first tarsal joint covered with
silvery pile ; second and third joints black, tip of the latter and

claw-joint reddish testaceous; the third joint fulvous beneath. The prosternum is simply rounded; the mesosternum vertical in front, and bituberculated.

Taken in the forest on the banks of the Cupari, Tapajos region. One example.

12. *Acanthoderes pigmentatus*, n. sp.

A. elongatus, subparallelus, depressus, tenuiter tomentosus, violaceo-brunneus, flavo nigroque variegatus: antennis crassis corpore multo longioribus, articulis 3-4 fortius sulcatis: elytris apice truncatis, angulis externis productis. Long. 8 lin. ♂.

Head plane in front, punctured, dusky brown, vertex paler and ornamented with two round black spots. Antennæ thick, brown, each joint from the third with two pale rings. Thorax with the lateral tubercles prominent, their apices produced and acute; the dorsal tubercles strongly raised, but obtuse, interstices punctured; violet-brown, with fulvous patches. Elytra gradually but very slightly narrowed posteriorly, the centro-basal ridges very feebly raised, granulated; their whole surface sparingly but coarsely punctured; the colour is violaceous brown, varied on each elytron with three discoidal angular dark-brown patches, viz. one near the base, one behind the middle, and the third near the apex; there is, besides, a transverse bowed yellowish streak behind the basal patch; the apical spot is also broadly margined with yellow; each elytron has a short white streak on the disk; the punctures are each covered with a dusky spot; the suture is tessellated with black and grey. Beneath and femora black, slightly shining, but clothed with yellowish-grey pile; tibiæ brown, with three pale rings; tarsi fulvous, second joint dusky. The pro- and mesosterna are both simply rounded, their surfaces closely punctured. The fore tarsi of the ♂ are slightly dilated; the first joint of the middle and hind tarsi are remarkably elongated for this genus.

One individual, taken at Tabatinga, on the Peruvian frontier. In the slenderness of the tarsi this species differs greatly from the rest of its congeners. This, however, is evidently merely a specific character, as the species is extremely nearly allied to *A. cylindricus* (*Ægomorphus* id., Dj. Cat.)* of Rio Janeiro, which possesses tarsi constructed as in the rest of the genus.

* *Acanthoderes cylindricus*, n. sp.—Elongatus, parallelus, depressus. Caput fuscum; fronte plana, punctata, opaca; vertice punctis duobus ocellaribus nigris iridibus flavis. Antennæ corpore longiores, piceæ, griseo-maculatæ. Thorax punctatus, fuliginosus, sericeus fulvo-variegatus, tuberibus lateralibus magnis, spiniferis, dorsalibus conicis acutis. Elytra perperum attenuata, apice sinuato-truncata, angulis internis acutis, externis productis, passim granulato-punctata

13. *Acanthoderes phasianus*, n. sp.

A. modice elongatus, depressus, tenuiter tomentosus, fulvo, flavo-griseo, cano nigroque multifariam variegatus: thoracis tuberculis dorsalibus fortissime elevatis: elytris apice late sinuato-truncatis, angulis externis in spinas longas productis. Long. 6 lin. ♂.

Head piceous, varied with paler shades, two rounded fulvous spots on the vertex, front uneven. Antennæ much longer than the body; basal joint black, with grey pile, remaining joints piceous, third and fourth with two greyish rings, the rest pale at their bases. Thorax with the lateral tubercles large, acute, the dorsal ones very large, conical, the interstices punctured; dusky brown, with fulvous streaks and spots. Elytra very slightly narrowed posteriorly, very sparingly granulate-punctate near the base; the centro-basal ridges produced forwards at the base, thence gradually subsiding towards the middle: the colour towards the base is brownish-black, varied on each elytron with four rounded fulvous spots odged with grey; the apical third is fulvous, edged in front with greyish, but near the apex with brownish black; there is a comma-shaped whitish mark on the disk of each before the middle, a large dark-brown V-shaped mark behind the middle; the rest is a combination of dusky brown, light grey, and fulvous minutely commingled. Body beneath shining black, clothed with grey pile. Femora dusky, each with a large fulvous spot: tibiæ and tarsi spotted with fulvous and dusky, the third and fourth tarsal joints being clear fulvous. The fore tarsi of the ♂ are strongly dilated and fringed. The prosternum is simple and rounded, the mesosternum subvertical and bituberculated.

One example taken at S. Paulo, on the Upper Amazon.

14. *Acanthoderes meleagris*, n. sp.

A. modice elongatus, depressus, postice attenuatus, tenuiter tomentosus, griseo, fulvo nigroque læte variegatus, capitis thoracisque lateribus griseis: elytris trigonis, breviter truncatis, angulis externis in spinas longas productis. Long. 5½-6½ lin. ♂ ♀.

Head sparingly punctured, varied with black and fulvous, sides clear pale grey. Antennæ rather short, black, the base of each joint from the second grey. Thorax with the lateral tubercles prominent and acute, the dorsal ones large, but only slightly elevated, interstices with a few large punctures; black varied with fulvous, the sides clear pale grey. Elytra briefly truncate at the

densius prope basin, utrinque obsolete bicarinata; fulvo-brunnea, apices versus griseo-varia, utrinque pone medium macula angulata nigra notata. Subtus niger, pilis cinereis vestitus. Pedes fusci, griseo-maculati. Sterna ut in *A. pigmentato*. Long. 8 lin. ♀.

apex, the outer angle armed with a long spine; centro-basal ridges prominent at the extreme base, each prolonged posteriorly as a smooth flexuous carina to the apical spine; sparingly punctured near the base and on the sides; pale grey, numerous small spots and three larger transverse patches brownish black, varied also with fulvous spots, chiefly near the scutellum, at one-third and at two-thirds the length. Body beneath, and legs black, spotted with light grey; the third and fourth tarsal joints fulvous. The opposing faces of the pro- and mesosterna are steeply inclined and bituberculated.

Taken at Ega and S. Paulo, on dead branches of trees.

15. *Acanthoderes Swederi*, White.

Acanthoderes Swederi, White, Cat. Long. Col. in Brit. Mus. ii. p. 360, pl. 9, fig. 6.

This is a common species near Pará, on dead trees; it is also found on the Upper Amazons and at Cayenne. In most collections it stands as *A. Daviesii* of Swederus and Olivier; but the descriptions of these authors, according to Mr. White, apply to a distinct Columbian species. The excellent description and figure quoted above are sufficient to make the insect perfectly well known. I will only add that the opposing faces of the pro- and mesosterna are steeply inclined and bituberculated, and that the fore tarsi of the ♂ are widely dilated and densely fringed.

16. *Acanthoderes chrysopus*, n. sp.

A. parum elongatus, valde depressus, postice paulo attenuatus, tenuiter tomentosus, rosaceo-fulvus maculis pallidioribus variegatus: elytris subtrigonis, apice late truncatis, angulis externis modice productis: tarsis aureo-fulvis. Long. 6 lin. ♂.

Head silky fulvous. Antennæ twice the length of the body, ferruginous, silky, base of each joint (from the third) greyish. Thorax with the lateral and dorsal tubercles equal in size and shape, large, conical, produced at their apices, rusty brown, punctured only on the fore and hind margins. Elytra with the centro-basal ridges much produced at the base, prolonged behind as smooth flexuous carinæ, to the apex; sparingly granulate-punctate on the ridges and on the sides near the base; the colour is fulvo-ferruginous, with a rosy tinge; near the base of each are two fulvous-yellow spots, and near the apex a large spot of the same colour, all encircled with dark brown; there are, besides, a few hoary-white specks scattered over the surface. Beneath, the body is black clothed with hoary pile, the apical half of the abdomen being yellow spotted with white pile. Femora black at the base, rusty-yellow on their apical halves; the tibiæ and tarsi are silky orange-yellow, the former ringed with dusky, the latter shining. The fore tarsi of the ♂ are broadly dilated and densely fringed. The opposing faces of the sterna are vertical and bituberculate.

At Ega, on severed and hanging woody lianas in new clearings. I consider it a local variety of the following, from which it differs in the more vivid coloration.

17. *Acanthoderes lotor*, White.

Acanthoderes lotor, White, Cat. Long. Col. in Brit. Mus. ii. p. 362.

The shape, sculpture, form of sterna, &c., are precisely the same as in *A. chrysopus*. I met with it only at Carepi, near Pará.

18. *Acanthoderes lateralis*, n. sp.

A. modice elongatus, subdepressus, postice attenuatus, tomentosus, cinereo-brunneus, thoracis lateribus, maculisque duobus elytrorum, altera magna triangulari pone medium, altera parva prope apicem saepe obsoleta, fuscis: elytris apice truncatis, angulis externis spina longa armatis. Long. 6–8 lin. ♂ ♀.

Head sooty-black, front and vertex ashy-brown. Antennæ about as long as the body, dusky, the third joint ringed with grey, 4–11 joints at the base testaceous grey. Thorax with the lateral tubercles prominent and acute, the dorsal ones prolonged into ridges, the dorsal line also forming a narrow ridge generally denuded; interstices punctured, ashy-brown, the sides sooty-black. Elytra narrowed to the apex, which is briefly truncate, the external angles being produced into long spines; the centro-basal ridges are feebly raised at the base, but prolonged behind each as a flexuous carina, which subsides at two-thirds the length; the basal half is rather thickly granulate-punctate; in colour they are ashy-brown, with a large triangular spot on the side behind the middle, and a small irregular one near the tip, silky dark brown. Beneath and legs black, clothed with ashy pile; tibiæ with two pale rings. The fore tarsi of the ♂ are widely dilated and densely fringed. The prosternum is simply rounded, the mesosternum steeply inclined in front and bituberculated. The terminal joints of the male antennæ are moderately slender and ciliated beneath.

This is a common species throughout the Amazon region, on felled trees in the forest; it is also common, apparently, in French Guiana. I have seen it in collections under the name of *A. lateralis*, Dej., which appellation I have adopted. *A. Jaspideus*, Germar (Sp. Nov. p. 475), and *A. consentaneus*, Dej., according to specimens sent to me by M. Deyrolle of Paris, are closely allied to *A. lateralis*; but their pro- and mesosterna are strongly convex. Our species is also near *A. satellinus*, Erichs.

(Consp. Insect. Peru., p. 143); but the latter is described as having the apex of the elytra armed with a very short spine.

19. *Acanthoderes bivitta*, White.

Steirastoma bivitta, White, Cat. Long. Col. in Brit. Mus. ii. p. 354.

This species was placed by Mr. White, whose description otherwise is a very good one, in the genus *Steirastoma*. It differs from that group in having simple instead of complex lateral thoracic tubercles. It stands in certain French collections as *A. tardigradus* of Dejean's Catalogue. It is a common insect, in the Upper Amazons, on the trunks of felled trees of a certain species, to whose bark its colours are assimilated. It is found also in French Guiana. It is inactive in its habits, but appears to be extremely prolific. The opposing faces of the pro- and mesosterna are steeply inclined and bituberculated.

§ 2. Antennæ with the terminal joints thickened and ciliated in the males, or triangularly dilated in both sexes; the third or third and fourth joints often furnished with tufts of hairs. (*Pteridotelus*, White, in part.)

[Although so diversified in structure and ornamentation of antennæ, this group is homogeneous in the form of the body, sterna, and in other respects.]

20. *Acanthoderes spectabilis*, n. sp.

A. elongatus, subdepressus, postice attenuatus, niger, velutinus, maculis magnis albis ornatus: antennarum articulo tertio scopa magna nigra instructo. Long. 8 lin. ♂ ♀.

Head black, with a triangular white spot in the middle of the forehead, two between the antennæ of the same colour, and two rounded on the vertex. Antennæ black, base of the fourth and following joints grey, the third joint encircled by a thick brush of black silky hairs, which extends nearer to the base on one side than on the other. Thorax with the lateral and dorsal tubercles large, conical, and acute; velvety-black, with two large rounded spots on each side, two elongate ones on the fore part, which are divergent behind, and another in the middle of the hind margin, white. Elytra elongate-trigonal, depressed together down the suture, the apex obliquely sinuato-truncate, outer angles acute; the centro-basal ridges much raised at the base, projecting over the hind edge of the thorax, prolonged behind to the apex, and granulated at their commencement; the surface impunctate, except on the suture near the base; the colour is velvety-black, each elytron having two spots along the suture behind the scutellum, a broken sutural stripe from the middle

to the apex, a spot near the shoulders, a large rounded one near
the lateral margin before the middle, a small one near the apex,
and one in the middle of the disk, all of a white colour. Body
beneath black, sides of the sterna and abdomen having large
white spots. Femora and tibiæ black clothed with grey pile;
tarsi fulvous. The opposing faces of the prosternum and
mesosternum are subvertical and sharply bituberculated. The
four apical joints of the antennæ in the ♂ are shortened and
ciliated.

This extremely beautiful species occurred only at Caiçara, a
village near Ega, on the Upper Amazons, on the trunks of felled
trees in the forest.

Genus DRYOCTENES, Serv.

Serville, Ann. Soc. Ent. Fr. iv. 24.

As already remarked, there is no character to distinguish this
genus from *Acanthoderes*, with which it will eventually have to
be incorporated. The species have a much broader and more
depressed form of body, and the antennæ are much longer, than
is the rule in the genus alluded to. The proportions of the antennal joints, form of muzzle, legs, male tarsi, and thorax are
the same as in *Acanthoderes*. In the style of coloration and
markings the species resemble most *A. bivitta* and its nearest
allies.

Dryoctenes scrupulosus, Germar.

Lamia scrupulosa, Germ. Insect. spec. nov. 470, 619.

There appear to be two somewhat distinct forms or geographical races of this species. The example before me, taken on
the banks of the Tapajos, differs considerably in colours and in
the shape of the elytra at the apex from the form found at Rio
Janeiro. The description of Germar with reference to the elytra
("glauco-tomentosa, apice truncata, intus dentata") applies to
the Amazonian example better than to those I have seen from
the south of Brazil. I do not know whether the latter may not
be the form described by Serville as *D. caliginosus*, his description not being sufficiently exact to decide. *D. caliginosus*, however, is generally considered to be synonymous with *D. scrupu-
losus* of Germar.

Genus OZOTROCTES, nov. gen.

Head somewhat narrow, antenniferous tubercles raised and
oblique. Palpi obliquely truncated at their apices, the labial
more strongly so than the maxillary. Thorax obtusely uni-
tuberculate on the sides, furnished with two very distinct tubercles on the disk. Prosternum simply rounded; mesosternum

much narrowed behind, steeply inclined in front. Elytra very slightly truncated at the apex. Legs and tarsi constructed as in *Acanthoderes*: the male sex, however, is as yet unknown. The antennæ are simple, the basal joint pyriform-clavate, shorter than the third; the second and third joints slightly furrowed above.

The truncation of the palpi and the attenuation posteriorly of the mesosternum amply distinguish this genus. The shape of the palpi is an anomaly amongst the Lamiaires, the pointed terminal joints being one of the very few characters which distinguish the tribe from the Cerambycidæ. The facies of the insect composing this genus, however, is entirely that of a Lamiaire, the shape of the thorax and elytra being almost precisely that of certain abnormal species of *Acanthoderes*, e. g. *A. hebes*.

Ozotroctes punctatissimus, n. sp.

O. oblongo-ovatus, subdepressus, obscure brunneo-ferrugineus: corpore supra punctis rotundis innumerosis impresso. Long. 4½ lin. ♀ .

Head brown, punctured. Antennæ about as long as the body, rufo-piceous, all the joints ringed with a paler shade. Thorax with the dorsal tubercles very distinct, conical, the rest of the surface almost even, punctured. Elytra very slightly truncated at the extreme apex, the centro-basal ridges short, the whole surface covered with punctures of a uniform size, partly arranged in rows. Beneath ashy-brown, shining. Legs dull ferruginous, spotted with a paler shade.

One individual, beaten from dried twigs in woods near Santarem.

Genus ÆTHOMERUS.

Thomson, Class. des Céramb. p. 338.

Syn. *Macronemus*, Dej. Cat.; White, Cat.

Char. emend. Body subcylindrical. Muzzle moderately broad, quadrate; front plane; antenniferous tubercles short, prominent, widely separated at their bases. Antennæ naked, excessively elongated, in some species being five or six times the length of the body, capilliform; the joints slightly increasing in length to the apex, the eleventh joint generally the longest; the basal joint short, very slender at the base, abruptly enlarged into an ovate club. Palpi normal. Prothorax unituberculated on the sides. Elytra rounded at the tip. Femora clavate; tarsal joints short. Prosternum greatly constricted between the large anterior coxæ.

The sexes are not distinguishable, as in Longicornes generally, by the relative length of the terminal antennal joint in most of the species; there is a sexual character, however, in the

P

apical ventral segment, the ♀ having in that part a deeply impressed fovea. The genus was established on certain curious species which agreed in having greatly elongated and hair-like antennæ, and strongly bowed fore tibiæ. I have extended the definition so as to embrace the *Alphus Lacordairei* of Dejean's catalogue—an insect which differs from all other *Alphi*, including *A. tuberosus* of Germar, to which it has otherwise some resemblance, in the curiously abrupt dilatation of the first antennal joint—a feature characteristic of the genus *Æthomerus*. *Æ. Lacordairei* differs from the other species in having straight fore tibiæ, and in having rather less elongated antennæ, whose articulations are much shorter in the ♀ than in the ♂.

The species are nocturnal in their habits. They are of rare occurrence, and are found in the daytime crouched on leaves,— *Æ. Lacordairei*, however, being seen only closely adhering to decayed boughs. In those species which have strongly bowed fore tibiæ, the anterior femora are greatly enlarged and furnished on the inner side with a sharp ridge, which fits a corresponding groove along the tibiæ. In the crouching position, the fore legs are closely folded, the almost invisible antennæ laid backwards, and the whole insect assumes a rigid aspect, well calculated to deceive its enemies. *Æ. Lacordairei*, on the other hand, possesses passive means of defence of quite a different character: its colours and markings give it a deceptive resemblance to a dead pupa covered with a fungous growth, such as is often seen adhering to trees in damp climates. The deception is perfect, the insect having on each side of its body a large spot coloured and reticulated like a wing seen through the integument of a pupa. Thus we see here another instance of the widely different means Nature employs, within the same genus, to maintain the existence of her specific forms. Every species exists by virtue of some endowment which enables it to triumph over the infinite diversity of adverse circumstances that surround it at all stages of its life. This concerns us here, inasmuch as the general principle has an important bearing upon the systematic arrangement of species, a knowledge of the fact that structures are adapted to the ends just mentioned being necessary to avoid errors in estimating their affinities. Longicornes are greatly subject to these adaptations, those parts of structure being modified, from species to species, on which we depend for the establishment of genera, thus rendering, in this family, real generic definitions almost impossible.

1. *Æthomerus antennator*, Fabricius.

Lamia antennator, Fab. Syst. Eleuth. ii. p. 288. 36.

Æ. elongatus, tenuiter tomentosus, niger vel brunneus, variegatus :

elytris inæqualibus, lineis tenuibus argenteo-albis inscriptis, basi elevatis, apud medium subnudis nitidulis. Mas segmento ultimo ventrali simplici: fœmina eodem fovea magna impresso. Long. 3½–4½ lin. ♂ ♀.

Head dark brown. Antennæ pitchy brown, the apices of the joints paler. Thorax with two dorsal tubercles in a transverse line with the lateral ones, all four of equal size; the surface punctured; dark brown or blackish, varied with lighter brown. Elytra with short but strongly elevated and crested centro-basal ridges, the space between the two being also elevated and clothed with a silky fulvous-brown pile; the sides in the middle have each a very large depression: the surface of the elytra is punctate-granulate in rows, one of which runs straight along the disk on each side, continuous with the centro-basal ridge; others are diverted out of their course by the lateral excavations, within which the surface is extremely irregular; the disk near the suture is irregularly punctured; towards the apex are some elevated lines; the disk is naked and shining: the colour is generally nearly black, in some specimens silky brown of various shades; there are also numerous very slender silvery-white lines, two of which, more conspicuous, oblique on the disk, form an inverted V. Body beneath and legs dark brown, covered with a slight pile, and varied with paler shades. Anterior femora dilated; tibiæ curved and grooved on the inner side. Antennæ capilliform. In the male the apical ventral segment is simple; in the female it has a large deep transverse fovea near the apex.

This species I met with at Pará, at Obydos in Brazilian Guiana, and at Santarem; it is found also at Cayenne. I have received it from M. Depuiset, of Paris, as *M. ruficornis*, var. I think there can be no doubt it is the *Lamia antennator* of Fabricius; his description (somewhat better than the Fabrician descriptions usually are) seems to suit our insect sufficiently well. I have thought it better to give a more detailed description, for the sake of fixing the Fabrician name with more precision. The white lines are faint or wanting in some examples.

2. *Æthomerus rufescens*, n. sp.

Æ. elongatus, tomentosus, brunneo-ferrugineus: elytris inæqualibus, basi elevatis, omnino brunneo-tomentosis: antennis pedibusque ferrugineis. Mas segmento ultimo ventrali apice fortiter bisinuato; fœmina latet. Long. 4 lin. ♂.

Head rufous brown. Thorax tuberculated as in *Æ. antennator*, clothed with rusty-brown pile, faintly punctured. Elytra with short but strongly elevated and crested centro-basal ridges, the space between them being slightly elevated; the sides in the middle have each a very large depression; the surface of the

---

clytra along the discal portion is impunctate, being clothed with pile, and there is no line of granulations in continuation of the contro-basal ridge: the strongly flexuous line along the disk is present, the lateral ones are broken and confused within the excavation, as in *Æ. antennator*; the whole surface is rusty tomentose and opake; there are indications of white lines in the same position as in the preceding species. Body beneath, legs, and antennæ ferruginous red. The apical ventral segment in the male is strongly bisinuated at the tip. Anterior femora dilated; tibiæ curved and grooved within. Antennæ capilliform.

Taken at Santarem. The distinctness of this species from the foregoing depends more upon the structure of the ventral apical segment than on the general colour and clothing, which seem to be variable in these species.

8. *Æthomerus Lacordairei*, n. sp.

Æ. subcylindricus, cano-tomentosus, fronte, vertice et thoracis vitta dorsali violaceo-brunneis: elytris utrinque apud humeros macula magna lineata alam mentiente instructis, in medio prope basin fuscia, apices versus canis tuberosis. Long. 6¼ lin. ♂ ♀.

Head rather broader, and front more plane, than in the preceding species; epistome and cheeks hoary white, rest of the head dark brown; antennæ yellowish, partially clothed with fine hoary-white pile. Thorax somewhat rugose transversely; lateral tubercles acute, dorsal ones only slightly raised, hoary white, a broad stripe of a violet-brown colour down the centre. Elytra with the centro-basal ridges short, obtuse, punctate-granulate, chiefly in rows, but more confused in the middle towards the base; on each side near the shoulders is a large yellowish spot traversed by the rows of granulations, which are of a darker colour and varied by discoloured punctures in the interstices, the whole producing an imitation of a wing; the basal space between the two spots is blackish; the apical half of the elytra is hoary-white, tomentose, varied with dusky, and having white tubercles in rows continuous with the granulate punctures of the basal part. Body beneath and legs yellowish testaceous, clothed unevenly with hoary-white tomentum. Fore femora and tibiæ simple. The antennæ in the male are about three times, in the female about twice, the length of the body.

Taken at various places on the Lower and Upper Amazons, closely clinging to dead boughs. As I have before stated, this species is the *Alphus Lacordairei* of Dejean's Catalogue, according to French collections.

Genus MYOXINUS (Dej. Cat.?), nov. gen.

Head narrow across the vertex, the antenniferous tubercles

being very prominent and directed upwards. Antennæ simple, the basal joint pyriform-clavate, though somewhat slender, shorter than the third. Palpi with their terminal joints slender and pointed, as in Lamiaires generally. Thorax with the sides furnished with a short simple spine, without conical tubercle; the disk having three small acute tubercles. Elytra with short, strongly raised and abrupt, crested centro-basal ridges; their tips rounded. Mesosternum narrowed behind, but broader than long, its front oblique and bituberculated. Prosternum simply rounded.

The narrowness of the head across the vertex, and the consequent approximation of the antenniferous tubercles, which at the same time are very prominent, amply distinguish this genus from *Acanthoderes*, as well as from the following, *Alphus*. It has, in common with *Alphus*, the comparative slenderness of the basal joint of the antennæ; but this is more pyriform and shorter in comparison with the third in *Myoxinus* than in *Alphus*. The form of the thorax and the crested ridges of the elytra contribute to give the species a peculiar facies. The name was first given, in Dejean's Catalogue, to an undescribed species; the genus has never been characterized; the species to which the generic name was applied I have seen in collections, and it appears different from the one I took; both belong, however, decidedly to the same genus. M. Thomson (Classif. des Cérambycides, p. 337) unites the genus to *Alphus*. It is more nearly allied to *Alphus* than to any other genus; but I think the characters given above will show that it should be separated from it.

Myoxinus pictus, Erichson.

Acanthoderes pictus, Erichs. Conspect. Ins. Peruan. p. 144.

I took this species at Ega and St. Paulo. It is sluggish in its motions, and is found on dead branches of trees, to the bark of which the insect is assimilated in colours. I have nothing to add to the excellent description given by Erichson in the place quoted.

Genus ALPHUS, Thomson.

Thomson, Classif. des Cérambyc. p. 10.

M. Thomson notices the shape of the basal joint of the antennæ, but, I think, not with sufficient detail to show the difference in that respect between this genus and its allies. In *Alphus* this joint is very gradually thickened, and is nearly equal in size to the third; therefore it is not pyriform in shape, as is the rule in the Acanthoderitæ. The genus differs from *Myoxinus* in the greater breadth of the head across the crown; the head, however, is much narrower than it is in *Acanthoderes* and the allied genera;

the muzzle also is much more obtuse. The genus, in fact, forms a connecting link between the Acanthoderitæ and the Acanthocinitæ, the chief character of the latter group being the great length of the basal joint of the antennæ, which exceeds that of the third. The other characters of *Alphus* which require mention are the sockets of the fore haunches, which in most of the species are angulated exteriorly; the fore tarsi, which are not dilated in the male; and the mesosternum, which is much narrowed behind, as in *Myoxinus*. As the genus is very imperfectly known at present, I add a list of all the described species, including those introduced in the present memoir.

1. *A. leuconotus*, Thoms. Classif. p. 10, = *sellatus*, Dej. Cat. sec. Chevrolat. South Brazil.

2. *A. pubicornis*, Serville. South Brazil.
   Oreodera pubicornis, Serv. Ann. Soc. Ent. Fr. iv. p. 21.
   Ægomorphus pubicornis, White, Cat. Long. Col. Brit. Mus.

3. *A. centrolineatus*, n. sp. Amazons and Venezuela.

4. *A. senilis*, n. sp. Amazons.

5. *A. scutellaris*, n. sp. Amazons.

6. *A. canescens* (Dej. Cat.?), n. sp.* South Brazil.

7. *A. tuberosus*, Germar.
   Lamia tuberosa, Germ. Ins. Sp. nov. p. 477.

8. *A. subsellatus*, White.
   Alphus subsellatus, White, Cat. Long. Col. Brit. Mus. p. 375.

The *Ædilis griseofasciata* of Serville, included by White in this genus, does not belong to it. Its proper position, as shown by the length of the basal joint of the antennæ and other characters, is amongst the Acanthocinitæ.

1. *Alphus centrolineatus*, n. sp.

A. oblongus, mediee convexus, fusco-ferrugineus, tomentosus, pilis cervinis passim vestitus: thorace fusco bilineato: elytris punctatis, punctis setiferis, apice oblique truncatis, apud medium linea abbreviata, suturali, communi, fusca ornatis. Long. 5 lin. ♂ ♀.

Head moderately broad, tomentose. Antennæ in both sexes

---

* *Alphus canescens*, n. sp.—Oblongus, antice leviter attenuatus, tomento cinereo-olivascente vestitus. Caput parvum, fronte inter antennas concava. Antennæ corpore duplo longiores, infra dense ciliatæ, canescentes, articulorum apicibus nigris. Thorax grosse punctatus, supra breviter tricarinatus. Elytra grosse irregulariter punctata, olivascentia, apud medium canescentia, apicibus breviter truncatis, carinis contro-basalibus parum elevatis postice prolongatis. Subtus niger, pilis cinereis vestitus. Pedes cinereo-pubescentes. Long. 7 lin. Rio Janeiro.

half as long again as the body, dull ferruginous, spotted with hoary tomentum, pubescent, more densely so beneath than above; the terminal joints more slender and less hairy than the preceding. Thorax with large lateral tubercles, and two impunctate obtuse dorsal ones, the interstices coarsely punctured: on each side of the upper surface is a longitudinal dark brown line. Elytra punctured throughout; the punctures closer and granulated towards the base, each furnished with a short blackish bristle: the centro-basal ridges are scarcely indicated: the surface is dull ferruginous, tomentose, with a few streaks of hoary colour; in the middle of the suture is a short, abruptly limited, dark-brown line. Body beneath black, thinly clothed with hoary pile. Legs ferruginous, clothed with similar pile and also with long pale hairs.

The elytra in the male taper towards the apex, which is obliquely truncated, the outer angle being slightly produced; in the female, the elytra are of equal breadth, and are obtusely rounded towards the tips, which are simply truncated obliquely.

This species, which is nearly allied to *A. pubicornis*, Serv., of Rio Janeiro, I found at Obydos, in Brazilian Guiana, on decayed branches. I have a specimen, ♀, also from Venezuela.

2. *Alphus senilis*, n. sp.

A. oblongus, tomento cano-olivascente vestitus: thorace punctato, tuberibus lateralibus productis, dorsalibus tribus acutis: elytris granulato-punctatis, fasciculis pilorum ornatis, apice singulatim rotundatis, regione scutellari fusca. Long. 8 lin.

Head punctate, tomentose, slightly depressed between the antennæ. Antennæ half as long again as the body, ashy; the tips of the joints blackish. Thorax with very acute prolonged lateral tubercles, and three acute and prominent dorsal ones arranged in a triangle; the surface closely punctured. Elytra oblong, moderately convex, rounded at the tips; the centro-basal ridges prominent, crested with tubercles, the scutellar space between them very thickly impressed with large, regular, oblong punctures; this space is of a dusky or brown colour; the rest of the surface is olive-ashy, coarsely granulate-punctate; each elytron has three indistinct incomplete longitudinal ribs, and along each of these is an interrupted row of small fascicles of hair. Body beneath and legs clothed with hoary tomentum.

On dead branches, Obydos and Pará.

8. *Alphus scutellaris*, n. sp.

A. oblongus, tomentosus, cinereus, thorace brunneo, spatio triangulari apud scutellum violaceo-brunneo: thorace punctato, tuberibus lateralibus productis acutis, dorsalibus tribus obtusis: elytris gra-

nulato-punctatis, fasciculis parvis pilorum ornatis, apice singulatim
rotundatis. Long. 4½ lin.

Head punctured, tomentose. Antennæ half as long again as
the body, ashy; tips of the joints blackish. Thorax with very
acute-pointed lateral tubercles, two obtuse dorsal ones, and a third
behind, smaller, also obtuse; the surface coarsely punctured,
pubescent and brown in colour. Elytra with moderately raised
crested centro-basal ridges, the scutellar space between them
densely and regularly punctured, violet-brown in colour; the
rest of the surface is ashy-white, sparingly punctured; each
elytron has two or three incomplete raised lines, along each of
which is a row of very small linear pencils of dark-coloured hair.
Body beneath and legs black, clothed with ashy pile.

This species I found at Caripí, near Pará. It is closely allied
to the preceding, and is probably a variety of it; but its much
smaller size, different coloration and punctation, give it so distinct a character that, in the absence of connecting links, I am
obliged to treat it as a separate species.

The present genus terminates the succession of generic forms
which lead from the *Acanthoderes* type to that of *Acanthocinus*
and *Leiopus*. I shall now return to a series of forms which appear to have branched off from *Acanthoderes*, especially from
those species resembling *Pteridotelus* in general structure.

Genus STEIRASTOMA, Serv.

Serville, Ann. Soc. Ent. Fr. iv. 24.

This is a well-defined genus, not very closely allied to any of
the preceding. Its nearest relationship seems to be with those
species of *Acanthoderes* which have slender fore tibiæ, rounded
anterior acetabula, and closed acetabular sutures, tricarinate
thorax, and prominent centro-basal ridges continuing as smooth
carinæ to the apex of the elytra. It differs, however, from
Acanthoderes in the complex tuberculation of the sides of the
thorax. Instead of a simple lateral conical tubercle or spine, as
is usual in the Longicorn family, the thorax presents, on each
side, an irregular prominence furnished with three tubercles.
In some species this prominence is very strongly developed, and
then two of the tubercles are carried to the apex, giving it a
bifid appearance, the third remaining at the base beneath. All
the species have, besides this tricuspid prominence, an acute
tubercle on each side near the fore margin of the thorax; and
some present, in addition to this, a similar pointed wart on the
side, some distance above it. The muzzle, although similar in
shape to that of *A. bivitta*, is considerably longer and more

broadened anteriorly than in any species of *Acanthoderes*, and the fore angles are more strongly pronounced. The mandibles are long, very slightly bowed, and much flattened. The centro-basal ridges of the elytra are curved outwards and prolonged behind as more or less flexuous smooth keels to the apex. In the males of some species the basal joint of the antennæ forms an irregular many-angled club, and is longer in proportion to the third than is the rule in the section to which the genus belongs.

1. *Steirastoma depressum*, Fabr.

Cerambyx depressus, Fabr. Ent. Syst. i. ii. 260. 32.

St. breve, depressum, postice valde attenuatum, nigrum, tenuiter griseo tomentosum: thoracis lateribus quinquetuberculatis, dorso depresso tricarinato: elytris subtrigonis, carinis centro-basalibus valde curvatis ramulum suturam versus emittentibus, apice spinosis: pedibus nigris, cinereo obscure variegatis: corpore subtus nigro, nitido, lateribus ochraceo tomentosis. Long. 7–11 lin. ♂ ♀.

Head black, scantily clothed with grey tomentum, sparingly punctured with three raised longitudinal lines, the lateral ones flexuous, the central one straight and running from the vertex to the edge of the epistome. Antennæ half as long again as the body in the ♂, a little longer than the body in the ♀, black, the bases of the joints ashy; the first joint in the ♂ of an irregular clavate shape, rugose, tuberculated at the apex. Thorax punctured near the fore and hind margins; the sides have each five tubercles, two anteriorly and three on the moderately produced lateral prominence; the dorsal carinæ are smooth, and shining black, the lateral ones being flexuous, interrupted, and tuberculate. The elytra are clothed with thin ashy tomentum, streaked and spotted with black; the centro-basal ridges are granulated and strongly curved, the posterior end of the curve emitting a short branch towards the suture; afterwards each is continued as a flexuous and smooth keel to the apex: the apex itself is produced into a strong tooth or spine, which varies in length in different individuals. The fore legs of the ♂ are much elongated.

This is a common Guiana species, and is generally distributed throughout the Amazon region; being found everywhere in new clearings, sometimes under the loose bark of trees. Like all the other species of the genus, it is sluggish in its motions, and feigns death when touched, bending its legs in a rigid position, and falling to the ground. As the Fabrician description is insufficient, and his name has been referred to a nearly allied but distinct South Brazilian species, I have thought it necessary to give a lengthened diagnosis. According to the British Museum collection, the *C. depressus* of Fabricius is the same as the *C. brevis* of Sulzer, an earlier author, and applies to the larger species of South Brazil above mentioned. I think, however, the description of Fabricius quoted above cannot apply to any other than the one I have described. It is probable, also, that Sulzer and the other old authors had the Guiana species in view in their *C. brevis*; for the productions of Brazil were not known in Europe at the time they wrote. I do not adopt Sulzer's name, however, because it is likely that the *C. depressus* of Linnæus, since Mr. White applies it to the Guiana insect, is the same species as the Fabrician; and therefore the name *depressus*, again having the priority, would stand. I have no means of deciding this point. Linnæus gives Coromandel as the locality of his *C. depressus*; and Fabricius does not quote his name in the synonymy. I have received a pair of *Steirastoma depressum* from M. Deyrolle of Paris, as coming from Venezuela, under the name of "*St. difformis*" of Dejean. It is considerably modified from the Guiano-Amazonian type, being more closely tomentose, and ochreous rather than grey in colour.

2. *Steirastoma melanogenys*, White.

Steirastoma melanogenys, White, Cat. Long. Col. in Brit. Mus. ii. p. 355.

The male in this species has a strong tooth or spine on the inner edge of the fore tibiæ near the middle. This was overlooked by Mr. White; otherwise his description, as cited above, leaves nothing to desire. This insect is the "*St. aculeata*" of Dejean's catalogue, according to specimens I have received from Paris. Cayenne examples do not differ at all from those found in the Amazon region. I met with the species only in the central parts of the Lower Amazon, at Obydos and Santarem.

3. *Steirastoma cænosum*, n. sp.

St. modice elongatum, postice attenuatum, depressum, tomento cervino-fusco vestitum : capitis thoracisque lateribus et plagis magnis duabus elytrorum fuscis : elytris apice valde spinosis. Long. 10 lin. ♀.

Head and labrum densely clothed with ashy-brown pile, the former punctured in front and marked with three fine longitudinal raised lines on the epistome, the central one extending to the vertex; the sides black. Antennæ brown; the third joint beneath with three very fine spines placed widely apart. Thorax quadrituberculate on each side, the lateral prominence very large, trituberculate, and the tubercle near the fore angle prominent; the dorsal surface depressed, punctured, tricarinate; the central keel very faint, the lateral ones prominent, flexuous; densely clothed with light-brown pile, the sides with a stripe of a coffee-brown colour between the lateral keel and the tubercles. Elytra depressed, elongate, subtrigonal; the centro-basal ridges prolonged behind to the apex, gently curved outwards and granulated from the base to two-thirds the length, then flexuous and smooth to their termination: the surface faintly and sparingly granulate-punctate, with a few large granulations, besides, on the shoulders; light or tawny-brown in colour, with a silky gloss; the sides have each two large irregular patches of a coffee-brown colour, one covering the shoulder and extending in a short streak to the disk, the other placed obliquely a little behind the middle; the apex of each elytron is produced into a spine. Body beneath thickly clothed with ashy-brown pile; the middle of the abdomen shining black. Legs thinly covered with ashy-brown pubescence.

One example, taken at Oyayá, banks of the Curuá, below Santarem.

The species is nearly allied to *St. melanogenys*, but differs in the shape and position of the spines beneath the third antennal joint, in its clothing and markings, and in the apex of the elytra not being squarely truncated with a spine at the external angles, but produced on each elytron into a stout spine.

4. *Steirastoma æthiops*, n. sp.

St. modice elongatum, depressum, supra tomento atro-griseo vestitum: capitis thoracisque lateribus et plagis magnis duabus elytrorum nigris: elytris apice valde spinosis. Long. 8–10 lin. ♂ ♀.

Head and labrum clothed with very dark-grey pile, the former with three longitudinal raised lines on the front, the central one extending to the nucbus, the lateral ones short and very prominent; sides black. Antennæ black; bases of the joints greyish, the third with a few very fine spines or bristles beneath, placed wide apart; the basal joint in the ♂ pyriform-clavate, smooth. Thorax quadrituberculate on each side, the lateral prominence very large, trituberculate, the tubercle near the fore angle prominent; the dorsal surface depressed, punctured, tricarinate; the central keel very faint, the lateral ones strongly pronounced; thickly clothed with very dark-grey pile, the sides each with a black stripe between the lateral keel and the tubercles. Elytra depressed, elongate, subtrigonal; the centro-basal ridges prolonged behind to the apex, strongly curved outwards, and granulated to two-thirds the length, then flexuous and smooth to their termination; the surface faintly and sparingly granulate-punctate, with a few large granulations, besides, on the shoulders; very dark grey, the sides having each two large, irregular, black patches—one, which is sometimes broken into smaller spots, covering the shoulder and extending in a slender streak to the disk, the other placed obliquely a little behind the middle; the tips of the elytra are briefly sinuate-truncate, the external angles produced into stout spines. Body beneath and legs clothed with very dark-grey pile; the middle of the abdomen naked, shining black. The fore tibiæ of the ♂ are untoothed.

This species occurred only at Ega and St. Paulo, on the Upper Amazons. It differs very little, except in colour, from *St. cœnosum*, and may be considered a geographical variety or race of that species. I have not seen either form in collections from other parts of South America.

Genus PLATYSTERNUS (Dej.), Blanch.

Blanchard, Histoire des Insectes, ii. 166.

The few words given by Blanchard as generic characters, in the place above quoted, have little or no meaning; the genus, however, is well known to entomologists from the figure given by Olivier of the only described species. It is a singular form of Lamiaire, partaking of the characters of *Steirastoma* and the Anisocerinæ—two widely different groups. The shape of the thorax, the closed acetabular sutures, and the direction of the centro-basal ridges of the elytra show a near affinity with the *Steirastomata*; whilst the form and smoothness of the muzzle, the broadly rounded apices of the elytra, and the depression of the fore edge of the metasternum are so many points of resemblance to the Anisocerinæ. The lateral prominences of the thorax are not simple, but bicuspid, the anterior cusp, however, being very much smaller than the posterior one. The antennæ are slender, one-fourth shorter than the body, and the eleventh joint, as in most of the Anisocerinæ, is much shorter than the tenth.

Platysternus hebræus, Fabricius.

Cerambyx hebræus, Fabr. Mant. Ins. i. 131.
——, Oliv. Ent. iv. p. 62, t. 15. f. 106.

I met with this rare and magnificent insect only at Caripí, near Pará. It was there found in some numbers, gnawing the bark of living Guariúba trees—a lofty tree of the order Leguminosæ, whose bark is thick, smooth, and friable, and much frequented by bark-feeding insects, especially Curculionides of the group Cryptorhynchini. Cicindelidæ of the rare genus *Iresia* are sometimes seen on the same tree, coursing over the trunk and preying upon the vegetable feeders; in fact, I never met with *Iresia* except on Guariúba trees. The large *Cratosomi* sometimes abound, and gnaw large holes in the bark. These insects do not seem to breed in the wood of the standing trees, but merely to resort to them for the purpose of gnawing the bark.

I have seen a second and undescribed species of *Platysternus* in the collection of Count Mniszech, at Paris.

Genus POLYRHAPHIS, Serv.

Serville, Ann. Soc. Ent. Fr. iv. 26.

From the seemingly capricious way in which the various parts of structure that, in other Coleoptera, furnish signs of affinity are modified from genus to genus in the Longicorn family, it is difficult to decide on the true position and relationship of the present group. In the general shape of the body, as well as in the form of the muzzle, thorax, and apex of the elytra, it seems to approach the genus *Acrocinus*. The antennæ, however, are quite glabrous beneath, instead of being ciliated partially or wholly as in *Acrocinus*; and the fore tarsi of the ♂ are dilated and ciliated, instead of being simple. In the proportions of the apical joints of the antennæ there is a great similarity between *Polyrhaphis* and the Anisocerinæ, the terminal joint in both sexes being extremely short compared with the penultimate. This seems to be a significant character. The form of the muzzle, too, is not greatly different from that of the Anisocerinæ; but the general form, the shape of the elytra and of the sterna, reveal no affinity with that group. The genus seems to have no close relationship with any other group of Lamiaires: it shows some resemblance to *Acrocinus* and the Anisocerinæ; but many intermediate links are wanting to prove a genealogical relationship. The prosternum is extremely narrow in this genus, and the mesosternum is contracted in the middle between the haunches. The anterior acetabula gape widely on the sides, the sutures being opened along their whole length. The genus is a very natural or well-defined one, comprising a cluster of species which agree with each other in facies as well as in structural characters. They are all of large size, have greatly elongated, filiform, rather stout antennæ, long and acute lateral thoracic spines, sometimes directed forwards, and ample oblong elytra, whose apices are broadly truncated and spined.

1. *Polyrhaphis spinosa*, Drury.

Lamia spinosa, Drury, Illustr. ii. p. 60, pl. 31. f. 3 (1773).
Cerambyx horridus, Oliv. Ent. iv. 66, pl. 4. f. 29 (1789–1808).
Lamia horrida, Fabr. Ent. Syst. i. ii. 273. 25 (1792).

The figures given by Drury and Olivier agree well in shape and form of the spines with the insect I have before me, taken at Villa Nova, on the Lower Amazons. My example, however, appears to be of a lighter colour. The general hue of the tomentum is hoary or ashy, the elytra, with the exception of the basal and apical parts, being of a violet-brown colour. The shape of the elytra in this species is elongate-quadrate, being only slightly narrowed posteriorly, with the base and apex rectangular, and the sides nearly straight. The spines on the elytra are as follows:—a row of small ones placed close to the suture, but deficient near the base and the apex; three large ones on the centro-basal ridges, two on the shoulders, and five or six very long ones on the disk. It occurs in Guiana as well as the Amazon region, and appears to be a rare insect. I met with only one example, which was found closely adhering to a dead branch, and scarcely distinguishable from it on account of the colours resembling those of the lichens with which the wood was covered.

2. *Polyrhaphis hystricina.*

P. brevis, subconvexa, spinosa, tomentosa, cervino-fusca, postice cinereo variegata: thoracis spinis antrorsum valde curvatis: elytris truncatis, angulis internis acutis, externis valde productis. Long. 12 lin. ♂.

Head scantily punctured, dull black, clothed with tawny-brown pile. Antennæ dark brown. Thorax punctured near the fore and hind margins, clothed with tawny-brown pile clouded with dusky; dorsal tubercles very large, obtuse; lateral spines strong, elongated, and more strongly curved forwards than in *P. spinosa.* Elytra rather short, subquadrate, slightly but gradually narrowed from the base, the sides nearly straight to three-fourths the length, and then gradually rounded to the apex, which is broadly truncated; the sutural angle very slightly produced, and the external one armed with a stout spine: the surface is studded with stout but not long or acute spines; there are five or six in a row on the strongly-raised centro-basal ridge, three or four along the suture near the middle, several smaller ones on the shoulders, and a short series of three or four between the shoulders and the centro-basal ridge, and, lastly, five on the disk, namely, two in the middle and three on the posterior part; the interspaces are studded with large, deep, and shining punctures, the apical portion of the elytra behind the spines alone being entirely smooth. Under surface of the body and legs black, thinly clothed with brownish pile; tarsi and a ring at the tips of the femora bright fulvous.

There is a specimen of this species in the British Museum, ticketed " *P. hystricina,* White," which name I have adopted; it is larger and paler in colour than my example, but agrees with it in all other respects. It appears to be a rare species. My specimen was taken near Pará.

3. *Polyrhaphis angustata*, Buquet.

Polyrhaphis angustatus, Buquet, Ann. de la Soc. Ent. de France, 1859, p. 445.

This species has been described at length by M. Buquet in the place quoted. It is an elongated parallel-sided species, 14 lines long; the elytra are free from spines or tubercles, being simply granulate and punctate partly in rows, but smooth towards the apex. The spines of the thorax are long and straight. The general colour is dull-reddish brown, varied with small specks and clouds of a dark-brown hue. The fore tarsi in the ♂ are feebly dilated and fringed, and the antennæ in the same sex are nearly twice the length of the body.

I met with the species on the banks of the Tapajos and at Ega. The examples found do not differ from the Cayenne specimen which I saw in M. Buquet's collection. The insect is found on the trunks of fallen trees in the virgin forest. Like many other large species of Longicornes, it comes abroad at night, and flies over broad rivers. I once found an individual along with many other dead or half-dead insects on a sand-bank in the middle of the Tapajos, which had been cast ashore after falling into the water during a squall in the night.

4. *Polyrhaphis gracilis*, n. sp.

P. elongata, angustata, subconvexa, tomentosa, violaceo-fusca: thoracis lateribus elytrisque postice flavo variegatis : elytrorum apicibus rotundato-truncatis, angulis externis spinosis. Long. 8 lin. ♀.

Head clothed with reddish pile, sides black; front coarsely punctured; muzzle short. Antennæ the length of the body, dull brown. Thorax punctured, reddish in colour, the sides behind varied with yellowish; the two dorsal tubercles small; the lateral spines long, slender, and slightly bent forwards. Scutellum yellowish. Elytra narrow, much elongated, and somewhat convex, gradually increasing in breadth from one-third to two-thirds their length, then slightly narrowed to the apex, which is obliquely and obtusely truncated, the external angle of the truncation produced into a spine ; the basal half of the surface is thickly granulate-punctate, the apical portion entirely smooth ; the colour is a dull-reddish or violet brown, the smooth posterior portion being varied with ashy yellow. The body beneath and legs are black, thinly clothed with ashy pile.

I only obtained one example of this small and elegantly shaped species, which was taken at Ega, on a dead branch.

5. *Polyrhaphis populosa*, Olivier.

Cerambyx populosus, Oliv. Ent. iv. 72, pl. 20. f. 156.

This fine species is found at Cayenne and, according to Erichaon (Consp. Ins. Peruana) in the forest region of Eastern Peru. My only example, a ♀, 15 lines in length, was taken on a slender dead branch in the forest at Ega—a locality midway between the two regions.

6. *Polyrhaphis Jansoni*, Pascoe.

Polyrhaphis Jansoni, Pascoe, Trans. Ent. Soc. Lond. v. n. s. pt. 1.

Mr. Pascoe, in the description referred to above, likens this species to the common *P. spinipennis* of Laporte, a native of South-east Brazil. It does not seem very closely allied, however, to that species. The elytra are less depressed, more thickly and deeply punctured on the base and disk, and less parallel-sided, being broad at the base and more tapering to the apex. In general outline it more nearly resembles *P. papulosa*. The colour above is fulvescent or tawny brown, the apical third of the elytra variegated with fine longitudinal streaks of a darker-brown hue. The bright-fulvous tarsi and the fulvous apical ring of the femora, contrasted with the deep-black legs, are features it possesses in common with *P. hystricina* and the following form, *P. Paraensis*. The surface of the elytra, except the apical portion, is studded with short obtuse spines, or, rather, conical tubercles; these vary in number in different examples, as they do in most species of *Polyrhaphis*; but, as is usual in the genus, they are constant in position. There is a row along the prominent centro-basal ridge, a series of three or four along the suture near the middle, and two oblique rows along the middle of the disk, the inner one of which extends in a flexuous direction to the base of the elytra. Besides these spines, the elytra on the sides and shoulders are thickly studded with tubercles arranged in rows, each accompanied, as the spines also are, by a large and deep puncture. The disk of the elytra towards the suture is much depressed, and, with the interspaces of the base, is thickly punctured; the apical third of the surface is smooth and impunctate. The apex of the elytra is truncated, the sutural angle has a very small projecting point, the external one being produced into a spine. The length varies from 9 to 15 lines.

This species is rather common at Ega, on the trunks of fallen trees in the forest. It is also found on the banks of the Cupari, an affluent of the Tapajos.

7. *Polyrhaphis Paraensis*, n. sp.

P. oblonga, tomentosa: capite fuliginoso: thorace fulvescente: elytris fuliginosis, basi et pone medium cervino variegatis: elytrorum tuberculis ut in *P. Jansoni* dispositis. Long. 10 lin. ♀.

Head and antennæ sooty black, the former punctured in front.

Thorax fulvous, the disk clouded with dusky; the lateral spines straight, the dorsal tubercles acute. Elytra broad at the base, then gradually narrowed to three-fourths their length, whence they are more abruptly narrowed and rounded to the apex, which is truncated; the sutural angles simply pointed, the external ones produced into spines; the tubercles and punctures on the surface are arranged precisely as in *P. Jansoni*, but the colour is different; the base is of a tawny-brown hue, the central parts and the apical third sooty brown, the interval between these darker patches being of a paler tawny colour. Legs black, a ring at the apex of the femora and the tarsi bright fulvous.

This species, which is no doubt a local modification of *P. Jansoni*, is found at Pará.

Group *Anisocerinæ*.

Genus TRIGONOPEPLUS, Thoms.

Thomson, Class. des Céramb. p. 339.

This genus is an aberrant form in the group Anisocerinæ, differing from most of the other genera in having the terminal joint of the antennæ, compared with the penultimate, of normal length, and the elytra obtusely truncated at the tip, instead of rounded. It resembles the genus *Chalastinus* so much in general form that I have thought it better to place it in this group. The third and three following joints of the antennæ are slender and slightly thickened at the tips; this indicates an affinity with the Anisocerinæ, where the thickening of the tips of the antennal joints is a very general character. The typical species of *Trigonopeplus* (*T. signatipennis*, Thoms., a native of South-east Brazil) has a deep semioval notch in the middle of the epistome—a singular peculiarity of structure, of which it is difficult to guess the purpose, especially as the labrum beneath it remains entire. The species of this genus which I found in the Amazon region has the epistome of the usual shape, the muzzle, in fact, being of exactly the same form as in the next genus, *Chalastinus*. It differs also from the type of the genus in the shape of the head, the antenniferous tubercles not being at all salient, whilst they are strongly raised in *T. signatipennis*. It agrees, however, so closely with the typical species in all other characters that I think it cannot be separated from it generically. It will be convenient, nevertheless, to treat it as a group or subgenus, which may be named *Anepsius*.

Trigonopeplus (*Anepsius*) *bispecularis*, White.

Trigonopeplus bispecularis, White, Cat. Long. Col. in Brit. Mus. ii. p. 403, pl. 10. f. 1.

Found occasionally on foliage in the forest at Ega.

Genus CHALASTINUS, nov. gen.

Head narrow; antenniferous tubercles prominent. Antennæ
slender, elongated, 11-jointed in both sexes, the terminal joint
shorter than the preceding, filiform; the third and five following joints thickened at the tips and curved, especially in the ♂.
Thorax much narrowed anteriorly, scarcely perceptibly tuberculated on the sides. Elytra subtrigonal, depressed, rounded together at the tips. Mesosternum bituberculated, hind margin
sinking behind into a fovea in common with the fore edge of the
metasternum. The fore tarsi of the ♂ not dilated, but fringed
with fine hairs.

This genus has a great resemblance in general figure to *Thryallis* (Thomson, Class. p. 31); but the antennæ in *Thryallis* have
only ten joints, and the sterna are broad and plane, the prosternum especially being remarkably broad. The Anisocerinæ
vary to such a degree in these and other parts of structure, that
almost every species might be made into a separate genus, if we
attached the same importance to those characters in this as in
other groups of Coleoptera.

Ch. Egaensis, White.

Anisocerus Egaensis, White, Cat. Long. Col. in Brit. Mus. ii. p. 408.

The typical form of this species seems to be confined to the
neighbourhood of Ega, on the Upper Amazons. It has on each
elytron behind the middle a short, oblique, ochreous belt, commencing on the sides near the middle, and not reaching the
suture, near which, in a line with the belt, is a round ochreous
spot. This is common at Ega on decaying branches of trees in
the forest. The following slight local modification was found
only at Fonte Boa, 120 miles in a straight line north-west of
Ega.

Local var. *Ch. postilenatus.* Like the type, except that the
oblique ochreous belt of the elytra is continuous from the sides
to the suture, and is also prolonged as a stripe along the sides
to the shoulders.

This variety wholly replaces the Ega form at Fonte Boa. At
Cayenne a nearly allied undescribed form * occurs, which, although apparently very different from *Ch. Egaensis*, I believe to
be a local modification of it. A species which varies in a small
degree from locality to locality a short distance apart becomes
modified in a greater degree in a more remote district and under
more greatly changed local conditions; at least, the distribution

* This is extremely rare in collections; and I regret being unable, from
having no specimen at command, to give a description of it.

of closely allied species and varieties, when carefully studied, seems to point to this conclusion.

Genus PHACELLOCERA.

Castelnau, Anim. Artic. ii. p. 468.

Char. emend. : Antennæ long and slender, eleventh joint about as long as the tenth, and filiform in both sexes; first joint slender at the base, and enlarged about the middle into a thick pyriform club; third joint thickened at the tip; fourth also sometimes dilated at its apex, and furnished with a small brush of hairs. Body elongate, parallel-sided, depressed. Thorax narrow, the lateral tubercles small, acute. The mesosternum plane or bituberculated, its hind margin depressed in conjunction with the fore edge of the metasternum, as is the rule in the Anisocerinæ. The fore tarsi in the ♂ are not dilated, and scarcely fringed.

I think the following species may be comprised in this genus:—

1. *P. plumicornis*, Klug, Entom. Bras., specimen alterum, pl. 42. f. 5. South-east Brazil.
2. *P. Buquetii*, Guér. Icon. R. A. p. 240. Cayenne.
3. *P. Batesii*, Pascoe, Trans. Ent. Soc. Lond. 1858, n. s. iv. Upper Amazons.
4. *P. limosa*, n. sp.* Venezuela.

Phacellocera Batesii, Pascoe.

Phacellocera Batesii, Pascoe, Trans. Ent. Soc. Lond. 1858, n. s. iv.

This species much resembles *P. Buquetii* of Cayenne, but it is considerably smaller. Its colour is light-greenish grey dusted with black; the sides of the head and thorax have a broad dusky stripe, and a narrow dusky zigzag belt runs across the elytra behind the middle, but does not reach the suture. The antennæ are three times the length of the body in the ♂, and not much shorter in the ♀; the joints are slender, almost capilliform, the basal one forms a very large and thick club, the third is thickened at the tip, the fourth simple like the rest; the apical joints are the longest : the colour is black, except a broad grey ring round the third joint.

This very curious insect is found at Ega, on the trunks and

---

* *P. limosa.* Corpus parum elongatum, depressum, fuliginosum, pilis minutis squamiformibus cinereo-fuscis vestitum. Antennæ (♀) corpore vix longiores, articulo quarto simplici. Caput et thorax punctata, hujus tuberculis lateralibus brevibus acutis, dorso trituberculato. Elytra passim granulato-punctata, singulis prope basin tuberculo magno cristato, et postice fasciculis duobus pilorum munitis. Antennæ fuscæ, articulo tertio apice valde dilatato. Long. 6 lin. ♀. *Hab.* Venezuela.

e 2

larger branches of fallen trees in the virgin forest. In crawling over the bark, it holds its antennæ straight forwards, and has a most striking resemblance to a greenish-coloured species of *Ptychoderes* belonging to the family Curculionides, which swarms at times on the same trees. I have a specimen from Yurimaguas, on the Huallaga, near the Andes, which differs (as all the other examples do which I have seen from the same place) from the Ega type only in being of a dull-grey colour without any greenish or olivaceous tinge.

Genus ANISOCERUS, Serv.

Serville, Ann. Soc. Ent. Fr. iv. p. 79.

This genus was founded by Serville on the *Lamia scopifera* of Germar, apparently the only species known at that time. Since then, a number of species have been added which do not belong to the genus, or at least would render its definition almost impossible were they to be included. I think it better to restrict it to those species which present the following characters :—

Body oblong, compact, subdepressed. Head broad; antenniferous tubercles slightly raised. Antennæ 11-jointed in the ♂, the terminal joint about half the length of the penultimate; 10-jointed in the ♀; the third joint in both sexes furnished at the tip with a compact rounded brush of short silky hairs. The mesosternum is very short, deeply depressed in the middle and on the hind edge in conjunction with the fore margin of the metasternum. The ligula is narrow at the base, then abruptly dilated, the lobes widely divergent. The palpi are gradually and obtusely pointed.

Anisocerus Onca, White.

Anisocerus Onca, White, Cat. Long. Col. Brit. Mus. ii. p. 405, pl. 10. f. 4.

Local var. a. *A. Fonteboensis*. Head and thorax as in *A. Onca*. Elytra at the base reddish brown and granulated, each with two rounded spots, and the humeral callus black; the rest of the surface has four rows of quadrate black spots divided only by narrow lines of a reddish-brown colour, and before the apex is a transverse black streak; the spaces between these black spots are quadrate in shape, and of a pale ochreous hue. Abdomen beneath varied with black. This variety diverges from the type only in the increased size and squared shape of the black spots, and the pallid hue of the equally squared interspaces. It is intermediate both in character and in geographical position between the type and local var. b, and is found near Fonte Boa, on the Upper Amazons.

Local var. b. *A. Olivencius*. Much larger than the type, being 7-7¼ lines in length. The occiput is black, with two pale ochreous lunate spots behind the eyes. Elytra at the base dingy ochreous and granulated, each with two rounded spots, and the humeral callus black; the rest of the surface is dark violet-brown, with four rows of angular spots, and the tip pale ochreous; the black spots in the same positions as in the type appear faintly through the violet-brown ground-colour. The rest as in the type. In this variety the pale spots of the elytra, already indicated in var. a, are strongly marked, and the ground-colour has become obscure. This change in the dress, added to the markings of the head and the size and robustness of the whole body, give the variety an aspect totally different from the typical form. Taken sparingly at St. Paulo de Olivencia: all the individuals found were conformable to the description here given.

This pretty insect seems very susceptible of local modification. The typical form is confined in its range to a very limited area around the town of Ega, on the Upper Amazons. It is there found in plenty on the trunks and branches of fallen trees in the virgin forest. At Fonte Boa, 120 miles above Ega, it occurs under a slightly modified local form (*A. Fonteboensis*), which would be scarcely worthy of remark were it not intermediate between the Ega type and the strangely transformed local variety or race, *A. Olivencius*, found at St. Paulo, 180 miles further west, or 300 miles in a straight line over a uniform country undivided by physical barriers from the home of its type. As before remarked, when a species varies in this way from district to district not far apart, it often happens that several closely allied but more distinct forms or species present themselves in districts further removed; these may be fairly suspected of being also modifications, considering the proof already obtained of the variability of the species. Several of these nearly allied forms occur in the present case. Thus I have no doubt, on perusing the excellent description, that the *A. stellatus* of Guérin-Méneville (Cat. Ins. Coléop. recueillis par Osculati, p. 27), found in Ecuador, probably on the banks of the Napo, is a further modification of the *A. Onca*, in the direction of our var. *Olivencius*. There is also an undescribed species found at Cayenne (*A. multiguttatus*, Laferté, MS.)*, which diverges from *A. Onca* in another direction; and this may with great probability be referred to the same type. It is the custom of naturalists, when they subordinate varieties to a species, to fix upon one of the forms as the original, to which the rest are referred: this original is generally the one first described or best known. In accordance with this usage, I have said that such and such forms are varieties of *A. Onca*; but, strictly speaking, no form can be said to be a variety

* I regret being unable, not having a specimen at command, to give a description of this species.

of *another existing* form unless it can be proved or shown to be highly probable that the one *descended* from the other, this other itself remaining meanwhile unchanged. It is necessary, therefore, to guard against the error of supposing that the arbitrarily chosen forms we see placed as species, with varieties subordinated to them are the true parents of those varieties; for whilst the varieties were being formed the parents themselves may have been undergoing modification, and therefore the so-called species and their varieties may be all equally varieties of some common possibly extinct form. In the present case, all that I mean to convey is, that, reasoning upon the fact of much local modification in *A. Onca*, we are constrained to infer that other closely allied forms have been derived from a pre-existing one nearly resembling them; and this might have been either *A. Onca* or the common parent of *A. Onca* and its subordinates*.

Genus GYMNOCERUS, Serv.

Serville, Ann. Soc. Ent. Fr. 1835, p. 84.

In this genus both sexes have eleven joints to the antennæ. According to Serville, the ♂ has the terminal joint very long. Amongst the species which I propose to include in the genus, some have this joint as long as the tenth, others much shorter; and it is always relatively shorter in the ♀ than in the ♂. All the joints are naked; but the third in some species, and the fourth in others, are more or less thickened at the tip. The body is convex and rather broad, and the elytra somewhat more gradually rounded to the apex than in *Anisocerus*.

This genus was omitted in Mr. White's 'Catalogue of the

---

* Some entomologists, however, believe that a local variety is an original creation equally with a species. Dr. Schaum, an author of high reputation, says, in discussing a case of local variation similar to the present one (Berliner entom. Zeitschr. 1861, p. 398), that many pairs of a species were originally created, and that, as there would be original differences amongst the individuals according to locality, so we have, at present, local varieties. This view will recommend itself to some minds by its extreme simplicity; for the excessive complexity of the relationships between existing varieties and species, on the other view above stated, repels by its difficulty of unravelment. In no case does the remark of Bacon so well apply, to the effect that the subtlety of nature far exceeds the subtlety of man's intellect. But Dr. Schaum's view ignores the fact that many local varieties shade off into mere individual variations or differences, such as we see occurring amongst the offspring of the same parents, making it extremely probable that local varieties or races have been derived by ordinary generation, with modification, from pre-existing forms. The hypothesis of the persistence, under the same conditions, of a local variety from the time of its creation is also quite at variance with the great mass of evidence, supplied by geology, of great migration and dislocation of species during the glacial and other epochs.

Longicorn Coleoptera of the British Museum;' and some of the species were included by him under *Anisocerus*, from which they are distinguishable by the naked antennæ.

1. *Gymnocerus capucinus*, White.

Anisocerus capucinus, White, Ann. Nat. Hist. xviii. t. 1. f. 7; Cat. Long. Col. Brit. Mus. ii. p. 406.

This remarkably beautiful species, which in its colours and markings resembles some kinds of *Doryphora* of the Chrysomelidæ group, I found only at Caripí, near Pará. It occurred sparingly, in January, on dead branches of trees in the forest. The third antennal joint is considerably thickened at the apex.

2. *Gymnocerus dulcissimus*, White.

Anisocerus dulcissimus, White, Cat. Long. Col. Brit. Mus. ii. p. 406.

I met with this species only on one occasion, in the forests on the banks of the Cuparí, a branch of the Tapajos, in 4° S. lat. and 55 W. long. It is still more beautiful in colours than *G. capucinus*; but I believe, with Mr. White, that it may be only a modification of that species. The third antennal joint is less thickened at its apex than in *G. capucinus*.

Three individuals only occurred, on a decaying branch in the depths of the forest.

3. *Gymnocerus cratosomoides*, n. sp.

G. ovalis, convexus, tomentosus, ochraceo-fulvus: thorace lævi: elytris seriatim punctato-granulatis, singulis prope basin tuberculo magno glabro instructis et apud medium breviter quadricarinatis, fascia lata undulata pone medium et marginibus posticis fuscis. Long. 9½ lin. ♀.

Head dull-greenish yellow, tomentose, smooth. Antennæ slender, shining black, about the length of the body, the third joint scarcely thickened at the apex. Thorax about half the breadth of the elytra; lateral tubercles rather small, acute, the surface quite impunctate; the disk tawny brown; the sides yellowish. Elytra very broad at the base, the breadth at that point being three-fourths the length; they are very gradually narrowed to two-thirds the length, thence more rapidly narrowed to the apex; each has in the middle, near the base, a very large naked obtuse tubercle; behind this, on the disk, are four short raised longitudinal lines, the one nearest the suture only being strongly elevated; the basal two-thirds of the surface is scantily covered with granulated punctures, mostly arranged in lines, each of which is accompanied by a dark-brown speck; the colour is ochreous brown or tawny; the posterior part of the suture and the discal ridges are finely streaked with grey; behind the middle is a broad irregular dark-brown belt, preceded by a yellow line; the dark-brown colour runs from the belt along the margins and suture to the apex. The body beneath and legs are clothed with greenish-yellow pile, which is denser on the sides of the breast and on the tarsi. The fore tibiæ are dilated and compressed.

One example, taken on the trunk of a tree at Tunantins, on the Upper Amazons. This and the following have a most deceptive resemblance to species of Curculionidæ of the genus *Cratosomus*, which occur in numbers on the trunks of certain trees. The general colour is exactly the same, and the resemblance is made more perfect by the large, glossy, basal tubercles of the elytra, which are merely modifications of the ordinary centro-basal ridges existing in this section of the Lamiaires. The shortness and slenderness of the antennæ, rendering the organs almost invisible at a short distance, also assist in perfecting the disguise, which completely deceived me when I saw the insect *in situ*. *G. scabripennis* (Serville), a native of Cayenne, belongs to this same group, all the forms of which appear to be excessively rare.

4. *Gymnocerus crassus*, n. sp.

G. ovalis, convexus, tomentosus; thorace elytrisque fulvis, his fascia latissima et macula subapicali canis. Long. 8½ lin. ♀.

This species very much resembles the preceding, and might be treated as a variety of it, although it seems more convenient to deal with it as a separate form. The punctures of the elytra, with their granulations, are much more strongly developed; otherwise the only differences observable are those of colour. The head is greenish yellow, with the crown and occiput grey. The thorax does not differ from that of *G. cratosomoïdes*. The base of the elytra is occupied by a narrow belt of a fulvous colour, and a much broader belt of the same hue crosses the elytra behind the middle; the rest of the surface is hoary grey, with the exception of the margins and suture near the apex, which are blackish.

I found one individual only of this form at Ega, on the trunk of a tree.

5. *Gymnocerus monachinus*, White.

Anisocerus monachinus, White, Cat. Long. Col. in Brit. Mus. ii. p. 406, pl. 10. f. 3.

This magnificent species varies in size from 7¼ to 11 lines. The fourth antennal joint is gradually and slightly dilated at the apex, the third is simple. The ground-colour of the upper surface in the ♂ is chalky white, in the ♀ rose-red, the latter being very bright during life. I found the species only within a radius of twenty miles of Ega, on the Upper Amazons. At Nauta, 540 miles to the west of Ega, the species recurs in a modified state; the modification is one of colour only, but is remarkable for its distinctness and its occurring in both sexes. The following is a short description of it:—

Local var. A. *Nautensis*; 8 lines, ♂ ♀.

The white fascia of the elytra is very much broader; the second black belt extends posteriorly along the suture, and the tooth-shaped black streak near the apex is replaced by a distinct isolated round spot.

I received one pair of this variety from Nauta, on the banks of the Upper Amazons, in Peru.

Genus ONYCHOCERUS, Serv.

Serville, Ann. Soc. Ent. Fr. iv. 83.

In this genus the antennæ have eleven joints in both sexes, but the terminal joint is much thinner than the others, and claw-shaped. In the males several of the apical joints are fringed beneath with long hairs; the second joint in both sexes is remarkably elongated. The sterna are in some species simple, and in others tuberculated, showing that this character has no generic value; for this genus is one of the most natural of the whole tribe. The tarsi of all the legs are strongly dilated; the fore tarsi of the males are more widely broadened than the others, but they are not fringed with long hairs. The ligula is elongated, not dilated on the sides, but simply rounded; the two lobes approximate, but are not united on their inner edges.

1. *Onychocerus scorpio*, Fabricius.

Lamia scorpio, Fabr. Mant. Ins. i. 131. 8.
——, Fabr. Ent. Syst. i. ii. 273. 26.

This well-known and common species is always found on the trunk of a particular kind of wild fruit tree called by the natives of the Amazon region Tapiribá; and the strange sculpture, shape, and colours of the body and limbs of the insect give it a most wonderfully exact resemblance to the bark. It is not possible to distinguish the insect, although a very large one (sometimes an inch long and a third of an inch broad), unless the tree is carefully examined. The tree is planted in fences very commonly in and near towns, on account of its rapid growth, and the insect accompanies it everywhere with the pertinacity of a parasite. It sometimes swarms on felled logs of Tapiribá.

2. *Onychocerus concentricus*, n. sp.

O. ovalis, postice paulo dilatatus, cinereo-fuscus: elytris multituber-

H

culatia, violaceo tinctis, lineis pallidioribus curvatis quasi concentricis ornatis. Long. 7 lin. ♂ ♀.

Head and antennæ dull black. Thorax ashy brown, the sides below the lateral tubercles black; the disk punctured and tri-tuberculate. Elytra widest behind; the sides near the base thickly granulate-punctate; the upper surface furnished with about four rather distinct rows of large and small acute dusky tubercles; ashy brown, the disk tinged with purplish, with several rather indistinct curved belts alternately of a darker or paler hue; the first is dark, and near the base on the outer side of the centro-basal ridge; the second, pale, is exterior to the first, and strongly curved outwards; the third, more distinct, of a dark colour and strongly curved, touches the suture, and forms a semicircular belt common to both elytra. The body beneath and legs are black, and clothed with ashy-brown pile. The three terminal joints of the antennæ in the ♂ have a few hairs beneath. The pro- and mesosterna are simple, the latter sloping from the hind to the fore margin.

I found this species on one occasion only, in great plenty, on a felled tree at Caripí, near Pará. The colour and sculpture of the insect gave it a deceptive resemblance to the bark on which it adhered.

Genus XYLOTRIBUS, Serv.

Serville, Ann. Soc. Ent. Fr. iv. 80.

The antennæ are short and mis-shapen; the third and fourth joints are dilated on one side at the tip, the fourth much more broadly so than the third. The eleventh joint is long and slender in the ♂, short and subuliform in the ♀. The head is narrow on the vertex, broadening below to the end of the muzzle, which is elongated. The terminal joint of the palpi is obtusely truncated at the tip. The ligula is short, slightly dilated in the middle, and the lobes joined together along their whole length, the two being conjointly and obtusely rounded at the apex. The prosternum is simple, the mesosternum short and bituberculated. The body is oblong and somewhat depressed. The thorax is short, transverse, and the lateral tubercles are small. The fore tarsi of the males are simple, and have only a slightly denser fringe of hairs than those of the females.

This genus is very closely allied to *Acanthotritus*, White (Cat. Long. Col. Brit. Mus.), and I think the latter might be with great advantage united to it. It appears to have been overlooked by Mr. White.

Xylotribus simulans, n. sp.

X. castaneo-rufus, thorace flavo trilineato: elytris pone basin minute granulatis et flavo sparsim irroratis, fascia lata sericeo-brunnea pone medium et prope apicem maculis oblongis carneis ornatis. Long. 5¼ lin. ♂ ♀.

Head dull red; front with four longitudinal yellow lines, the two outermost running obliquely down the muzzle from beneath the eyes. Antennæ about the length of the body, shining dull red. Thorax with two short transverse raised lines on the disk, dull reddish, with three fine, interrupted, dorsal, yellow lines. Elytra with the basal half minutely, densely, and evenly granulated; the colour of the surface is reddish brown, the basal half sprinkled with small yellow specks of different sizes and shapes; close behind the middle is a rather broad, silky, brown fascia, not touching the suture, its fore margin dentate and speckled with yellow; behind the fascia are a few oblong flesh-coloured spots, which are placed longitudinally at first, and then towards the apex transversely.

The body beneath and legs are dull reddish; the fore coxæ and a few spots on the sides of the breast are yellow, and there is a large, round, bright orange-yellow spot on each side of the post-pectus; the abdomen has on each side two rows of round whitish spots.

I have adopted for this species the name under which it stands in White's Catalogue; but it has not before been described. It seems to be peculiar, like nearly the whole of the *Anisocerinæ* I have here enumerated, to the Amazon region. I found it on the Lower Amazons only, at Obydos, Santarem, and Pará. It occurs on woody sipós or lianas, especially those which have been severed with knives or axes, on the borders of new clearings. It is closely allied to the *X. heterocerus* of Serville, to which in fact it should stand in the relation of a geographical form or race.

Genus HOPLISTOCERUS, Blanch.

Blanchard, Voyage de D'Orbigny, Ins. p. 210 (not characterized).

The antennæ in this remarkable genus are short and thick as in *Xylotribus*, and the eleventh joint in the ♀ is slender and claw-shaped. The basal joint is thickened from the base; the second, third and fourth are each produced at their tips into a very sharp and rather long spine. The body is oblong and depressed; the thorax cylindrical and unarmed. The species are adorned with brilliantly metallic colours. The terminal joints of the palpi are gradually and sharply pointed.

Hoplistocerus gloriosus, n. sp.

H. castaneo-rufus, glaberrimus, antennis pedibusque violaceo-cupreis: elytris alutaceis, confertim punctulatis, rubro-cupreis, vitta angusta suturali apicem haud attingente, altera lata marginali viridi-cyanea. Long. 5 lin. ♀!

Head and front tumid, very finely rugose-punctate; occiput and thorax marked with fine transverse striæ. The cheeks have a spot of brilliant green; the rest of the head and thorax is of a dark-chestnut hue. The elytra are oblong, broadly rounded behind, even on their surface, and uniformly punctured; their colour is red or orange-copper, with the exception of a narrow sutural stripe not extending to the apex, and a broader marginal one, which are of a greenish-blue lustre. The antennæ and legs are of a brilliant violet copper hue; the underside of the body is chestnut-red, and, with the legs and the other portions of the body, glabrous.

I took one individual only of this extraordinary and beautiful Longicorn, flying over a mass of dried twigs in an open place in the forest at Ega. It has a near resemblance to the *Hoplistocerus refulgens* of Blanchard (Voy. de D'Orbigny, Ins. p. 210, pl. 22. f. 9); but that species is described as having the body, with head and thorax, of a green colour. D'Orbigny's species was taken in the province of Santa Cruz de la Sierra (Bolivia), which region is connected with the Ega district by an uninterrupted stretch of low wooded country over 14° of latitude. I have seen several undescribed and distinct species of this genus in the collections of Count Mniszech at Paris and Messrs. Bowring and Pascoe in London.

Genus CYCLOPEPLUS, Thomson.

Thomson, Classif. des Cérambyc. p. 32.

In this genus, which is still more extraordinary in form than the preceding, the second and third antennal joints have an elongated and very slender spine at their tips; but the fourth, instead of being armed with a spine, is dilated on one side of the apex into a large, thick, rounded knob, clothed with a velvety pile. The antennæ differ also from those of the preceding two genera in being greatly elongated and slender, in the ♂ being twice the length of the body. The basal joint is very thin at its origin, and is dilated beyond the middle into a pyriform club; in length it departs from the almost universal rule in the subtribe Acanthoderitæ by being as long as the third. The fore tarsi in the ♂ are strongly dilated and fringed.

Cyclopeplus Batesii, Thomson.

Cyclopeplus Batesii, Thoms. Class. de Céramb. p. 32.

Ega, Upper Amazons, on dead branches on the margins of small tobacco plantations in the forest. The form of the insect is quite an exception to the prevailing character of the Longicorn family, the elytra being excessively dilated—in fact, as near as possible hemispherical in shape, instead of elongated as is the almost universal rule. When I first met with it, I was deceived by its great resemblance to a common insect of the family Eumorphidæ (*Corynomalus discoideus*) which swarms at times on the same decaying branches of trees on which the Longicorn is found. It is true the size is much larger than that of the *Corynomalus*, but this is not noticed when they are *in situ*. The very curious black knob on the fourth antennal joint assists greatly to complete the disguise; for this mimics the terminal club of the antennæ of the *Corynomalus*; and as the remaining joints in the Longicorn are very slender and imperceptible when the insect is on the tree, the organs in motion resemble precisely those of the *Corynomalus*. It is further remarkable that the Longicorn mimics especially a pale variety of *Corynomalus discoideus*, which is the prevailing form of the species at Ega.

A second species of this genus is known from Cayenne,—the *Cyclopeplus cyaneus* (Thoms. *l. c.*). I do not know whether this has its analogue in the same country, in a species of *Corynomalus*. Both species are excessively rare. The Anisocerinæ furnish many instances of adaptive mimetic resemblances; and to this peculiarity of the group is no doubt attributable the strange divergences or aberrations of form which it contains. In addition to the clearer cases which I have noticed, there are others not quite so evident. For instance, I think our *Hoplistocerus* is the mimetic analogue of a species of *Stenochia*, a Heteromerous genus, and the *Acanthotritus dorsalis* of South-east Brazil appears to resemble much a species of *Heilipus*, belonging to the family Curculionidæ. The *Onychoceri*, instead of mimicking other insects, have deceptive resemblances to the bark of trees on which they live. A tendency to mimetic resemblances seems to run in certain groups; and these groups are remarkable for the aberrations from their types in minor points of structure or in facies, and for the rarity and diversity of the specific forms which they contain.

Subtribe ACANTHOCINITÆ.

Group *Lagocheirinæ*.

Genus LAGOCHEIRUS (Dej. Cat.), Thomson.

Thomson, Classif. des Cérambyc. p. 9.

Body of large size, broad, oblong, slightly convex. Antennæ stout, half as long again as the body, and of nearly equal length in both sexes; the sixth joint in the males having a tubercle beneath its apex, surmounted by a pencil of stiff hairs; the basal joint is as long as the third, gradually thickened from the base, and in both sexes toothed beneath at the apex. Thorax obtusely tuberculated on its disk, and with large conical lateral tubercles. Elytra very broad at the shoulders, gradually and slightly tapering to the apex, which latter is briefly truncated. Thighs abruptly clavate; basal joint of the tarsi not much longer than the second.

The females have not elongated ovipositors and sheaths; the terminal abdominal segments, however, are much longer in the females than in the males. In one of the two species which I have examined (*L. araneiformis*) both the ventral and dorsal segments have their apical edges excised, whilst in the other (*L. fasciculatus*) they are entire. The males have their anterior tarsi ciliated.

1. *Lagocheirus araneiformis*, Linnæus.

Cerambyx araneiformis, Linn. Syst. Nat. ii. p. 625; Drury, Illustr. ii. t. 35. f. 4.
Acanthoderes araneiformis, Serv. Ann. Soc. Ent. Fr. iv. p. 30.

L. oblongus, postice modice attenuatus: thoracis tuberculis lateralibus acutis: elytris nigro fasciculatis, olivaceo-griseis, macula magna laterali triangulari fusco-nigra lineisque transversis pallidis ornatis: tarsis articulis duobus basalibus griseis, duobus apicalibus nigris nitidis. Long. 7–11 lin. ♂ ♀.

This is a well-known and widely distributed insect. I found it occasionally at most stations on the banks of the Amazons, from Pará to Peru: it is also a native of Guiana, the West Indian Islands, and the Island of Tahiti, where, according to M.Vesco*, it is common, the larva inhabiting the trunks of *Spondias dulcis*. It is not stated whether the Tahitian examples differ from those of America; those of the West Indian Islands form a tolerably distinct local variety. The species, however, has probably been introduced by the agency of man into the distant Polynesian island.

2. *Lagocheirus fasciculatus*, White.

Trypanidius fasciculatus, White, Cat. Long. Col. Brit. Mus. ii. p. 377, pl. 9. f. 9.

L. oblongus, postice valde attenuatus: thoracis tuberculis lateralibus obtusis: elytris nigro fasciculatis, olivaceo-griseis, maculis duabus lateralibus triangularibus (altera magna, altera parva) fasciaque lata pallida ornatis: tarsis ochraceis, articulo ultimo apice nigro. Long. 8–9½ lin. ♂ ♀.

Not uncommon at Ega, Upper Amazons, on dead branches in the forest, in company with *Acrocinus trochlearis* and other wood-

* Léon Fairmaire, Coléoptères de la Polynésie, p. 88.

eating Coleoptera. The tubercle at the tip of the sixth antennal joint of the males is much larger in this species than in *L. araneiformis*. The figure given in White's Catalogue represents a female.

Genus LEPTOSTYLUS.

Leconte, Journ. Acad. Nat. Sc. Philad. n. s. ii. p. 168.

Syn. *Annisous*, Dej. Cat. (part.).

The chief characters given by Leconte as distinguishing this from the allied genera are the shortness of the basal joint of the posterior tarsi and the tuberculose surface of the thorax, whose sides are simply prominent instead of being armed with a tooth or spine. The genus consists of a number of small-sized species more nearly allied to *Lagocheirus* than to *Leiopus* and *Acanthocinus*, being of compact, oval, convex form, and having short legs with thighs abruptly clavate. The basal joint of the posterior (as well as the other) tarsi is scarcely longer than the second; the thorax is very much narrower than the elytra, and its surface is studded with obtuse tubercles, the lateral tubercles in some of the species being scarcely visible, and in none spiniform: the elytra are also tuberculated or uneven, and are not spined at the apex. Most of the species which I have examined have the basal joint of the antennæ much flattened beneath; and in all, the apex of the same joint is produced beneath into a short tooth. The elytra are generally fasciculated, but have not very distinct centrobasal ridges.

Leptostylus appears to be closely related to *Erphæa* of Erichson (Consp. Ins. Peruana, p. 144), differing chiefly in the absence of acute lateral thoracic tubercles.

1. *Leptostylus pleurostictus*, n. sp.

L. oblongo-ovatus, subconvexus, tomento cinereo-brunneo vestitus : thoracis dorso quinquetuberculato : elytris multifasciculatis, lateribus macula magna nigro-fusca ornatis. Long. 4½ lin.

Head clothed with tawny-brown pile. Antennæ not much longer than the body, brown; basal joint (except the tip) and base of the remaining joints grey. Thorax with five distinct dorsal tubercles; the lateral tubercles short, conical, obtuse, and accompanied, near the front angle on each side, by a smaller one : greyish or hoary, a lateral spot behind the tubercle dark brown. Elytra ovate, not narrowed before three-fourths of their length ; apex very briefly, obtusely, and obliquely truncate : surface coarsely punctured (except near the tip), and furnished with numerous small tubercles arranged in three irregular rows, and surmounted each by a pencil of short bristles pointing towards the apex : the colour is ashy or greyish brown, a large dark brown patch occu-

pying each side from the base to the middle, and an indistinct
oblique whitish belt traversing the middle of each elytron. Underneath and legs brownish, varied with grey. The sterna are
all plane.

Occurred sparingly at Ega on slender dead branches.

2. *Leptostylus cretatellus*, n. sp.

L. oblongus, subconvexus, tomento canescente vestitus: elytris linea
laterali nigra, macula magna apicali fusca: thoracis dorso indistincte tuberculato. Long. 3½ lin.

Head clothed with grey pile. Antennæ grey, spotted with
brown. Thorax uneven above; tubercles indistinct, the lateral
ones conical, obtuse, placed behind the middle; the colour is
hoary white, the fore part of the disk having two small dark
brown spots. Elytra oblong, sharply and obliquely truncated at
the apex; surface punctured, and furnished with three faint
raised lines, on which rise a few small elevations, surmounted
each by a minute pencil of black hairs; the colour is hoary
white, except at the apex, which has a large brown spot re-
mounting in an angle on the suture; the sides near the base
have also a thick blackish line. Legs and underside greyish,
varied with brown.

One example taken at Obydos.

3. *Leptostylus ovalis*, n. sp.

L. curtus, ovatus, convexus, nigrinus: thoracis dorso trituberculato,
tuberculis lateralibus obtusis. Long. 3 lin.

Head olive-grey, with minute black spots. Antennæ with the
three basal joints dark grey, speckled with black; the remainder
grey, with the tips blackish. Thorax with an elevation on the
front part of the disk, surmounted by three obtuse tubercles;
the lateral tubercles very obtuse: punctured, scantily clothed
with dark grey pile irrorated with black. Elytra short, ovate,
very briefly truncated at the tip, coarsely punctured, and furnished with rows of small tubercles, each surmounted by a short
pencil of black hairs; the colour is sooty black, with scanty dark
grey pile, but towards the apex the grey pile forms a patch
speckled with black. Beneath iron-grey, slightly shining. Legs
grey, speckled with black.

Found at Obydos and Pará, on slender dead twigs.

4. *Leptostylus obscurellus*, n. sp.

L. elongato-ovatus, fuliginosus: thorace brevi, dorso inæquali, tuberculis lateralibus prominentibus. Long. 3 lin.

Head clothed with sooty pile; antennæ of the same hue, with
the bases of the joints (after the third) pallid. Thorax small

compared with the elytra; disk very uneven, the depressed parts
coarsely punctured, the lateral tubercles prominent; colour sooty.
Elytra elongate-ovate, the broadest part being about two-thirds
their length, the tip not perceptibly truncate; their surface is
thickly punctured, and is furnished with a few small tubercles
or ridges crested with hairs, as in the allied species; the colour
is sooty, with a few spots of white pile on the disk, sometimes
forming a patch near the apex. Beneath grey; legs sooty,
varied with grey; base of the thighs pallid.
Taken on slender dry twigs, in the suburbs of Santarem.

Group *Leiopodinæ*.

Genus AMNISCUS (Dej. Cat.).

Besides the species taken to form the genus *Leptostylus*, the
vague group standing in collections under the yet uncharacterized
name of *Amniscus*, Dejean, comprises others which might con-
veniently bear this title, as they differ in many respects from
the types of *Leptostylus*. These have an elongated and subdepressed form, with the basal joint of the posterior tarsi equal
to the two following united. They form a connecting link between *Leptostylus* and *Alcidion*, differing from the latter in having
the elytra oblong without prominent shoulders, instead of the
triangular form, broad and elevated at the base, which so well
distinguishes *Alcidion*. The thorax is tubercular on the disk,
as in *Leptostylus*, and its sides are simply prominent in the
middle, without acute or spiniform lateral tubercles. The elytra
are briefly truncated or rounded at the apex. The thighs are
abruptly clavate. In the only species which I have been able
to examine closely, the apical segment of the abdomen is conical
and somewhat produced in both sexes; but in the male both
dorsal and ventral segments are truncated or slightly emarginated
at the tip, whilst in the female the dorsal segment is obtusely
pointed.

Alcidion polyrhaphoides of White (Cat. Long. Col. Brit. Mus.
p. 394, pl. 10. f. 6) may be cited as the type of the genus *Amniscus* as here defined. In this and the species I have to describe the basal joint of the antennæ is abruptly clavate near the
tip; but it is doubtful whether this will prove to be a generic
character, as some species of *Alcidion* also have the same feature,
whilst their nearest allied species have the joint of the same
shape as the generality of the Acanthocinitæ. The joint, although abruptly clavate, is of the same relative length as in the
rest of the allied genera, and it presents also, near the tip on
the underside, the small dentiform process which is characteristic
of the subtribe.

I

Amniscus pictipes, n. sp.

A. oblongus, testaceo-rufus, nigro canoque variegatus : thoracis dorso trituberculato, tuberculis anticis fortiter elevatis: elytris prope basin bifasciculatis. Long. 3½ lin. ♂ ♀.

Head yellowish, spotted with black. Antennæ reddish, joints tipped with black; basal joint swollen beneath near the apex, the latter toothed. Thorax with three tubercles in a triangle on the disk, the two anterior very prominent; lateral tubercles obtuse; the colour is brown testaceous, with two black dorsal stripes. Elytra oblong, gradually narrowed from the middle to the tip, which latter is not truncated; the surface is thickly punctured, especially towards the base, and in the place of the centro-basal ridge there is a large pencil of black hairs; the rest of the surface even; the colour is testaceous brown, with the base and a few scattered marks blackish, an indistinct whitish line obliquely crossing the disk. Body beneath testaceous, clothed with pile of the same colour. Legs and tarsi reddish, spotted with grey and black.

One example, taken at S. Paulo, Upper Amazons. The species also inhabits South-eastern Brazil, specimens from Rio Janeiro (taken by Mr. Squires) not differing from the Amazonian example except in being rather duller in colour.

Genus ALCIDION (Dej. Cat.), Thomson.

Thomson, Classif. des Cérambyc. p. 12.

Char. emend. Thorax free from tubercles on the disk, or at most but slightly uneven, its sides unarmed. Elytra broad and convex at the base, thence narrowing in a nearly straight line to the apex, with the surface sloping equally in that direction; the apex truncated and toothed or spined, and the centrobasal ridges more or less prominent. Apical segments of the abdomen and ovipositor not produced in the female. Thighs abruptly clavate; basal joint of the tarsi generally longer than the two following united.

As above defined, the genus *Alcidion* will comprise a considerable number of species distinguished from *Amniscus* by the peculiar shape of the elytra, and from other allied genera by the thorax wanting the lateral spines. It is divisible into two groups, —one of which is distinguished by the species having a raised line along the whole length of the elytra on each side, from the centro-basal ridge to the external apical angle; and the other by the absence of these lines, the centrobasal ridges at the same time being very prominent. The Amazonian species belong wholly to the second group*.

* *A. latum* (Thomson, *l. c.*), of Mexico, seems to belong to the first group;

1. *Alcidion oculatum*, n. sp.

A. oblongum, postice modice attenuatum, tomento cervino subsericeo vestitum : thorace maculis duabus nigro-fuscis, albo marginatis : elytris lateribus acute carinatis, dorso lævibus, utrinque fascia discali interrupta nigro-fusco ornatis. Long. 3½ lin.

Head and thorax tawny brown ; disk of the latter with two short blackish lines, narrowly margined with whitish. Antennæ dusky, base of each joint from the fourth pallid ; basal joint gradually clavate, the outline waved beneath. Elytra prominent at the shoulders, gradually narrowed to three-fourths of their length, then more quickly so to the apex, which is briefly and very obliquely truncate, without spines : the sides are acutely carinated ; the dorsal carina is effaced, but the centro-basal ridge is very prominent, and crested with hairs ; the surface is punctured (except near the apex) ; the colour is tawny or violaceous brown, with a slight silky gloss, having on the disk behind the middle a short blackish-brown fascia, bordered on the basal side with pale ashy ; a small linear mark of the same colour is seen also near the suture towards the apex, and the suture, disk, and lateral margins have rows of small dark spots. Beneath tawny ashy. Legs testaceous ; thighs varied with ashy ; tibiæ black at the base and apex ; tarsi with the middle joints black.

Ega ; on slender dead branches in the forest.

2. *Alcidion triangulare*, n. sp.

A. breve, postice valde attenuatum, fulvo-griseum : thorace fusco bimaculato : elytris medio irregulariter cinereo fasciatis, apicem

---

also *Leiopus emeritus* (Erichson, Conspectus Ins. in Peruam, p. 147) of Eastern Peru. The two following should also be added :—

A. bispitum (Dej. Cat.). Modice elongatum, apud humeros latum, postice declivum et attenuatum, fusco-testaceum griseo olivaceoque variegatum. Caput nigrum. Antennæ fusco-testaceæ, articulo basali subiter elavato, clava infra barbata. Thorax griseo-olivaceus. Elytra apice sinuato-truncata, angulo interno acuto, externo longe mucronato, supra utrinque bicarinata, carinis lævibus, interstitiis punctatis, carina centrobasali nigro penicillata ; grisea olivaceo varia, fascia pone basin nigricante. Subtus testaceum, sternis nigricantibus. Pedes testacei, griseo maculati. Long. 3¼ lin. *Hab.* Rio Janeiro.

A. lineatum, n. sp. Elongatum, apud humeros minus latum, olivaceofulvum, olivaceo-fusco varium. Caput olivaceum, vertice fusco bipunctato. Antennæ fusco-testaceæ, articulo basali sensim clavato, infra planato et barbato. Thorax medio fusco bivittatus. Elytra valde elongata, punctata, apice sinuato-truncata, angulo interno acuto, externo obtuse dentato ; carina dorsali tenui et acutissima, controbasali parum prominente, lateribus discoque etiam carinatis, carinis obtusis et abbreviatis. Subtus nigricans ; pedibus fuscis. Long. 5¼ lin. *Hab.* Venezuela.

versus cinereo strigosis, apice breviter oblique sinuato-truncatis. Long. 3¼ lin.

Head dusky; antennæ testaceous brown, apices of the joints from the third black, basal joint waved beneath. Thorax tawny brown; disk with two round blackish spots, sometimes wanting. Elytra gradually narrowed from shoulders to apex, which latter is very obliquely sinuate-truncate; the sides acutely carinated from the shoulder, the carina effaced before the apex; surface even and punctured, the centro-basal ridge extremely prominent and destitute of hairs: the colour is tawny brownish, with a very indistinct, waved, ash-coloured fascia across the middle, the apical part being silky brownish, streaked with ashy. Beneath and legs testaceous brown; base and apex of tibiæ and middle joints of tarsi blackish.

Var. *Paraënse*, rather more robust; the surface and sides of elytra more thickly punctured, and the apex simply truncated, without sinuation.

This was rather a common insect at Ega, on dead twigs. The variety was found at Pará.

3. *Alcidion latipenne*, n. sp.

A. crassum, postice modice attenuatum, apice transverse sinuato-truncatum: elytris humeris valde productis lateribusque carinatis. Long. 4–6 lin. ♀ ♂.

Head dusky; antennæ testaceous brown; base of all the joints from the third pallid. Thorax much broader than long, surface uneven, tawny brown, silky, with a V-shaped dusky mark behind, joining two dusky spots on the sides of the scutellum. Elytra very broad and convex at the base, gradually narrowed to the apex, which is broadly and transversely sinuate-truncate, with the external angles somewhat produced; the shoulders are very prominent, and from the acute edge of each commences the lateral carina, which extends nearly to the apex; the surface is even and moderately punctured, the colour being reddish or tawny brown, slightly streaked here and there with ashy, especially in the middle, and having rows of small dusky specks, a large violet-brown spot lying on the deflexed margin beneath the shoulders. Beneath and legs testaceous brown.

Ega, and on the banks of the Cupari, a branch of the Tapajos.

4. *Alcidion interrogationis*, n. sp.

A. valde elongatum, postice parum attenuatum, carneo-griseum sericeum: elytris apice longe mucronatis, litura nigra signum interrogationis simulante ornatis. Long. 5¼ lin.

Head brown; antennæ dusky, base of joints from the third pallid, the first joint strongly waved. Thorax tawny, sides black, and disk with a black mark shaped like a horse-shoe. Elytra elongated, quickly narrowed behind the shoulders, then widening slightly, afterwards towards the apex again narrowed, the apex itself being rather broad and sinuate-truncate, the internal angle dentiform, the external one produced into a lengthened spine; the shoulders are acute, and from their edge commences the lateral carina, which is very prominent, but is so placed as to leave the deflexed portion of the elytra beneath it visible from above; the surface is rather thickly punctured in the middle towards the base, sparingly so in other parts; the centro-basal ridges are short, but extremely elevated, hooked posteriorly, and surmounted by a crest of black hairs; the colour is grey, with a rosy tint in some lights, and towards the apex there is on the disk of each a black curved line and spot resembling the note of interrogation. Legs (especially the posterior thighs) elongated; like the under surface of the body, they are of a dusky hue.

This elegantly shaped and curiously marked insect occurred only at Ega, on dead branches in the forest.

5. *Alcidion olivaceum*, n. sp.

A. oblongum, postice modice attenuatum, tomento olivaceo signaturis obscurioribus variegato vestitum: elytris apice breviter oblique truncatis, femoribus crassissimis. Long. 5¼ lin.

Head dusky; antennæ dusky reddish, base of joints from the third pallid. Thorax olivaceous grey, with a short black streak in the middle of the hind margin. Elytra rather broad at the shoulders, and narrowed curvilinearly thence to the apex, which is briefly and obliquely truncate; the lateral carina is less pronounced than in the last species, but the centro-basal ridges are very prominent; they are hooked behind, although not crested with hairs; the surface is rather uneven, having two faint and obtuse dorsal carinæ, which, however, are effaced shortly behind the middle; the colour is olivaceous grey, varied with small dusky spots which accompany the carinæ, and two oblique discal streaks placed behind the middle. Beneath and legs dusky, varied with grey; base of thighs and tip of tarsi testaceous; the thighs are short and very thickly clubbed.

Ega, Upper Amazons.

6. *Alcidion minimum*, n. sp.

A. parvum, depressum, postice modice attenuatum, fusco-cinereum, fuliginoso maculatum: elytris apice oblique sinuato-truncatis, angulis obtusis, carina centrobasali modice elevata. Long. 2 lin.

Head sooty brown; antennæ with the basal joint strongly flexuous beneath, sooty brown. Thorax smooth, sides rounded, hinder part punctured. Elytra subtrigonal, depressed, with lateral carinæ; centro-basal ridges moderately elevated, and not abrupt posteriorly; apex obliquely sinuate-truncate, angles not produced; the surface (except towards the apex) has a number of large punctures, and is of an ashy-brown colour, with a number of oblong dusky spots on the apical portion. Body beneath and legs dusky; base of thighs and of the first tarsal joint and a ring on the tibiæ pale testaceous.

Taken flying in the evening, banks of the river, S. Paulo, Upper Amazons.

The species, from its small size and general appearance, would consort well with those I have placed in the genus *Osines*; but the absence of lateral thoracic spines compels us to treat it as a member of the *Alcidion* group*.

Genus Lophopœum, nov. gen.

Head, antennæ, and general shape of the body as in *Alcidion*. The thorax differs from that and the allied genera in being armed near the centre of each side with an acute tubercle or

* The following species also belong to the second section of *Alcidion* :—

A. bicristatum. Elongatum, postice sensim attenuatum, olivaceo-griseum, fusco varium. Caput olivaceum. Antennæ infra parce setosæ, articulo basali angulo inferiore apicali producto, olivaceo-brunneæ, articulis basi pallidis. Thorax dorso obtuse tuberculatus, antice et postice punctatus, olivaceo-griseus, subsericeus. Elytra elongata, depressa, postice sensim attenuata, apice oblique sinuato-truncata, angulis externis productis; lateribus et disco obtuse carinatis; carinis centrobasalibus parum prominentibus, singulis cristis duabus pilorum nigrorum ornatis; elytris utrinque penicillis minutis atris in duplici serie (altera suturali, altera obliqua discoidali) notatis: olivaceo-grisea, tertia parte posteriore saturatiore griseo lineata, marginibus nigro punctatis. Corpus subtus griseum, podibus olivaceo-griseis nigro punctatis, femoribus basi testaceis. Abdomen fœminæ segmento ultimo dorsali apice attenuato producto, maris rotundato medio emarginato. Long. 5 lin. ♂ ♀. *Hab*. Rio Janeiro. (Coll. Squires.)

A. trivittatum. Elongatum, depressum, postice sensim attenuatum, brunneo-sericeum, vittis et maculis fusco-atris ornatum. Caput fuscum. Antennæ rufescentes, articulis basi pallidis, articulo basali aequaliter clavato. Thorax rotundatus, dorso æqualis, olivaceo-brunneus, sericeus, vittis latis tribus atro-fuscis velutinis (una centrali, alteris lateralibus) marginem anticum haud attingentibus ornatus. Elytra elongata, depressa, apice sinuato-truncata, angulis internis acutis, externis valde productis, humeris parum prominentibus, lateribus utrinque acute flexuoso-carinatis; disco æqualis; carina centrobasali valde prominente, brevi, nuda; rufescenti-brunnea, certo visu carnea nitentia, passim punctata et fusco maculata; lateribus prope basin fuliginosis; disco pone medium plaga atro-fusca antice utrinque linea obliqua grisea marginata ornato. Corpus subtus nigricanti-sericeum. Pedes nigricantes, tibiis et tarsorum articulo primo cano annulatis. Long. 5 lin. *Hab*. Venezuela.

spine. The surface of the thorax is generally smooth on the disk, but is in some species slightly uneven. The elytra are subtrigonal in shape and depressed as in *Alcidion*; the shoulders are moderately prominent, and in most species a lateral carina extends thence towards the apex, but, as in *Alcidion*, this becomes very obtuse or almost obliterated in some of the species. The apex of the elytra is more or less truncated and spined, and the centro-basal ridges are always prominent, although unconnected posteriorly with a dorsal carina, the disk of the elytra being always even. The ovipositor is not exserted in the females, nor is the apical segment of the abdomen produced. The thighs are thickly and abruptly clavate, and the tarsi very moderate in length.

1. *Lophopœum carinatulum*, n. sp.

L. curtulum, minus depressum, postice rotundato-attenuatum, fusco-ferrugineum, nigro-fusco maculatum: elytris lateribus haud carinatis, carina centrobasali parum elevata nigro setosa, disco plagis tomentosis ochraceis variegato. Long. 3½ lin.

Head rusty brown. Antennæ rust-coloured, base of each joint (from the third) paler; the basal joint somewhat evenly clavate, the upper side being convex, and the lower scarcely flexuous, but tubercled at the apex. Thorax rusty brown, varied with dingy ochreous; the lateral spine quite central and acute. Elytra less depressed than in allied species, and attenuated curvilinearly to the apex, which is briefly and obliquely truncated, and without acute angles to the truncation: there is no lateral carina, and the centro-basal ridge is only moderately raised, but is crested with black hairs; the surface is thickly punctured, except near the apex, and is without raised lines or inequalities, the colour being rusty brown, sprinkled over with darker spots, and varied behind the middle with two or three linear patches of decumbent hairs of an ochreous hue. The legs are rusty brown, with paler rings; the posterior tarsi, with their basal joints, are almost as short relatively as in the genus *Leptostylus*.

One example; Ega, Upper Amazons.

2. *Lophopœum fuliginosum*, n. sp.

L. oblongum, depressum, postice angustatum, fuliginosum, cano nigroque parce variegatum: elytris apice breviter truncatis, lateribus obtuse carinatis, carina centrobasali valde elevata et nuda. Long. 3¼ lin.

Head dingy brown. Antennæ (in ♂) greatly elongated and, towards the apex, fine as a hair, sooty brown, the extreme base of each joint (from the third) pallid; basal joint strongly flexuous beneath. Thorax dingy brown, punctured on the disk, which also is marked with four indistinct black spots. Elytra depressed, sides nearly straight, apex briefly truncated, angles of the truncation obtuse; humeral angles moderately prominent, the lateral carina extending thence towards the apex, obtuse, but the centro-basal ridges are strongly elevated, acute, and naked; the surface is somewhat evenly punctured throughout, and of a sooty hue, like the rest of the body, but the apical half has a few whitish specks, and on the disk behind the middle on each side is a short oblique black line. Body beneath and legs olive-brown, base of thighs pallid; tibiæ and basal joint of tarsi also each with a pale ring. The basal joint of the posterior tarsi is longer than the two following taken together.

Neighbourhood of Santarem, Lower Amazons.

3. *Lophopœum circumflexum*, n. sp.

L. curtulum, depressum, postice valde attenuatum, ferrugineo-fuscum: elytris singulis pone medium linea abbreviata cinerea maculaque laterali notatis, apice breviter transverse truncatis, carina centro-basali magna acutissima. Long. 3½ lin.

Head dull brown. Antennæ greatly elongated, fine as a hair, most of the joints furnished with a short bristle at the apex beneath; rust-coloured. Thorax with the lateral tubercles spiniform, disk punctured, rusty brown, the middle with two broad dusky stripes, flanked by spots of the same colour. Elytra trigonal, shoulders prominent; lateral carina obtuse, apex briefly truncated, with the angles rounded; surface thickly punctured towards the base; centrobasal ridges strongly raised and acute, naked, or nearly so; ashy or rusty brown, spotted with darker brown, the disk behind the middle having on each side a short, oblique pale ashy line, accompanied on the side nearer the base with a smaller line of the same hue, lying at right angles to it, the pale lines margined with dark brown. Body beneath and legs rusty brown.

Ega; on dried twigs.

4. *Lophopœum bituberculatum*, White.

Leiopus bituberculatus, White, Cat. Long. Col. Brit. Mus. ii. p. 382.

"*L.* punctulatus, cinereo-fuscus: antennis subferrugineis: oculis supra approximatis: elytris supra subplanis, arcu postmediano cinereo, singulis ad basin medio tuberculo parvo uncinato, ad mediam et posticam partes cinereo variegatis; femoribus basi pallidis; tibiis pallido uniannulatis; tarsorum articulo primo pallido. Long. 3¼ lines. Ega." (White, *l. c.*)

This species very closely resembles *L. circumflexum* in shape and colours; but the centro-basal ridges are reduced to a tubercle. The lateral thoracic spines are placed near the middle of the thorax—the position they occupy in the genus *Lophopœum*; but the sides behind them are deeply sinuated, which gives the thorax a similar shape to that possessed by the typical *Leiopi*.

5. *Lophopœum acutispine*, n. sp.

L. latiusculum, depressum, brunneum, postice cano marmoratum: thoracis disco obtuse tuberculato: elytris apice sinuato truncatis, angulo interno acuto, externo longe mucronato. Long. 5½ lin.

Head and antennæ dingy brown, basal joint of the latter flexuous beneath. Thorax wide; disk obtusely tuberculated, sides with the lateral tubercle very prominent and acute; colour dingy olivaceous brown, silky. Elytra rather broad; shoulders prominent, lateral carina proceeding thence, strongly marked and acute, apex rather broadly and transversely sinuate-truncate, the internal angle acute, the external produced into a long tooth or spine; the centro-basal ridge not much raised, but surmounted by a very high crest of hairs; the surface is coarsely punctured only near the base; the colour is the same as that of the thorax, but the posterior half is marbled with light grey. Body beneath silky brownish; legs the same, varied with grey. The basal joint of the tarsi is a little longer than the two following taken together.

Pará; on dead branches in the forest.

6. *Lophopœum cultrifer*, White.

Ægomorphus cultrifer, White, Cat. Long. Col. Brit. Mus. ii. p. 374.

"*Æ.* griseus, fusco variegatus: thorace supra subtuberculato, dorso medio maculis duabus subtriangularibus fuscis, lateribus tuberculo apice acuminato et elevato: scutello fusco-griseo subcincto: elytris singulis basi subtuberculatis, medio tuberculo supra acuto et apice postice producto: elytris singulis apice fasciis duabus fuscis: abdominis segmentis subtus lateribus fusco maculatis: pedibus griseis; femoribus intus fusco punctulatis; tibiis apice late fusco. Long. 6 lin." (White, l. c.)

The elongated form and grey colour of this species give it some resemblance to *Ægomorphus*, in which genus Mr. White placed it; but its depressed body would seem to suggest rather a relationship with the *Oreodera*. The great length and flexuous shape of the basal joint of the antennæ show, however, that its true place is amongst the Acanthocinitæ; and its strongly raised centro-basal ridges and acute lateral thoracic tubercles point out an affinity with the species I have grouped under the genus *Lophopœum*. Its form is elongate-oblong and depressed; the elytra have not very prominent shoulders, and do not taper to the apex; they therefore have not the trigonal shape which is usual in *Lophopœum* and *Alcidion*: as the species of these genera, however, vary in general shape, this is of less importance. The elytra are sinuate-truncate at the apex, and have both angles of the truncation slightly produced; there is no lateral carina, and the dorsal surface, with the exception of the strongly raised and naked centro-basal ridge, is free from raised lines. The thighs are strongly clavate, and the basal joint of the tarsi elongated.

Taken at Pará, and also at Ega on the Upper Amazons; the species has therefore a wide range.

Genus OZINEUS, nov. gen.

Body small, slender, depressed, and posteriorly attenuated. Antennæ as in *Alcidion* and the allied genera. Thorax with the lateral spines short, placed much behind the middle—in some species close to the hind angles, and in others coincident with them, but remaining always distinct. Elytra narrowed to the tips, which are truncated and toothed or spined; the centro-basal ridges prominent, but generally much smaller than in *Lophopœum* and *Alcidion*. Legs moderate in length; thighs abruptly clavate; tarsi slender, with the basal joint elongated.

This genus seems to form a connecting link between *Lopho-pœum* and the well-known group *Anisopodus*. Some of the species are almost as much flattened as the Anisopodi, but their hind legs are never elongated as in *Anisopodus*; the possession of prominent centro-basal ridges on the elytra is also a good distinctive character.

The species are all small and fragile; they are found, like most of those of the allied genera, on the bark of broken and decaying branches of trees in the forest, undergoing their transformations beneath the bark.

1. *Ozineus elongatus*, n. sp.

O. angustatus, elongatus, postice parum attenuatus, carneo-cinereus, ferrugineo-fusco maculatus: thoracis spinis lateralibus pone medium positis. Long. 3¼ lin.

Head dusky: eyes large, nearly touching above; labrum yellow. Antennæ much elongated, capilliform, rusty brown, base of joints pallid. Thorax rather elongated, the lateral spines placed behind the middle, but leaving a considerable space between them and the hind angles; surface punctured, pinkish ashy, with dark-brown spots. Elytra narrow and elongate, tapering posteriorly to the apex, which is briefly truncate, the angles not produced; sides and disk without raised lines, the centro-basal ridges rising in the form of small tubercles crested with hairs; the surface (except the apical part) is punctured, and is of a pinkish-ashy hue, with numerous darker-coloured spots, some of which are collected into a transverse belt a little beyond the middle. Body beneath and legs pale teataceous; thighs with a large dusky spot; base and apex of tibiæ and tips of tarsi also dusky.

Ega. The position of the lateral thoracic spines is almost the same as in *Lophopœum bituberculatum*; the species would therefore seem to belong to the last genus rather than to the present one; but in the general shape of the body it agrees better with the species placed in the group *Oximeus*, the distance of the thoracic spines from the hind angles being probably due to the general elongation of the thorax and the rest of the body.

2. *Oximeus mysticus*, n. sp.

O. subelongatus, depressus, cinereo-fuscus, lineis angulatis canis ornatus: thoracis spinis lateralibus ab angulis posticis paulo distantibus: elytris postice modice attenuatis, apice perobliquo sinuato-truncatis, angulis externis productis, carina centrobasali paulo elevata, elongata, pilis nigris cristata. Long. 3 lin.

Head ashy brown; eyes distant above. Antennæ greatly elongated and hair-like, rusty brown; tips of joints (from the third) blackish. Thorax with the lateral spines placed at a short distance from the hind angles; surface rusty brown, with several curved ashy lines. Elytra depressed, curvilinearly narrowed to the apex; sides with an obtuse carina; tips very obliquely sinuate-truncate, with the outer angles produced; the centrobasal ridge is elongated, very slightly raised, but fringed with black hairs; the surface is moderately punctured in some parts, and is of an ashy-brown hue, with whitish ashy markings, which are in the form of lines near the base, but united in a very oblique angulated belt near the middle, two curved letters remaining close to the suture near the apex. Body beneath and legs pallid; thighs and tibiæ dusky near their apices.

Ega, Upper Amazons.

3. *Oximeus doctus*, n. sp.

O. oblongus, modice depressus, postice rotundato-angustatus, fuliginosus, lineis curvatis albo-cinereis notatus: elytris apice oblique sinuato-truncatis et dentatis, carina centrobasali modice elevata, nigro penicillata. Long. 3 lin.

Head sooty black. Antennæ the same, with the bases of the joints (from the third) pallid. Thorax with the lateral spines placed near the hind angles, sooty brown, with a few whitish specks. Elytra oblong, curvilinearly narrowed to the apex, which is very obliquely truncated, both angles pointed; sides with an obtuse lateral carina; centro-basal ridges slightly raised, fringed with black hairs; the surface thickly punctured nearly to the apex, sooty brown, with a number of ashy-white letter-like marks, some of which are united near the middle of the disk of each elytron so as to form the letter V. Body beneath rust-coloured : legs dusky, base of thighs pale testaceous, the tibiæ in the middle and the basal and claw-joints of the tarsi ringed with pale testaceous.

On dried twigs in the forest at Obydos, on the Guiana side of the Lower Amazons ; also at Pará. A closely allied species, having, however, different markings on the elytra, is found at Cayenne, and exists in several collections*.

4. *Osinous cinerascens*, n. sp.

O. subellipticus, olivaceo-cinereus fusco variegatus : thoracis spinis lateralibus prope angulos posteriores sitis : elytris oblique sinuato-truncatis ; carinis centrobasalibus parvis, convexis, penicillatis : antennis pallide annulatis. Long. 3¼ lin.

Head and thorax olivaceous ashy. Antennæ brown, the third, fourth, sixth, and eighth joints with their basal halves pale testaceous. Thorax partially punctured, its small acute lateral tubercles placed very near the posterior angles, a minute notch only appearing between them and the hind margins. Elytra subelliptical and depressed, the tips obliquely sinuate-truncate ; the centro-basal ridge slightly elevated and crowned with blackish hairs ; the basal half of the surface is punctured, and the colour is the same as that of the head and thorax, but varied with small olive-brown specks placed partly in rows, and an undulated fascia of the same behind the middle interrupted at the suture. Body beneath and legs yellowish testaceous, thinly clothed with ashy pile ; apex of femora and tibiæ and middle part of the tarsi blackish.

Taken on dead slender branches at Santarem and on the banks of the Tapajos. At Villa Nova, on the banks of the Lower Amazons, I obtained one example, which differs considerably in colour from the Tapajos form. This may be called

Local var. *O. pallipes.* Same size and shape as the type. General colour tawny ashy. Antennæ with the basal joint reddish, the remaining joints dusky, with the basal portions of the

---

* *O. strigosus.*—Oblongus, modice depressus, postice rotundato-angustatus. Caput fuliginosum. Antennæ obscure rufescentes, articulis apice nigricantibus. Thorax spinis lateralibus prope angulos posticos sitis, dorso fuliginoso-cinereo notato. Elytra apice peroblique sinuato-truncata, angulis acutis, lateribus obtuse carinatis ; carina centrobasali modice elevata, pilis nigris cristata, punctata, fuliginosa, singulis lineis quinque fusco-cinereis basin band attingentibus et plus minusve confluentibus ornatis. Corpus subtus obscure rufescente. Pedes fuliginosi, femoribus basi rufescentibus, tibiis tarsisque rufescenti annulatis. Long. 3 lin. *Hab.* Cayenne.

third, fourth, sixth, eighth, and following joints clear pale testaceous. The disk of the thorax has several olive-brown spots. The elytra are of a reddish-tawny hue, with tawny-ashy pile; they are spotted as in the typical *O. cinerascens*, but instead of a waved fascia behind the middle, they have simply a short oblique stripe on the disk of each. Body beneath and legs pale testaceous; knees, tips of tibiæ, and middle of the tarsi dusky.

O. cinerascens resembles much in markings a small allied species from Rio Janeiro, which is as yet unnamed in collections*.

Genus ANISOPODUS, White.

White, Cat. Long. Col. Brit. Mus. ii. p. 349.

Syn. *Anisopus*, Serville, Ann. Soc. Ent. Fr. iv. p. 30 (name preoccupied).
Leptostetis, Erichson, Consp. Ins. Peruana, p. 145 (name preoccupied).

Char. emend. Body elongated and extremely flattened. Prothorax even on its upper surface, its lateral spines placed near the hind angles. Elytra oblong-oval, flattened, *without centro-basal ridges;* their apices sinuate-truncate and mucronate, their sides each furnished with a sharp lateral carina extending from the humeral angle to the tip. Thighs abruptly clubbed; the hind legs elongated, in the males of some species excessively so. Ovipositor of the female not apparent.

* *O. ignobilis*, n. sp.—Parvus, subellipticus, olivaceo-cinereus fusco variegatus. Caput fuscum. Antennæ rufo-piceæ, articulis (duobus basalibus exceptis) apice nigris. Thorax olivaceus, spinis lateralibus longioribus acutissimis, prope angulos posteriores sitis. Elytra sub-depressa, apice peroblique sinuato-truncata, angulis externis longe mucronatis, carinis centrobasalibus longe nigro penicillatis; dorso punctata, cinereo-olivaceo, maculis minutis fasciaque lata pone medium fuscis. Corpus subtus nigrum. Pedes nigricantes; femoribus, tibiis tarsisque dimidiis basalibus testaceis. Long. 2 lin. *Hab.* Rio Janeiro. Coll. Bakewell, Bates.

A very pretty species of this genus, captured near Rio Janeiro, by Squires, in some number, is

O. rotundicollis.—Oblongus, latiusculus, depressus, cinereus, griseus vel fulvus, cano fuscoque variegatus. Antennæ rufescentes, articulis (duobus basalibus exceptis) apice nigris. Thorax brevis, latus, lateribus ante medium dilatato-rotundatis, utrinque spina minuta distincta ab angulo postico distante; dorso punctato, nigro bimaculato, lateribus (infra spinam) plaga fusca cum vitta basali elytrorum conjuncto notatis. Elytra subtrigona, depressa, apice peroblique sinuato-truncata, angulis externis longe mucronatis, carinis centrobasalibus brevibus, carinis lateralibus acutis; dorso punctato, griseo vel fulvo, medio late cano fasciato, fascia fusco maculata, postice sinuata et fusco marginata. Corpus subtus testaceum. Pedes rufescentes, tibiis apice tarsisque nigro maculatis; tibiis anticis medio intus tuberculo conico instructis. Long. 3½ lin. *Hab.* Rio Janeiro. Coll. Bakewell, Bates, &c.

The last genus (*Ozineus*) seems to form the connecting link between the groups allied to *Alcidion* and the present genus, some of the species of *Ozineus* (e. g. *O. mysticus* and *O. rotundicollis*) having very much the general appearance of *Anisopodi*. The absence of the centro-basal ridges of the elytra, and the elongation of the hind legs, however, amply distinguish *Anisopodus* from the four preceding genera.

1. *Anisopodus phalangodes*, Erichs.

Leptoscelis phalangodes, Erichson, Consp. Ins. Col. Peruana, p. 145.

A. "oblongus, planus, badius, dense cinereo pubescens, infra lateribus nigro vittatis: elytris seriatim fusco punctatis, apice mucronatis: pedibus posticis fortiter elongatis, femoribus abrupte clavatis. Long. 5½ lin." (Erichs. *l. c.*) Eastern Peru.

This species is distinguished from its congeners, to some of which (*A. arachnoides*, *A. cognatus*) it is very closely allied, by the sides of the breast and the abdomen being marked each with a streak of a sooty-brown hue (extending to the deflexed margin of the elytra), which, from the silky nature of the pile that clothes the under surface of the body, is fainter in some lights than in others, and in small examples is scarcely perceptible. The site of the centro-basal ridges of the elytra is marked by a small rounded tubercle coloured black. The male is much larger than the female, reaching 5½ lines in length, the female being seldom longer than 4½ lines. The hind legs in the male are sometimes 10 lines long. Besides the black spot on the elytra over the centro-basal tubercle, there is, in the males, a larger and irregular spot on the disk of each towards the apex, and in most specimens a small lateral streak on the edge of the lateral carina a little behind the middle; this latter, however, never extends towards the disk of the elytra, as does the similarly placed spot in *A. arachnoides*, *A. cognatus*, and *A. sparsus*. *A. phalangodes* also differs from its relatives in the shape and direction of the lateral thoracic spine, this being large, robust, and prominent, directed obliquely towards the edge of the humeral angle of the elytra, and not standing at right angles to the side of the thorax, as in *A. arachnoides*, or having its point directed in continuation of the thoracic outline, as in *A. cognatus*. Both angles of the truncation of the elytra are mucronate.

The species occurred, sometimes abundantly, on the boughs of fallen trees, in moist hollows of the forest at Ega, Upper Amazons. I could not ascertain the special use of the elongated hind legs of the male. Like all the species of this and the allied genera, the insects pass their lives on the bark, their larvæ feeding and undergoing their transformations between the bark and the wood, and the perfect insects rambling on the outside of the fallen boughs, on which, after copulation, the females deposit their eggs.

In the collection of Mr. Bakewell there is a specimen of this species from Cayenne, differing from Upper-Amazons examples only in the dark-brown points of the elytra being a little more distinct and encircled with grey.

2. *Anisopodus arachnoides*, Serv.

Anisopus arachnoides, Serville, Ann. Soc. Ent. Fr. iv. p. 31.

A. oblongus, griseus, sericeus, fusco punctatus: thoracis spinis latc-ralibus distinctis porrectis: elytris apice modice spinosis; tuberculis centrobasalibus hirsutis; lateribus atro-fuscis, maculis tribus adja-centibus, una pone medium, irregulari, majore: pedibus posticis vix elongatis. Long. 4½–6 lin.

Head brown. Antennæ pitchy red, tips of joints (from the third) dusky. Thorax greyish, with two black spots on the fore part of the disk; sparsely punctate; the lateral spines standing out at right angles from the sides. Elytra transversely sinuate-truncate at the apex, the inner angle of the truncation pointed, the outer moderately prolonged as a spine; the surface is grey, with scattered black points and a black pencil of hairs over the centro-basal tubercles: the deflexed sides are dark brown and silky, and, above, this colour extends in three irregular spots— one near the base, one after the middle reaching to the disk of the elytron, and one smaller near the apex. The underside of the body is silky ashy. The legs are dusky, ringed with grey. The hind legs of the male are only slightly elongated.

The above description applies to a species which I have seen in collections at Paris under the name of *A. arachnoides* of Serville, and to which the description of Serville applies, as far as it goes. I found it on the same trees with *A. phalangodes* at Ega, and also met with it at Pará.

3. *Anisopodus cognatus*, n. sp.

A. subellipticus, carneo-griseus, sericeus, nigro punctatus: thoracis spinis lateralibus unciformibus: elytris absque tuberculis centro-basalibus, apice breviter mucronatis, pone medium macula laterali nigra obliqua notatis: pedibus posticis maris elongatis, fortiter clavatis. Long. 4 lin.

Head brownish. Antennæ pitchy red, tips of the joints dusky. Thorax greyish, fore part of the disk with two black spots, punctured; the lateral spines pointing to the hind angles, their anterior sides (nearest the head) being continuous with the lateral outline of the thorax, and therefore giving them a hooked shape. Elytra sinuate-truncate at the apex, both angles of the trunca-tion pointed, but neither prolonged into a spine; the surface is grey with a pinkish shade, irregularly spotted with blackish, and having each, behind the middle, an oblique spot or fascia extending from the side to the middle; the deflexed margins are of a light silky-brown hue, like the under surface of the body. Legs dusky, base of thighs and tarsi pale testaceous. The hind legs of the male considerably elongated; the thighs in both sexes abruptly clavate.

Ega and S. Paulo, Upper Amazons, in the same situations as, and sometimes in company with *A. phalangodes.*

4. *Anisopodus sparsus,* n. sp.

A. subellipticus, carneo-griseus, sericeus, nigro-punctatus: thoracis spinis lateralibus brevissimis, antice a lateribus vix distinctis: elytris apice oblique sinuato-truncatis, angulis vix productis, absque tuberculis centrobasalibus, pone medium atro-fusco fasciatis. Long. 4½–5½ lin.

Head greyish silky. Antennæ reddish, tips of joints dusky. Thorax pinkish grey, punctured, bimaculate; lateral spines on their anterior sides scarcely distinct from the sides of the thorax, their minute points only being directed outwards; they have, therefore, not the hook-like shape of those of *A. cognatus.* Elytra with the angles of their truncation scarcely produced; their surface is grey, with a pinkish tinge, finely spotted with blackish, and with a largish black spot marking the site of the centro-basal tubercles (which are quite absent), besides an angled fascia of the same hue crossing behind the middle. The deflexed sides and under surface of the body are light brown and silky. The legs are reddish, with the tips of the thighs, tibiæ, and tarsi dusky. The hind legs of the males are greatly elongated, but their thighs are not very abruptly clubbed.

Santarem and Obydos; generally on severed and hanging sipós, or woody climbers, near the borders of clearings. It appears to be not uncommon in Cayenne, examples from which country have been sent to me from Paris as *A. sparsus* of Dejean's catalogue.

5. *Anisopodus pusillus,* n. sp.

A. oblongus, griseus, fusco maculatus: thoracis spinis lateralibus antice a lateribus vix distinctis, retrorsum spectantibus: elytris apice breviter sinuato-truncatis; angulis internis acutis, externis productis; carinis lateralibus obtusis: femoribus modice clavatis. Long. 3 lin.

Head brown. Antennæ reddish, tips of joints dusky. Thorax thinly clothed with grey pile, sides with rufous-brown patches, and disk with two large rounded blackish spots; lateral spines large and acute, scarcely distinct anteriorly from the sides of the thorax, and pointing obliquely towards the humeral angles of the elytra. Elytra oblong, the lateral keels not sharp, and hence the surface apparently less flattened than in the preceding species; apex moderately mucronated; surface greyish, with a moderate number of rather large round brownish spots more or less confluent, one on the site of the centro-basal tubercles (which are quite absent), and another, lateral, near the middle of the elytra, being larger than the others. Body beneath and legs reddish. The legs are rather slender, and the thighs not abruptly, although distinctly clubbed.

This small and delicate species was found only at Pará.

6. *Anisopodus elongatus*, n. sp.

A. elongatus, ellipticus, fulvo-griseus, cano fuscoque punctatus: thoracis spinis lateralibus acutissimis, retrorsum curvatis, basi tumidis: elytris apice utrinque bimucronatis : pedibus posticis femoribus modice clavatis. Long. 5½ lin. ♀.

Head clothed with shining tawny pile. Antennæ reddish, tips of joints dusky. Thorax with the sides rather dilated and tumid at the base (on the fore side) of the lateral spines, which appear curved posteriorly, and have very acute points; disk punctured, tawny brown, with a short polished dorsal line and two discal black spots edged with light grey. Elytra narrow and greatly elongated, both angles of the truncation produced into spines; lateral carinæ rather obtuse, surface punctured, except near the apex, and of a tawny-brown hue, with three rows of alternate whitish and brownish spots, besides a sutural row of brown and grey specks. Body beneath and legs pale brown; hind thighs moderately clubbed in the female.

Found only at Ega, Upper Amazons.

7. *Anisopodus macropus*, n. sp.

A. elongatus, planus, griseus, fusco punctatus : elytris dimidio apicali nigro, cano notato, apicibus utrinque bispinosis : pedibus posticis maris maxime elongatis, tenuibus, femoribus abrupte clavatis. Long. 4¼ lin.

Head dusky. Antennæ rust-coloured. Thorax rather long, narrow in front, sinuated on each side, and then abruptly dilated, the dilatation terminating at the apex of the spine, which is obtuse, and points towards the hind angle; the surface is punctured, and has an impressed dorsal line, the colour being obscure greyish, with four indistinct oblong sooty spots not reaching the hind margin. Elytra elongate, narrow, flattened, although having the lateral keels obtuse ; the apex transversely sinuate-truncate, with both angles spiniform; their surface is covered with equidistant punctures, and is, on the basal half, of a dull pinkish-grey hue, spotted with dark brown, whilst the apical half is dull black, with a few greyish marks. Body beneath clothed with silvery ashy pile. Legs slender; all thighs abruptly clubbed, the fore and middle pair having on their under side near the base a small tooth; they are of a dusky hue, with the base of the femora pallid. The hind legs in the male (the only sex known) are extremely long and slender.

Of this elegant species I found only a single example at S. Paulo, Upper Amazons.

8. *Anisopodus gracillimus*, n. sp.

A. oblongus, gracilis, olivaceus, nigro punctatus: thoracis spinis lateralibus retrorsum curvatis, basi tumidis: elytris apice utrinque bidentatis, pone medium nigro undulato-fasciatis: pedibus posticis (♂) fortiter elongatis, femoribus abrupte incrassatis. Long. 2¾–4 lin.

Head olive-green. Antennæ blackish, base of joints reddish. Thorax with the sides dilated and tumid before the spines, which latter consequently appear curved, and are placed close to the hind angles; the pile of the surface has an olive-green hue, leaving three blackish streaks in the middle—two touching the front margin, and one, lying between the anterior two, the hind margin. Elytra with both angles of the truncation produced, but not to a notable length, the external one longest; lateral keels obtuse; surface olive-green, spotted with blackish, an undulated fascia of the same colour crossing the middle, and two small patches lying on the sides, namely, one near the base, and one near the apex. Body beneath of an olive-ashy tinge. Legs dusky, base of thighs pale. Hind legs of the male greatly elongated; the thighs much more thickly clubbed in the male than in the female.

Taken once abundantly on dried twigs in the forest at Ega.

9. *Anisopodus ligneus*, n. sp.

A. oblongo-ovatus, fulvus, strigosus: thoracis spinis lateralibus conicis, prope angulum posticum sitis: elytris postice valde attenuatis, apice peroblique truncatis, angulis externis mucronatis; femoribus abrupte clavatis. Long. 4¾–5 lin.

Head tawny brown, vertex spotted with dark brown. Antennæ reddish, tips of joints dusky. Thorax rather short, coarsely punctured, the lateral spines conical, and placed very close to the hind angles; colour tawny brown. Elytra rather oval in shape, rapidly attenuated from three-fourths their length to the tip; the tip is consequently pointed, and the truncation so short that each elytron may be said to end in a spine notched on the inner side, instead of being obliquely sinuate-truncate;

the lateral keel is sharply marked, the surface is marked with several (seven or eight) slightly raised lines extending from the base to near the apex, but most of them bent near the base, and with as many corresponding depressed lines between them, the latter of which are thickly punctured, whilst the raised lines are impunctate; the colour is of a tawny-brownish hue, the base being dusky, and the apical third of a deeper tawny hue—the whole giving to the insect a striking resemblance to a chip of wood. Body beneath and legs reddish. Thighs abruptly clubbed; hind legs of the male greatly elongated.

Taken in the forests of the Tapajos and at Ega. Rare.

10. *Anisopodus lignicola*, n. sp.

A. oblongo-ovatus, cinereo-ochraceus, humeris fulvescentibus : thoracis spinis lateralibus magnis, acutis, obliquis : elytris postice valde attenuatis, apice peroblique sinuato-truncatis, subplanis, punctatis. Long. 3 lin.

Head reddish. Antennæ reddish, tips of joints dusky. Thorax ochraceous or yellowish ashy, with two obscure dusky lines on the disk; the lateral spines large, thick, directed obliquely rearwards, the thorax behind the spines being much narrowed. Elytra narrowed to the tips, which are obliquely sinuate-truncate, the inner angles pointed, the outer spiniform; the lateral keels are obtuse, but distinct; the surface is plane, but not notably depressed, minutely punctured, ashy-ochraceous in hue, with obscure spots and oblique fasciæ (on the sides) of a darker colour, a triangular spot on each shoulder, extending over the scutellum, being of a ruddier ochreous tinge. Body beneath and legs of a tawny colour; thighs moderately clavate.

Pará and the banks of the Tapajos.

11. *Anisopodus humeralis*, n. sp.

A. oblongo-ovatus, niger : thoracis lateribus humerisque fulvescentibus; spinis lateralibus magnis, acutis, obliquis. Long. 3 lin.

Head dusky, with the sides bright tawny. Antennæ pitchy red, tips of joints dusky. Thorax with the sides shining tawny, the middle portion dusky, with two more distinct black dorsal stripes; the spines as in *A. lignicola*—namely, large, acute, obliquely directed rearwards, and followed by a narrowing of the thorax to the base. Elytra oval, narrowed to the tips, which are obliquely sinuate-truncate, the inner angles pointed, and the outer spiniform; the lateral keels are indistinct; the surface is closely punctured, the colour sooty black, varied with a few ashy marks, the shoulders having each a triangular tawny spot, which does not cover the scutellum. Body beneath and legs dusky; hind femora (in the ♀) but slightly clavate.

K 2

One example, S. Paulo, Upper Amazons. It is possible, notwithstanding the great difference in colour, that it may be but a local variety of *A. lignicola.*

Two other species of *Anisopodus*, in addition to the eleven here enumerated, have been described, namely, *A. curvilineatus* (White, Brit. Mus. Cat. ii. p. 350, pl. 9. f. 1) of South Brazil, and *L. prolixus* (Erichson, Consp. Ins. Peruana, p. 145) of Eastern Peru. The latter is the largest species at present known, and seems to be closely allied to *A. arachnoides.* I add a description, at the foot, of a fourteenth species*.

Genus LEPTURGES, nov. gen.

Body depressed, oblong, elliptical or elongate, free from irregularities or tubercles on its surface, and clothed with fine, prettily variegated tomentum. Antennæ long and hair-like, sparsely clothed with short, stiff hairs; the basal joint greatly elongated, gradually thickened from the base, the club thus formed being waved or not in its outline beneath; the remaining joints (except the second) very slender. Thorax trapezoidal, depressed, the lateral spines placed close to the hind angles, or at a short distance from them. Elytra free from centro-basal ridges or tubercles, more or less truncated at the tip, except in rare instances, where they are entire. Abdomen with the terminal segment slightly elongated in the females, the dorsal plate obtusely pointed at the tip, the ventral truncated or scarce perceptibly emarginated; in the males the same terminal segment has both its ventral and dorsal plates entire at the tips. Legs moderate in length, the thighs moderately clavate, and the basal joints of the tarsi elongated.

This group, which comprises a large number of small Leiopodine Longicorns of Tropical America, is so closely allied to the European genus *Leiopus* that I have great hesitation in separating it. All the species, however, differ from the European *Leiopus nebulosus* (the type of the genus) in the shape of the thorax, and in the antennæ having very slender and elongated joints more or less clothed with stiff hairs. The thorax has, in nearly all the species, a trapezoidal outline, the lateral spines being placed very near to, or coincident with, the hind angles, the surface depressed, and the sides widening from the head

* *A. canus.*—Oblongus, planus, tomento denso canescente vestitus. Antennæ rufescentes, articulis apice nigris. Thorax punctatus, antice nigro bivittatus, spinis lateralibus tenuibus porrectis. Elytra lateribus parallelis, prope apicem subito attenuata, dorso inæqualia, medio fortiter depressa, carinis lateralibus acutissimis, apicibus longe mucronatis; canescentia, maculis minutis nigris sparsa, quarum duabus distinctioribus prope apicem. Pedes nigricantes, femoribus tibiisque dimidiis basalibus rufis. Femora postica (maris?) elongata, subito clavata. Long. 2¼ lin. *Hab.* Brasilia meridionalis. Coll. Bakewell.

towards the base. In one section, however, the spines are more or less distant from the hind angles, and they then have the acute tips and recurved shape of the thoracic spines of *Leiopus*; so that this character is not wholly to be relied on. The flatness of the thorax and the great slenderness of the antennæ are perhaps distinctive characters of more value. The species are prettily variegated in the hues of the fine pubescence with which they are clothed; and the group, whether treated as a section of *Leiopus* or as an independent genus, appears to me a very natural one [*].

§ 1. Thoracic spines very near to, or coincident with, the hind angles; small, not curved posteriorly.

1. *Lepturges elegantulus*, n. sp.

L. subellipticus, depressus, carneo-fulvus, fusco variegatus : elytris oblique et obtuse truncatis : femoribus posticis vix clavatis, tarsis maxime elongatis. Long. 3¼ lin. ♂.

Head pinkish tawny. Antennæ the same, with the extreme tips of all the joints dusky; they are filiform, or rather stout, and nearly three times the length of the body (♂). Thorax with the lateral spines nearly coincident with the hind angles, porrect

* The genus *Leiopus* is represented by three European species, one only of which (*L. nebulosus*) I have been able to examine. Leconte enumerates several North-American species, and, according to the characters he gives of the genus, these seem to agree generically with the European forms ; but one (*L. angulatus* of Georgia) would appear rather to belong to our new genus *Lepturges*. The chief features enumerated by Leconte as distinguishing *Leiopus* from the many allied genera are—(1) the shortness and conical shape of the ovipositor of the females (to which may be added the uncleft tip of the apical ventral segment which forms part of it), (2) the rounded apex of the dorsal plate of the apical abdominal segment in the males, (3) the naked antennæ, and (4) the elongation of the basal joint of the posterior tarsi. I propose to limit the genus to those species which have, in addition to the above characters, the thorax of quadrate outline and of more or less convex shape, with the lateral spines placed at a distance from the hind angles, long, acute, and curved posteriorly. I did not meet with a single species answering to this definition in the Amazons region : the following, however, found in South-east Brazil, seems to be a true *Leiopus*, with the exception of the antennæ being long and slender, and furnished with stiff hairs :—

L. amoenulus. Oblongus, convexiusculus, tomento carneo-griseo læte variegatus. Caput nigrum, vertice rufo. Antennæ elongatæ, tenues, setiferæ, rufo-piceæ, articulis (duobus basalibus exceptis) apice nigris. Thorax subquadratus, convexus, spinis lateralibus pone medium sitis, acutis, recurvis ; tomento carneo-griseo vestitus, maculis duabus dorsalibus claviformibus nigris. Elytra apice breviter et obtuse truncata, modice convexa, punctata, nigricantia, utrinque plaga irregulari ab humero usque ad apicem extensa grisea, nigro quadrimaculata, apud humeros roseo tincta ornata. Corpus subtus rufo-piceum. Pedes picei, femoribus omnibus valde clavatis. Long. 2 lin. ♂. *Hab.* Rio Janeiro Brasiliæ. Coll. Bakewell, Bates.

or standing out at right angles to the body; surface pinkish grey, with a large irregular brownish blotch in the middle, and a stripe of the same colour on each side beneath, above the sockets of the haunches. Elytra depressed, tapering from base to apex, the latter obtusely and obliquely truncated; base near the scutellum slightly convex; surface punctured, pinkish fulvous or grey, silky, with a few brown spots and patches, namely, one on the convex part near the scutellum; a second, kidney-shaped, on the margin near the humeral angle; a third, behind the middle, extending as a large angulated blotch towards the suture; and a fourth, small and oblique, near the apex. Body beneath and legs pinkish fulvous or grey; front and middle thighs with dusky patches. Hind thighs gradually thickened. Tarsi greatly elongated, the hind pair nearly as long as the tibiæ, the basal joint especially being of excessive length.

This handsome little species was only once met with, namely, flying in the evening twilight on the banks of the river at S. Paulo, Upper Amazons. It differs from all other species of the genus in the length of its tarsi and the slenderness of its hind thighs, in which characters it approaches the genus *Parœcus*; but the depressed form and general facies make it consort better with *Lepturges* than with *Parœcus*.

2. *Lepturges linearis*, n. sp.

L. linearis, fuliginosus : elytris griseo bilineatis. Long. 4 lin. ♂ ♀.

Head sooty, with a shining olivaceous pile. Thorax with the lateral spines placed near the hind angles, and forming each a large acute tubercle separated by an impressed line from the body of the thorax; surface with an impressed dorsal line, sooty, varied with silky greyish-olivaceous pile. Elytra greatly elongated, almost linear, sinuate-truncate at the tips, the external angle of the truncature produced and acute; surface coarsely but somewhat evenly punctured, sooty brown, each elytron with two olive-grey vittæ united before reaching the apex. Scutellum grey. Body beneath and legs clothed with iron-grey pile. Legs rather short; all the thighs clavate; tarsi slender and elongate.

Ega; not uncommon on dry twigs in the forest.

3. *Lepturges flaviceps*, n. sp.

L. elongatus, sublinearis, niger : capite, vittis duabus thoracis, annuloque antennarum flavis. Long. 4 lin. ♀.

Head shining testaceous yellow, with two black vittæ extending from the front of the eyes to the occiput. Antennæ twice the length of the body, the basal joint very greatly elongated; black, with the basal half of the fourth joint pale yellow. Thorax with the lateral spines placed near to the hind angle; surface behind with a transverse depression, testaceous yellow, silky, with a broad black vitta in the middle, and another still broader on each side above the sockets of the haunches. Elytra elongated, depressed, with the sides nearly parallel, obliquely truncated at the apex, with the external angle of the truncature produced and acute; surface thickly punctured, sooty black, with an indistinct pale streak in the middle of the base on each side. Body beneath black, with the exception of the pro- and mesosterna, which are yellow. Legs black, coxæ and base of thighs yellow; thighs slenderly clavate; tarsi slender, the basal joint elongated.

One example, taken at Pará.

4. *Lepturges complanatus*, n. sp.

L. oblongus, depressus, carneo-griseus, fusco maculatus: thoracis spinis lateralibus angulos posticos constituentibus; elytris apice singulatim rotundatis. Long. 3¼ lin. ♂.

Head black. Antennæ reddish, with the extreme tips of all the joints dusky. Thorax blackish, clothed with ashy changeable pile, the lateral spines coincident with the posterior angles. Elytra oblong, broadly rounded at the tips, plane above and thickly punctured, with a slight indication of two longitudinal raised lines on each, pinkish grey in colour, with five dark brown spots or patches,—namely, one, minute, under the humeral angle; one, linear-oblique, in the middle of the base; a third, subtriangular, on the side near the base; a fourth extending as a broad irregular fascia nearly to the suture; and a fifth, wedge-shaped, near the apex. Body beneath and legs dusky. Front and middle thighs thickly clavate; hind thighs more slender. Tarsi moderately elongated.

One example, taken at S. Paulo, Upper Amazons, flying in the evening.

5. *Lepturges amabilis*, n. sp.

L. oblongus, depressus, griseus: thoracis spinis lateralibus prope angulos posticos sitis: elytris griseo nigroque læte variegatis, apice breviter oblique truncatis. Long. 3¼ lin. ♂ ♀.

Head sooty black. Antennæ greatly elongated, pitchy black. Thorax grey; the disk occupied by two large square black spots, which leave only a central line and the margins of the ground-colour; the lateral spines are prominent and porrect, a small space only intervening between them and the hind margin. Elytra oblong, very slightly narrowing towards the apex, which latter is obliquely and briefly truncated; the surface is slightly depressed and closely punctured; the colour is clear grey, with (on each) four black spots,—namely, one, oval-oblique, in the middle of the base; a second, elongate-bilobed, on the side near the base; a third extending as a broad fascia behind the middle to the suture; and a fourth, small and transverse, near the apex: the lateral spots are sometimes united on the extreme margin. Body beneath and legs clothed with grey pile; club of hind femora slender.

Ega; on dry twigs in the forest.

6. *Lepturges inscriptus*, n. sp.

L. oblongus, subdepressus, griseus, fusco laete variegatus: thoracis spinis lateralibus prope angulos posticos sitis: elytris oblongo-ovatis, apice sinuato-truncatis, griseis, plagis maculisque fuscis notatis. Long. 3¼ lin. ♀.

Head reddish brown. Antennae reddish, extreme tips of the joints (from the third) black. Thorax regularly widened posteriorly; the spines situated near the hind angles, very acute and directed obliquely outwards; the surface finely punctured, greyish, with the sides and two dorsal vittae brownish. Elytra oblong-ovate, briefly sinuate-truncate, with the angles obtuse; the surface finely punctured, slightly convex, grey, with various patches of a reddish-brown hue,—the patches consisting of a spot in the middle of the base, an elongate hooked spot on the side at the base, a lateral twin spot behind the middle, a V-like spot in the middle near the suture, and an oblique zigzag fascia between these latter and the apex. Body beneath and legs reddish brown. Legs rather slender; all the thighs slenderly clavate.

S. Paulo, Upper Amazons.

7. *Lepturges candicans*, n. sp.

L. oblongus, subdepressus, canescens: elytris pone medium fasciis duabus fuscis ad suturam convergentibus ornatis. Long. 3¼ lin. ♀.

Head clothed with hoary pile. Antennae reddish, clothed with hoary pile. Thorax not much widened posteriorly, the spines placed near the posterior angles, the disk with a few scattered punctures, and clothed uniformly with hoary pile. Elytra oblong, the sides rounded and rather enlarged behind the middle, the tip sinuate-truncate, both angles slightly produced and acute; the surface punctured, hoary, with two irregular brown fasciae behind the middle converging on the suture; the anterior fascia is broken towards the sides, and a narrow line connects the two in the middle; besides these fasciae, there is a V-shaped brown mark in the middle of the base on each elytron and a streak on each side from the base to the middle. Body beneath and legs reddish, clothed with hoary pile. Thighs slenderly clavate.

Ega.

8. *Lepturges venustus*, n. sp.

L. subelongatus, griseus: thorace supra nigro trivittato: elytris maculis vittisque nigris. Long. 2¼–3¼ lin. ♂ ♀.

Head greyish. Antennæ black. Thorax grey or light brown, the upper surface having three broad and regular black stripes, and the sides each having a similar stripe above the insertion of the coxæ; the lateral spines placed close to the hind angles, short, obtuse. Elytra oblong, rounded, and somewhat widened behind the middle in the ♀, shorter and more tapering in the ♂, broadly sinuate-truncate at the apex, outer angle of the truncature produced into a tooth in the ♀, both angles produced and acute in the ♂; upper surface with punctures scarcely apparent through the tomentum, grey or light brown, with a black vitta over the suture, dilated about the middle and narrowed towards the apex, a similar vitta, of more equal breadth, on each side, beginning at the shoulder, detached from the margin of the elytron at one-third its length, and ending in a curve before the apex, and two elongate black spots in the middle of each elytron —one near the base and one behind the middle. Body beneath and legs blackish, with grey pile.

Ega and Pará, on dried twigs and branches. The lateral vitta of the elytra is sometimes interrupted near its termination, leaving a detached spot on the disk near the apex.

9. *Lepturges dilectus*, n. sp.

L. oblongus, depressus: thoracis spinis lateralibus magnis, acutis, subporrectis: elytris profunde sinuato-truncatis, fuscis, plaga communi irregulari ante medium maculisque posticis griseis. Long. 3¼ lin. ♂.

Head brown. Thorax brown, with grey pile, punctured on the disk and hind margin; lateral spines placed near to the hind angles, large, prominent, and acute, standing out somewhat from the sides of the thorax. Elytra deeply sinuate-truncate, both angles of the truncature produced and acute, the outer ones most so; surface closely punctured, brown, with a large common grey patch about the middle, which emits short lines towards the base and apex; at the base, on each side the scutellum, there is a small round grey spot, and behind the large grey patch there are, on each elytron, two short grey lines, followed by a transverse grey streak connected with the suture near the apex. Body beneath and legs reddish, with grey pile.

Ega, on dead branches. There is a Cayenne species* resem-

* *L. Barii*, n. sp. Oblongus, depressus. Caput et antennæ rufescentes. Thorax rufescens, griseo-sericeus, disco et margine posteriore punctatis; spinis lateralibus prope angulos posticos sitis, parvis, subporrectis, acutis. Elytra apice sinuato-truncata, angulis truncaturæ pro-

---

bling the present one greatly in markings, but differing in the smaller size of the thoracic spines and in other minor features.

10. *Lepturges perelegans*, n. sp.

L. parvus, oblongo-ovatus, griseus: thorace lituris duabus nigris : elytris sinuato-truncatis, griseo fuscoque lituratis. Long. 2¾ lin. ♀.

Head blackish. Antennæ pitchy red. Thorax grey, with two black vittæ on the disk, and an oblique spot of the same colour on each side; lateral spines short, not distinct anteriorly from the outline of the thorax. Elytra sinuate-truncate, angles of the truncature not produced, surface punctured, grey, with several flexuous black bands and spots,—namely, one basal, S-shaped, extending from the shoulder to the suture; a second, in the form of a large spot, on the side; a third extending as a broad zigzag belt across the elytra behind the middle; and a fourth, comma-shaped, near the apex. Body beneath and legs dusky.

One example; S. Paulo, Upper Amazons.

11. *Lepturges lineaticollis*, n. sp.

L. parvus, oblongus, minus depressus, griseus : thorace supra nigro quinquelineato : elytris nigro lineatis et plagiatis, apice sinuato-truncatis. Long. 2 lin. ♂.

Head greyish, vertex with two dusky stripes. Antennæ pitchy red. Thorax with the lateral spines short and conical, placed at a short distance from the hind angles; grey, with five black vittæ, the middle one much the broadest; the sides of the thorax are also blackish. Elytra oblong-ovate, narrowed near the apex (♂), sinuate-truncate, angles of the truncature slightly produced; surface slightly convex, punctured, greyish, with irregular black patches near the base; middle, sides, and apex partially connected with each other by indistinct lines of the same colour. Body beneath and legs dusky.

Santarem; on dried twigs.

12. *Lepturges fragillimus*, n. sp.

L. parvus, oblongus, minus depressus, griseus: thorace supra fusco bivittato : elytris maculis circa septem fuscis ornatis, apice leviter sinuato-truncatis. Long. 2¼ lin. ♀.

Head dusky or reddish. Antennæ very long and thin, reddish; apical halves of the joints (including the basal one) blackish. Thorax grey; disk with two black vittæ; sides deep black, with

ductis et acutis; dorso punctata, rufescentia vel brunnea, ante medium fascia antice posticeque dentata latera haud attingente, pone discum punctum minutum et prope apicem ad suturam macula unciformi griseo-albis. Corpus subtus rufoscens, griseo tomentosum. Pedes pallidiores. Long. 3¼ lin. ♀. *Hab.* Cayenne. Dom. Bar legit.

silky grey pile; lateral spines placed close to the hind angles, acute. Elytra oblong, rounded on the sides; apex sinuate-truncate, angles of truncature not produced; surface grey, with (on each) about seven angular blackish spots,—namely, one under the shoulder; a second, oblique, near the scutellum; a third, of large size, on the side near the middle; a fourth, elongated, near the apex; and, finally, three, more or less contiguous, on the disk behind the middle: some of the spots are partially confluent in some examples. Body beneath and legs reddish, clothed with grey pile.

Santarem, on dry twigs.

13. *Lepturges pulchellus*, n. sp.

L. parvus, elongatus, carneo-griseus: thorace supra fusco trivittato: elytris maculis magnis fuscis, apice late sinuato-truncatis. Long. 2 lin. ♀.

Head clothed with changeable grey pile. Antennæ dusky. Thorax pinkish or tawny grey, with a broad black vitta in the middle and one on each side, the latter varying in hue according to the light; lateral spines placed close to the hind angles, very small and obtuse. Elytra oblong, rather narrow, broadly sinuate-truncate, the external angles of the truncature produced; surface punctured, pinkish or tawny grey, with a large dusky spot close to the scutellum, a second, larger and rounded behind the middle, near the suture, and a third, smaller, near the apex; the sides also, except near the apex, occupied by an elongate stripe or spot of a dusky colour. Body beneath and legs tawny grey.

Santarem; on dry twigs in the woods.

14. *Lepturges delicatus*, n. sp.

L. parvus, oblongus, depressus, griseus: thorace vittis duabus rufescentibus: elytris punctis numerosis rufescentibus, utrinque macula magna posteriore nigra. Long. 2 lin. ♂ ♀.

Head reddish, clothed with grey pile. Antennæ reddish testaceous, each joint from the third tipped with black. Thorax reddish testaceous, clothed with grey pile, and with two abbreviated vittæ on the disk of a darker reddish-brown colour; lateral spines distinct, acute, placed very near the hind angles. Elytra oblong-oval, depressed; apex obliquely sinuate-truncate, both angles of the truncature produced; surface punctured, grey, sprinkled with brownish-red spots, and having on each elytron behind the middle a large black spot extending from the side to the disk: in some specimens there is also a dusky spot on the side towards the base. Body beneath and legs reddish testaceous; tips of tibiæ and tarsi black.

Upper and Lower Amazons, at S. Paulo and Santarem.

15. *Lepturges musculus*, n. sp.

L. parvus, oblongo-ovatus, minus depressus, postice apicem versus rotundato-attenuatus, fuliginosus: elytris obscure griseis, punctis fuliginosis sparsis: corpore subtus rufo. Long. 2¼ lin. ♂.

Head blackish; labrum hirsute. Antennæ dull black. Thorax sooty black, with obscure greyish pile, which leaves two abbreviated oblique vittæ on the disk, of the sooty ground-colour; lateral spines placed near the hind angles, short, porrect, or standing out from the sides of the thorax. Elytra oval, apex briefly and obliquely sinuate-truncate; surface dull grey, sprinkled with small soot-coloured spots, some of which unite to form patches; in some specimens there is also a whitish speck on the side of each elytron near the middle. Body beneath, coxæ, and base of the thighs reddish; legs dusky. Tarsi shorter than usual in this genus; the basal joint of the hind foot, however, is as long as the two following taken together.

S. Paulo, Upper Amazons; flying, in the evening, on the banks of the river.

16. *Lepturges deliciolus*, n. sp.

L. parvus, oblongus, carneo-griseus, fusco læte variegatus: antennis pedibusque testaceis, nigro maculatis. Long. 1¾ lin. ♂.

Head dusky or reddish. Antennæ reddish testaceous, tips of the joints (from the third) blackish. Thorax reddish (black on the sides), clothed with pinkish or tawny-grey pile, and varied with four arcuated streaks or vittæ of a reddish-brown hue; lateral spines placed a short distance from the hind angles, and bent posteriorly, as in *Leiopus*. Elytra oblong, narrowed behind towards the apex, briefly sinuate-truncate, angles of the truncature slightly prominent; surface punctured, pinkish or tawny grey, varied with numerous reddish-brown spots,—namely, one, angular, over the shoulder; a second, transverse, near the suture behind the scutellum; three, oblong-linear, in an oblique row across the elytron before the middle; a sixth, N-shaped, on the disk behind the middle; and a seventh, minute, near the apex. Body beneath and legs reddish testaceous; tips of femora, tibiæ, and tarsi dusky.

This very pretty little species occurred only at Santarem, on dry twigs on the borders of woods.

17. *Lepturges angustatus*, n. sp.

L. parvus, angustatus, postice attenuatus, nigricans: elytris maculis linearibus obscure griseis. Long. 2¼ lin. ♂.

Head and antennæ black. Thorax black, with obscure grey pile and a faint grey dorsal line; lateral spines acute, placed almost coincident with the hind angles. Elytra elongated, narrowed towards the apex, briefly sinuate-truncate, outer angle of the truncature much produced; surface punctured, black, clothed with olivaceous-sooty pile, and varied with a few short grey streaks arranged in lines from base to apex. Body beneath and legs pitchy black, clothed with dull greyish pile.

Ega.

18. *Lepturges inops*, n. sp.

L. parvus, angustatus, depressus, obscure rufescens : elytris griseo lituratis, apice truncatis : femoribus posticis vix clavatis. Long. 2¼ lin. ♂.

Head reddish, with scanty grey pile. Antennæ dull reddish. Thorax reddish, with a dusky tinge, and scanty silky grey pile ; sides reddish; lateral spines large, pointing backwards, and situated close to the hind angles. Elytra narrow, slightly widening towards two-thirds their length, truncated at the apex, with the outer angles of the truncature slightly produced; surface punctured, dull reddish, dusky on the sides near the base, variegated with dull greyish marks, there being a line on each side of the scutellum, an irregular, elongate, flexuous spot extending from the base to the middle near the suture, a small spot on the disk near the termination of the before-mentioned streak, and three oblong spots in a transverse row behind the middle of the disk ; besides these marks, the suture near the apex and the apex itself of the elytra are bordered with dull grey. Body beneath and legs dull testaceous red. Legs feeble; hind thighs scarcely clavate.

S. Paulo, Upper Amazons*.

19. *Lepturges griseostriatus*, n. sp.

L. oblongus, postice attenuatus, fuscus : elytris rufescenti-fuscis, utrinque lineis griseis octo, quarum tribus interioribus postice interruptis, notatis : pedibus validis, femoribus fortiter clavatis ; tarsis posticis maxime elongatis. Long. 3¼ lin. ♂.

Head and antennæ dull reddish. Thorax above blackish, thinly clothed with hoary pile; the disk with a few punctures ; lateral spines large and thick, placed very near to the hind

---

* A species inhabiting South-east Brazil closely resembles *L. inops* in general appearance and markings ; the following is a description of it :—

L. miser. Parvus, oblongus, subangustatus, depressus, obscure fuscus, griseo variegatus. Caput nigricans, tomento fulvo vestitum. Antennæ tenues, parce setosæ, rufescentes, articulo basali piceo, reliquis apice obscuris. Thorax nigricans, griseo parce tomentosus ; spinis lateralibus parvis, acutis, paulo ante basin sitis. Elytra apice integra, dorso punctata, obscure fusca ; fascia valde dentata ante medium liturisque subapicalibus griseis. Corpus subtus pedesque nigro-picea ; femoribus omnibus clavatis. Long. 2 lin. ♂. *Hab.* Rio Janeiro. Coll. Bakewell.

angles. Elytra rather elongate, narrowed from base to apex, sinuate-truncate, both angles of the truncature slightly produced; surface feebly convex, punctured, light brown, each with eight longitudinal lines (besides a short one near the scutellum) of an ashy-grey colour; the second, third, and fourth from the suture interrupted a little beyond the middle of the elytron, and leaving a considerable space free from lines; towards the apex these three lines are represented by a thick streak. Body beneath and legs dull reddish, clothed with ashy pile. The legs are rather long and stout, the thighs thickly clubbed, the hind tarsi greatly elongated, especially the basal joint, which is much longer than the remaining three taken together.

Forests of the Cupari, River Tapajos.

20. *Lepturges alboscriptus*, n. sp.

L. oblongo-ovatus, niger: elytris utrinque linea arcuata lineolisque duabus albis ornatis. Long. 3½ lin. ♀.

Head black, with a few silvery-grey hairs. Antennæ black, furnished with numerous bristles. Thorax black, with patches of silvery-grey pile; surface sparingly punctured; lateral spines prominent and acute, placed very near the hind angles. Elytra oblong-ovate, slightly convex, very briefly and obtusely truncated; surface punctured; each elytron with a distinct white line extending from the shoulder to the suture behind the middle, and then sharply bent, terminating on the lateral margin; besides this line, there are two short white streaks placed transversely,—namely, one on the side, at one-third the length of the elytron, and the other very near the apex. Body beneath and legs dusky. Legs moderately stout; hind tarsi moderately elongated.

One example, taken at Caripí, near Pará.

21. *Lepturges dulcissimus*, n. sp.

L. oblongus, depressus, testaceo-flavus: capite nigro, lineola flava: elytris fulvo-griseis, apice flavis; marginibus, sutura fasciaque subapicali nigris. Long. 3½ lin. ♀.

Head deep shining black; labrum and a short and broad line on the crown yellow. Antennæ black. Thorax testaceous yellow; disk clothed with rich golden pile; lateral spines reduced to mere tubercles, and placed near to the hind angles. Elytra oblong, slightly narrowed near the tip, depressed, broadly truncated, outer angle of the truncature slightly produced; surface punctured, clear tawny grey, with the sutura, lateral margins, and a fascia near the tip deep black; the apical space behind the fascia yellow. Body beneath reddish testaceous, except the tip of the terminal abdominal segment, which is shining black. Legs shining black; basal halves of the femora reddish testaceous. Thighs all somewhat abruptly clavate; basal joint of the hind tarsi moderately elongated.

I met with only one example of this charming species. S. Paulo, Upper Amazons.

§ 2. Thoracic spines placed at a distance from the hind angles: large, acute, curved posteriorly.

22. *Lepturges dorcadioides*, White.

Leiopus dorcadioides, White, Cat. Long. Col. Brit. Mus. ii. p. 382.

" *L.* punctulatus, brunneus, cano sublineatus: capite inter antennas linea impressa transversa et linea longitudinali ab ore ad verticem currente; oculis supra distantibus: thorace cinereo, fusco punctulato, vittis duabus medianis antice approximatis; scutello cinereo: elytris singulis apice oblique abruptis; margine, sutura et lineolis abbreviatis cinereis." (White, *l. c.*) Long. 3½ lin.

The lateral spines of the thorax are placed at some distance from the hind angles, and are long, acute, and directed obliquely outwards with a slight curve. The elytra are obliquely truncated in a waved line, and the external angle of the truncature forms a small tooth directed outwards; their colour would be better described as hoary or ashy, with (on each side) a broad, irregular, arcuated, blackish vitta, extending from near the scutellum to three-fourths the length of the elytra, and followed by a small, angular subapical spot of the same colour. The legs and antennæ are of the same shape as those of the many allied species.

Ega and Pará. In my own Collection and that of the British Museum.

23. *Lepturges obscurellus*, n. sp.

L. parvus, elongatulus, fuliginosus: elytris griseis utrinque medio macula magna triangulari nigricante, apice sinuato-truncatis, angulis obtusis. Long. 2¼ lin. ♂.

Head blackish. Antennæ reddish. Thorax dusky, with obscure grey pile; the lateral spines placed a short distance from the hind angles, acute, and directed posteriorly. Elytra oblong, apex briefly sinuate-truncate, angles of the truncature not produced; surface punctured, dull grey, with (on each side in the middle) a large triangular blackish spot, whose apex touches on the suture the apex of the corresponding spot on the other elytron. Body beneath and legs dull pitchy red, shining.

Ega.

24. *Lepturges minutissimus*, n. sp.

L. minutus, oblongus, rufescens, tomento rufescenti-griseo variegatus: elytris apice integris. Long. 1¼ lin. ♂.

Head rust-coloured. Antennæ twice the length of the body, reddish testaceous, naked. Thorax rusty red, clothed with dull grey pile, leaving the sides and two dorsal vittæ of the rusty ground-colour; lateral spines placed a little behind the middle, large, acute, slightly curved posteriorly. Elytra elongate-ovate, convex, entire at the apex; surface coarsely punctured, rusty red, clothed partially with dull grey pile, leaving the region of the scutellum, two short basal vittæ, and an irregular dentated fascia behind the middle, of the ruddy ground-colour. Body beneath and legs testaceous red. Thighs all clavate; basal joint of the posterior tarsi moderately elongated.

Santarem; on dry twigs*.

Genus PAROECUS, nov. gen.

Body elliptical, narrowed equally anteriorly and posteriorly, and slightly convex. Antennæ stout, filiform rather than setaceous, greatly elongated, two and a half times the length of the body in both sexes. Thorax of trapezoidal outline; lateral spines thick and conical, placed close to the hind angles. Elytra without prominences on the surface, apex of each sinuate-truncate and bispinose. Legs rather long and stout; front and middle thighs thickly clavate; hind thighs gradually thickened from base to apex; hind tarsi greatly elongated, the basal joint longer

---

* The following species also belong to section 2 of this genus:—

L. spinifer. Elongatus, modice depressus, cinereus, brunneo lineolatus et maculatus. Caput brunneum, oculis postice cinereo marginatis. Antennæ testaceæ, tomento cinereo parce vestitæ. Thorax cinereus, dorso maculis duabus brunneis cinereo marginatis; spinis lateralibus magnis, acutis, retrorsum oblique spectantibus, basi cinereis. Elytra angustata, apice peroblique et obtuse breviter truncata; dorso punctata, cinerea, vittis abbreviatis basalibus quatuor pallide brunneis, maculis et fascia irregulari pone medium obscurioribus. Corpus subtus et pedes testacea, tomento cinereo parce vestita; femoribus omnibus clavatis, tarsis posticis elongatis. Long. 2–2¼ lin. ♂. *Hab.* Rio Janeiro. Coll. Bakewell, Bates.

L. humilis. Oblongus, postice paulo ampliatus, deinde apicem versus attenuatus, fuliginosus, cinereo lineatus et fasciatus. Caput piceum, vertice linea cinerea. Antennæ rufo-piceæ. Thorax fuliginosus, dorso cinereo trilineatus; spinis lateralibus grossis, minus acutis. Elytra thorace latiora, pone medium paulo ampliata, apice vix truncata; dorso convexiuscula, punctata, fasciis duabus e maculis oblongis obscure cinereis, una ante, altera pone medium. Scutellum cinereum. Corpus subtus et pedes rufo-piceæ; femoribus omnibus clavatis; articulo primo tarsorum posticorum modice elongato. Long. 2–2¼ lin. ♂♀. *Hab.* Rio Janeiro. Coll. Bakewell, Bates.

than the three remaining taken together. Ovipositor of the female elongated (1½ line long), tubular; dorsal plate of the terminal abdominal segment pointed, ventral plate notched; ventral plate of the same segment in the males notched or sinuated, dorsal plate entire or sinuated.

The general appearance of the two species which I place in this genus resembles that of the *Anisopodi* and of the larger species of *Lepturges*; but the thickness of the antennæ and the length of the ovipositor of the females forbid their being associated with either genus.

1. *Parœcus ellipticus*, n. sp.

P. ellipticus, tomento carneo-cinereo vestitus: elytris plaga magna communi irregulari subtriangulari maculisque posticis nonnullis adjacentibus. Long. 3½–5 lin, ♂ ♀.

Head clothed with ashy-fulvous pile; forehead dusky. Antennæ reddish ashy; tips of most of the joints slightly thickened. Thorax clothed with pinkish-ashy pile, sparingly punctured on the disk and hind margin; lateral spines conical, oblique, placed very near to the hind angles, and separated from the body of the thorax by a deep fovea. Elytra sinuate-truncate at the tip, both angles of the truncature produced into a short spine; surface faintly punctured, thickly clad with pinkish-ashy changeable tomentum, and having a large, common, dark brown blotch of irregular triangular shape, the apex of which touches the scutellum, and the base (behind the middle of the elytra) broken into two or more elongate spots followed by an oblique spot (on each elytron) of the same hue lying nearer to the apex. Body beneath reddish; sides of breast dusky. Legs dull reddish, sparsely clothed with ashy pile. Apical ventral segment in the males deeply notched, dorsal entire.

Fonte Boa, Upper Amazons, on fallen trunks of gigantic trees of the order Leguminosæ. The pupæ were found in numbers, lying in oval chambers formed by the larvæ between the bark and the wood.

2. *Parœcus rigidus*, n. sp.

P. oblongo-ellipticus, parum convexus, tomento cinereo vestitus: thorace fusco notato: elytris lateribus fuscis, cinereo maculatis. Long. 4½ lin. ♂.

Head clothed with ashy-fulvous pile, forehead dusky. Antennæ reddish, clothed with ashy pile. Thorax rather strongly punctured on the disk; lateral spines conical, oblique, placed very near the hind angles; ashy, varied with small, oblong fuscous spots, two of which form an interrupted vitta on each side of the dorsal line. Elytra strongly sinuate-truncate at the tip, both angles of the truncature produced into spines, the external

L

one very long; surface punctured, ashy, the sides occupied by a dark-brown streak or elongate patch, of very irregular outline and broken throughout with short spots and lines of the ashy ground-colour of the elytra. Body beneath clothed with ashy pile. Legs reddish; hind tibiæ with rather long apical spurs. Ega.

Genus Baryssinus, nov. gen.

Body oblong, convex. Antennæ stout, furnished sparingly with setæ beneath. Thorax somewhat short and broad, widening from the front to the tips of the lateral spines, which are very thick, and placed near to the hind angles. Elytra furnished with centro-basal tubercles, surmounted each by a pencil of hairs; the rest of their surface naked; apices scarcely perceptibly truncated. Apical abdominal segments in the male short and obtuse, in the female slightly prolonged, so as to form a short sheath for the ovipositor, the dorsal plate being flattened and obtuse, the ventral bluntly truncated. Mesosternum depressed, not tuberculated. Legs stout; thighs clavate; basal joint of the tarsi short, not surpassing in length the second and third taken together.

This genus, which comprises a few small species resembling *Trypanidius* in facies, has some affinity with *Leptostylus*. We are therefore, after pursuing the line of affinities which leads through a series of depressed forms of Leiopodinæ from *Alcidion* to *Parœcus*, brought back again to the starting-point,—the present genus commencing a suite of genera of more convex form of body. The presence of hairy-crested centro-basal ridges or tubercles distinguishes *Baryssinus* from all the genera which follow, whilst the existence of a prominent ovipositor in the females, and the shape of the thorax, with the position of its lateral spines, separate it from *Leptostylus* and the allied groups.

1. *Baryssinus penicillatus*, n. sp.

B. oblongus, cinereo-brunneus, fusco obscure variegatus: thoracis dorso antice tumido: elytris utrinque tricostatis, apice rotundatis. Long. 4 lin. ♂.

Head ash-coloured. Antennæ stout, one and a half times the length of the body (♂), stout, setose beneath, ashy testaceous, tips of the joints (from the third) dusky. Thorax with the anterior part of its disk rising into a large obtusely conical elevation; lateral spines stout and curving posteriorly; surface ashy brown, with indistinct darker brown markings. Elytra oblong-quadrate, being but slightly narrowed to the tips, which are broadly rounded; the disk of each has three faintly marked ribs which do not reach either the base or the apex; centro-basal tubercles each with a thin pencil of black hairs; surface, to the tips, covered with largish punctures, ashy brown, with blackish-brown markings, which form two fasciæ beyond the middle, the anterior one oblique, the posterior one forming a curve on each elytron. Body beneath ashy. Legs pale reddish, clothed with tomentum, which forms rings alternately of an ashy and brown hue.

Beaten from dead branches; woods near Santarem.

2. *Baryssinus bilineatus*, n. sp.

B. oblongus, rufescenti-brunneus, cano fuscoque variegatus: thorace nigro birittato: elytris apice breviter obtuse truncatis, pone medium vittis duabus abbreviatis nigris. Long. 4½ lin. ♀.

Head ash-coloured. Antennæ twice the length of the body (♀), setose beneath, ashy reddish, with the tips of the joints (including the first) blackish. Thorax regularly and moderately convex; hind margin with a single row of punctures; surface smooth, and ornamented with two blackish stripes, a line of the same colour also encircling (above) the bases of the lateral tubercles. Elytra oblong, very slightly narrowed to the tips, which appear rounded, but are seen, on close examination, to be obtusely truncate; surface free from raised lines, with the exception of the centro-basal ridges, which are slightly elevated, but crested with black hairs; the basal part only of the elytra is punctured; the colour is reddish brown, with a light grey tinge near the middle and apex, and a number of small blackish spots, besides a short vitta, on the disk of each behind the middle, which has at each end a whitish spot. Body beneath pale reddish, clothed with ashy-brown tomentum; legs reddish, ringed with ashy and black.

Taken at Ega, on dead branches.

Genus CHÆTANES, nov. gen.

Body oblong, convex, setose. Antennæ one and a half times the length of the body, stout, furnished with a few setæ beneath. Thorax rather narrow, widening slightly from the front; lateral spines distinct, acute, standing out from the sides, and placed at a distance from the hind angles. Elytra with centro-basal tubercles, surmounted each by a crest of hairs; the rest of the surface hispid, with tufts of short bristles and longer setæ; truncated at the tip. Legs stout; thighs abruptly clavate; basal joint of tarsi equal in length to the second and third taken together. Apical abdominal segment in the males with both dorsal and ventral plates deeply notched, the angles of the ventral plate produced into spines: ovipositor elongated in the ♀;

dorsal plate acute lanceolate, ventral truncated. Mesosternum in the ♂ plane, in the ♀ tumid, as in *Trypanidius*.

The only species which I have at present seen belonging to this genus has the bulk and general form of the *Trypanidii*; but it differs from them by the presence of crested centro-basal tubercles on the elytra, and by the absence of tubercles on the mesosternum in the male sex.

Chætanes setiger, n. sp.

C. oblongo-ovatus, fuscus, fulvo cinereoque varius: antennis pedibusque squamis cinereis sparsis: elytris apice utrinque macula triangulari nigra velutina liturisque cinereis. Long. 5–6½ lin. ♂ ♀.

Head dull black, sprinkled with tawny-coloured hair-scales. Antennæ scarcely one and a half times the length of the body, even in the males, dull black, sprinkled with minute ash-coloured hair-scales; the bases of the joints (from the fourth) ashy; sparingly setose beneath, the second joint having a little tuft of stiff hairs. Thorax moderately convex, widened from the front to the bases of the lateral spines, which are small and acute, and placed at a short distance from the hind angles; the surface dull black, variegated with tawny; sides (below the spines) ashy-tawny, sprinkled with black. Elytra oblong oval, briefly and rather obliquely truncated at the apex, moderately convex; surface densely clothed with short, erect, black bristles, some of which arise from a little tuft of shorter bristles; moderately punctured; centro-basal tubercles surmounted each by a rather long pencil of hairs: the colour is blackish brown, with a few tawny specks; behind the middle is a short transverse ash-coloured line crossing the suture (in some examples almost obliterated), and close to the apex on each side is a triangular velvety-black spot, notched on its inner side and margined with ashy, the sutural space between the spots being sometimes wholly ash-coloured. Body beneath and legs tawny ashy, sprinkled with black; middle of abdomen black, with edges of segments tawny. The legs are stout; the thighs clavate, the basal joint of the tarsi fully equal in length to the two following taken together.

♂ The apical ventral segment in the male is semicircularly notched, the dorsal segment briefly and obtusely notched.

♀ The apical dorsal segment in the female is much elongated, lanceolate and acute, but not keeled above; the ventral segment semitubular and truncated at the tip.

Ega and S. Paulo, Upper Amazons, on dead branches in the forest. I have a specimen also from the interior of French Guiana, collected by M. Bar.

Genus ATRYPANIUS, nov. gen.

Body oblong-oval or elliptical, convex. Head with the front elongated; eyes oblong. Antennæ not much longer than the body, and nearly naked. Thorax as in *Trypanidius*—namely, slightly uneven on the surface, widening from the front to the tips of the lateral spines—which are short, conical, and acute, not curved posteriorly, and placed not much after the middle of the thorax. Elytra with centro-basal ridges not conspicuous; obtuse at the tip, naked. Feet very stout; thighs strongly clavate; basal joint of the tarsi short, scarcely longer than the second. Mesosternum simple. Dorsal and ventral plates of the apical abdominal segment obtuse in the male : ovipositor in the female very short, scarcely apparent beyond the tips of the elytra, the dorsal plate broadly rounded at the tip, the ventral truncated.

The present genus is founded on *Lamia conspersa* of Germar, a species which differs from all the allied genera, except *Trypanidius*, in the shortness of the basal joint of the tarsi. The obtuseness of the apical abdominal segment in both sexes, the shortness of the ovipositor in the female, and the elongation of the eyes and forehead, also distinguish it from most of the groups to which it is in other respects most nearly related. It differs from *Trypanidius* (besides the elongation of the eyes) in the mesosternum being plane instead of tumid, and also in the style of coloration, although agreeing in general form as well as in the shape of the tarsi. The genus *Trypanidius* is unknown in the Amazons region.

Atrypanius conspersus, Germar.

Lamia conspersa, Germar, Ins. Spec. nov. p. 474.

A. ellipticus, griseus, carneo nigroque conspersus : capite carneo-fulvo, fronte grisea : elytris plaga irregulari pone medium alteraque apicali griseis nigro maculatis, apice breviter oblique truncatis. Long. 4–6 lin. ♂ ♀.

The species has a wide range, being found near Rio Janeiro, on the Upper Amazons, and in Mexico. I see no difference in specimens which I have compared from all these widely distant countries.

Genus PROBATIUS (Dej. Cat.), Thomson.

Thomson, Classif. des Cérambyc. p. 16.

Body elliptical. Antennæ scarcely one and a half times the length of the body, furnished with short stiff hairs, some of which are arranged in whorls around the tips of the joints. Thorax slightly convex, its outline curvilinearly widening from the front to the tips of the lateral spines, which are placed near the hind angles : ovipositor not produced in the females; terminal dorsal plate truncate and bidentate in both sexes; corresponding ventral plate obtuse, and, in the females, slightly notched. Elytra setose, even, briefly truncated and generally spined at the apex. Legs moderate; thighs clavate; tarsi short; basal joint of the hind foot about as long as the two following taken together.

This is one of the best-defined and most homogeneous genera in the host of variable forms constituting the group Leiopodinæ. The character drawn from the apical abdominal segment is seen to be constant here, bringing together species agreeing in facies and many other points, but greatly diversified in colours and markings.

1. *Probatius Chryseis*, n. sp.

P. ellipticus : capite thoraceque auratis: elytris viridibus sericeis, nigro setosis, apice mucronatis ; abdomine testaceo-rufo. Long. 5–6 lin. ♂ ♀.

Head and thorax shagreened, of a rich golden colour shading into green with the play of light, naked. Antennæ black, setose; thoracic spines large, acute, pointing obliquely rearwards, and placed very near to the hind angles. Elytra of a breadth nearly equal to three-fourths their length, thence narrowing to the apex, which, in each elytron, is prolonged into an acute spine ; surface clothed with minute silky-green scales, and having a well-marked sutural stria with regular rows of black bristles, each proceeding from a puncture. Sides of breast rich golden green ; sternum clothed with hoary tomentum ; abdomen reddish testaceous, clothed (especially towards the base) with hoary pile. Legs black ; thighs beneath hoary.

♂ Apical abdominal segment much longer than the medial segments; ventral plate obtuse at the tip ; dorsal plate square, with the angles each produced into a stout tooth.

♀ Apical abdominal segment of the same relative length as in the ♂, but the ventral plate more convex ; dorsal plate narrowed towards the tip, and terminating in two stout teeth.

One pair, taken *in copulâ* on a dead branch, at Obydos, on the Guiana side of the Lower Amazons. The insect has a deceptive resemblance to species of the Cerambycideous genus *Chrysoprasis*, found in the same localities.

2. *Probatius humeralis*, Perty.

Acanthocinus humeralis, Perty, Delectus Anim. Art. Itin. Spix & Martius, p. 91, pl. 18. fig. 3.

P. oblongus, fusco-niger, sericeus : thoracis vitta centrali, scutello, macula elytrorum utrinque humerum circumdante lineaque abbreviata marginali aurantiacis. Long. 4¼ lin. ♂ ♀.

The apical abdominal segment is of nearly the same shape

in both sexes as in *P. Chryseis*; but the ventral plate in the ♀ is briefly notched at the apex.

This species has a wide range. I have specimens before me from Rio Janeiro, the Upper Amazons, and Cayenne. It is also found in Mexico, but exists there under the form of a well-marked local variety or race, the *P. mexicanus* of Thomson (Classif. des Céramb. p. 17). This differs from the South-American form by the orange-coloured marginal streak extending to the tip of the elytra, instead of halting halfway, and by the thoracic vitta extending over the crown of the head.

3. *Probatius partitus*, White.

P. ellipticus : capite, thorace (brunneo indistincte bivittato) articuloque basali antennarum testaceo-ochraceis : elytris postice valde attenuatis, nigris, nitidis, basi obscure testaceis, maculis minutis griseis in fasciis undulatis tribus dispositis, nigro setosis, striato-punctatis, postice unicarinatis. Long. 4–4½ lin.

The apical abdominal segment is of a similar form in both sexes to that of *P. Chryseis*. The antennæ, excepting the pale basal joint, are black, with the bases of the joints whitish. The thoracic spines are thick, conical, and obtuse. The apical spines of the elytra are very long and acute, the smooth posterior carinæ of the wing-cases continuing to their tips.

Found at Pará, on dead boughs in the forest. It has also been found by M. Bar in the interior of French Guiana, specimens collected by that gentleman having been sent to me from Paris under the MS. name of *P. ruficollis*.

4. *Probatius apicalis*, n. sp.

P. oblongo-ovatus, fusco-niger, sordide ochraceo canoque variegatus; thorace ochraceo nebuloso, medio fusco bivittato : elytris postice modice attenuatis, apice transverse sinuato-truncatis, angulis externis longe spinosis, dorso punctis setiferis in striis dispositis, dimidio basali maculaque apicali ochraceis nigro conspersis, fascia lata pone medium fusco-nigra lineolis canis variegata : antennis fusco-nigris, infra dense setosis, articulo quarto annulo lato pallido. Long. 4¾ lin. ♂ ♀.

Head dingy ochraceous ; vertex with two divergent blackish vittæ. Antennæ black ; basal joint dingy rufous, second and base of third joints hoary, basal half of fourth whitish ; they are furnished with long setæ, the first joint being fringed beneath with them. Thorax dingy ochraceous, partly lighter and partly darker in hue, the centre having two nearly parallel, distinct, dark-brown vittæ ; the lateral spines are large, conical, simply acute, and not prolonged at their points. Elytra gradually and moderately narrowed to their tips, which latter are transversely sinuate-truncate, the sutural angle of the truncature being advanced but obtuse, the external angle prolonged into a stout spine; the posterior carina, which in *P. partitus* is very long, is here reduced to a faint elevation close to the tip of each elytron, and the sutural stria is not strongly impressed; the surface has many rows of setiferous punctures; the basal half is dingy ochraceous, much speckled with black; the apical part has an ochraceous patch neatly limited on its anterior edge, and varied with dusky points, the extreme apex near the suture having a smaller opake ochreous spot; the rest of the elytra is dull blackish, which colour forms a broad fascia, varied only by minute grey linear specks, arranged in lines. Body beneath and legs dingy ashy; tarsi pale testaceous; apical segment of abdomen black; ventral plate obtuse, dorsal plate truncate and bidentate in both sexes.

This was rather a common insect at Ega, on branches of dead trees in the forest.

5. *Probatius ramulorum*, n. sp.

P. ellipticus, fusco-niger, fulvo-ochraceo variegatus: thorace dorso nigro, lineis duabus ochraceis antice convergentibus: elytris fusco-nigris, maculis ochraceis conspersis et fasciatis, apice oblique truncatis, angulis externis vix dentatis: antennis nigris, infra parce setosis, articulo quarto basi pallido. Long. 4¾ lin.

Head tawny ochraceous, vertex with two oblique flexuous dusky lines. Antennæ black, two basal joints dingy ochraceous, fourth with the basal third of its length pale testaceous; they are setose, but the basal joint beneath is free from setæ. Thorax on the sides tawny ochraceous, the middle blackish, with two oblique S-shaped streaks converging in front on the fore part of the disk; the dorsal line is also dotted with ochraceous. Elytra strongly narrowed towards the tip, which is not spined, but obliquely truncated, with the sutural angle rounded off and the external one simply acute; the surface is furnished with rows of setiferous punctures; there is no posterior carina, and the colour is shining brown-black, with a sprinkling of tawny specks, some of which collect to form two indistinct narrow fasciæ, one near the middle, the other near the apex. Body beneath dusky, with dingy ashy pile; margins of ventral segments whitish. Legs blackish; claw-joints of the tarsi testaceous. The apical dorsal plate has shortish and rather blunt teeth.

Valley of the Irurá, Santarem; on dead boughs.

Genus OXATHRES, nov. gen.

Body elliptical, moderately convex, setose. Antennæ furnished with numerous setæ. Thorax as in *Probatius*, its outline widening from the front to the tips of the lateral spines, which are small, conical, and placed behind the middle. Abdomen in the male with the apical dorsal plate notched or entire; ventral plate truncated and terminating in two stout teeth, like the dorsal plate of *Probatius*. In the female the apical abdominal segment is prolonged as a conical or short tubular sheath to the ovipositor, the dorsal plate tapering into a very sharp point, and in some species acutely carinated on its upper surface; the ventral plate is simply truncated in the female. Legs moderately short and stout; thighs abruptly clavate; tarsi short, even in the hind legs, much shorter than the tibiæ, but the first joint slender and longer than the two following taken together.

This genus, although closely allied to *Probatius*, is distinguished at once by the bidentation of the apex of the abdomen existing on the ventral instead of the dorsal plate; but this is seen in the males only, the females differing from the same sex in *Probatius* still more widely—namely, by having an exserted ovipositor and a prolonged pointed dorsal plate, instead of a bidentated one.

I. *Oxathres navicula*, n. sp.

O. ellipticus, rufescenti-fuscus vel obscure fuscus; thorace dorso convexo, lævi, spinis lateralibus brevibus conicis mox pone medium sitis: elytris granulato-punctatis, lineis interruptis cinereis apice peroblique truncatis, angulis interioribus obtusis, externis subacutis. Long. 3½ lin. ♂ ♀.

Head dusky. Antennæ reddish, with a few short stiff hairs both above and beneath the joints, especially at their apices. Thorax convex, smooth, reddish brown, sometimes with a faint ashy vitta on each side, the lateral spines forming simply conical protuberances on the sides, a little behind the middle. Elytra strongly narrowed posteriorly in the ♂, more ovate in the ♀, obliquely and obtusely truncated at the tips; surface with numerous punctures, each surmounted by a raised point, dull reddish or dark brown, with scanty ashy pubescence arranged partly in interrupted lines. Body beneath and legs dull reddish brown or blackish, shining, scantily clothed with ashy pile; legs hirsute; basal joint of the hind tarsi as long as the three remaining joints taken together.

♂ The apical abdominal segment in the ♂ is greatly elongated, at least the ventral plate, with the angles of the apex acute; the dorsal plate is much shorter and deeply notched.

♀ The ovipositor projects beyond the tips of the elytra; apical dorsal plate with raised margins and a sharp keel running into a prolonged point.

Pará and Santarem, on slender dead branches.

M

2. *Oxathres Erotyloides*, n. sp.

O. ellipticus, flavo-testacens: elytris cinereis, utrinque nigro decem-
maculatis. Long. 4¾ lin. ♂ .

Head testaceous yellow, with golden-yellow pile. Antennæ
black, with three white rings, sparingly clothed with long setæ.
Thorax small, yellow-testaceous; lateral spines forming simply
conical protuberances on the sides, a little behind the middle.
Elytra of equal breadth to about three-fourths of their length,
thence rapidly narrowing to their tips, which latter are briefly
and rather obliquely truncated, the sutural angle of the truncature rounded off, the external one produced into a short blunt
tooth directed outwards; surface densely clothed with dusky
setæ and punctured, with an obtuse smooth posterior carina on
each side running into the apical tooth. The colour is ashy,
with, on each elytron, ten rounded black spots—namely, three
on the margin near the shoulders, four in a line parallel to
and near the suture, and three on the disk. Body beneath and
legs yellow-testaceous; basal joint of tarsi equal in length to
the two following taken together. Apical dorsal plate of the
abdomen (♂) rounded and closely applied to the ventral plate,
which is truncated and strongly bidentate.

I found only one example of this singularly coloured species.
It was met with at Ega, on an old stump in the forest, and was
mistaken at first for an Erotylien, especially a species of *Prio-
telus*, which is almost identical in colours with this Longicorn,
and which was often seen in similar places when the fallen trees
were covered by fungi.

3. *Oxathres muscosus*, n. sp.

O. oblongus, cinereo-olivaceus, nigro conspersus: antennis, pedibus
elytrisque longe setosis, his apice sinuato-truncatis, angulis externis
breviter mucronatis. Long. 2½–2¾ lin. ♀ .

Head olivaceous, vertex with two dusky spots. Antennæ with
all the joints except the basal one furnished with a few longish
and straight bristles, placed both above and beneath; they are
blackish in colour, with the bases of the joints (from the fourth)
pallid. Thorax rather narrow; lateral spines thick, conical,
and placed near the hind angles; surface olivaceous, with dusky
marks and four rather darker spots on the disk. Elytra oblong,
apex sinuate-truncate; sutural angle obtuse, external one produced into a short tooth; surface clothed with long black bristles, punctured towards the base, partly in lines, ashy olivaceous,
with numerous black spots, some of which are united near the
middle and form an imperfect flexuous belt. Body beneath
ashy. Legs hirsute, dingy ashy; tarsi reddish testaceous.

♀ Apical abdominal segment with the dorsal plate lanceolate, the point prolonged and acute; ventral plate simply truncated. Ega; on dead branches in the forest.

Genus TRICHONIUS, nov. gen.

Body oblong, setose. Antennæ about twice the length of the body in both sexes, furnished with numerous longish setæ. Thorax broad, widening in a curved line from the front to the tips of the lateral spines, which form thick conical protuberances situated near to the hind angles. Elytra setose, not much narrowed behind, with their tips obtusely truncated or rounded. Legs stout, bristly; thighs clavate; tarsi much shorter than the tibiæ, but the basal joint slender and elongated. Apical abdominal segment in ♂ not elongated, with tips of both dorsal and ventral plates broadly rounded; in ♀ slightly prolonged as a sheath to the very short ovipositor, which scarcely passes the tips of the elytra.; its dorsal plate flattened and rounded at the tip, its ventral plate truncated.

In the shape of the thorax and setose clothing of the body and limbs this genus resembles *Probatius* and *Oxathres.* It differs from both in the obtuse apices of the ventral and dorsal plates of the terminal abdominal segment in both sexes. There is a somewhat close relationship between *Trichonius* and *Baryssinus,* but this latter genus is amply distinguished by its crested centrobasal tubercles.

1. *Trichonius quadrivittatus,* n. sp.

T. oblongus, subdepressus : thorace lato, cinereo, fusco quadrivittato : elytris apice obtuse breviter truncatis, cinereis, fusco multiguttatis. Long. 3¼ lin. ♀.

Head brownish ashy, vertex with two fuscous dots. Antennæ more than twice the length of the body, bristly both above and beneath, reddish, bases of the joints from the fourth pallid. Thorax short and broad; lateral spines conical, scarcely pointed, and placed very near to the hind angles; surface smooth, brownish ashy, darker in the middle, and with four distinct dark-brown stripes, besides one less distinct on each side below the lateral spines. Elytra rather depressed, slightly narrowed behind, obliquely and very obtusely truncated at the tips, clothed with long bristles, which arise from punctures placed in rows independent of smaller punctures lying rather thickly towards the base; brownish ashy in colour, with a large number of small dark-brown spots which cover the setiferous punctures. Body beneath and legs dingy ashy, the latter bristly.

Villa Nova; on dead branches.

2. *Trichonius fasciatus*, n. sp.

T. oblongus, subdepressus: thorace cinereo, fusco bivittato: elytris apice truncatis, angulis exterioribus prominentibus, cinereis, pone medium fusco fasciatis. Long. 3¼ lin. ♀.

Head brownish ashy. Antennæ more than twice the length of the body, blackish, with the basal joint dull reddish, bristly both above and beneath; bases of the joints, from the fourth, pallid. Thorax broad; lateral spines thick, pointed, and slightly curving behind; surface smooth, brownish ashy, darker in the middle, and with two distinct dark-brown vittæ, besides two dusky spots on each side above the lateral spines. Elytra rather depressed, very slightly narrowed behind, transversely truncated at the apex, with the exterior angles slightly produced; surface clothed with long bristles, which arise from punctures placed in rows independent of smaller punctures lying rather thickly towards the base; brownish ashy in colour, with a rather well-defined dark-brown fascia lying behind the middle, besides two small lateral marks of dark-brown colour, one near the base, the other near the apex. Body beneath and legs reddish testaceous, clothed with ashy down; the legs bristly and ringed with dusky.

Santarem.

3. *Trichonius picticollis*, n. sp.

T. oblongus, convexus: thorace spinis lateralibus crassis, leviter curvatis, dorso brunneo vittis posticis duabus nigris intus albo notatis: elytris breviter setosis, brunneis, postice nigro alboque guttatis, apice rotundatis. Long. 3¼ lin. ♂.

Head dingy brown. Antennæ scarcely twice the length of the body, setose, the bristles much longer and more numerous beneath than above; reddish, tips of all the joints blackish; bases of the joints, from the fourth, pallid. Thorax broad, lateral spines very thick, large, and pointed, slightly curving and placed at a short distance from the posterior angles; surface anteriorly clear brown, posteriorly dusky, with two indistinct black vittæ not reaching the front margin, and dotted on their inner sides with whitish; there are also other white specks on the base and sides. Elytra rather more convex than in the allied species, scarce perceptibly truncated at the tips; surface clothed with shortish bristles, which are decumbent, instead of suberect as in the preceding species, and which arise from punctures placed in rows independent of the smaller irregular basal punctures; the colour is purplish brown, with a few indistinct black streaks and dots, and, behind the middle, a number of white specks. Body beneath and legs reddish, clothed with ashy pile; legs hirsute, but not bristly.

S. Paulo, Upper Amazons.

Genus Sporetus, nov. gen.

Body elongate-oblong, free from inequalities on the surface. Antennæ twice the length of the body, setose both above and beneath. Thorax subquadrate, rounded on the sides, and not widened from the front, the lateral spines forming small obtuse tubercles situated some distance from the hind angles. Elytra oblong, clothed with stiff hairs, simply and briefly truncated at the tip. Thighs clavate; hind tarsi with the basal joint elongated. Apical abdominal segment with both dorsal and ventral plates notched in the male, forming a sheath to the more or less exserted ovipositor in the female, the dorsal plate of which is pointed, the ventral truncate.

The few species combined to form this group differ from all the other genera of setose Leiopodinæ in the oblong narrow form of their bodies and the subquadrate shape of the thorax, with the shortness of its lateral spines. In colour they are either blackish with ashy spots, or grey sprinkled with blackish.

§ 1. Ovipositor of the ♀ short, scarcely apparent beyond the tips of the elytra (*Chætissus*).

1. *Sporetus* (*Chætissus*) *porcinus*, n. sp.

S. elongatus, hirsutus, griseus, fusco conspersus: capite thoraceque elytris angustioribus. Long. 2⅔ lin. ♀.

Head dull greyish. Antennæ brownish, with the bases of the joints from the fourth pallid; setose both above and beneath. Thorax grey, with a few indistinct darker markings. Elytra rather wider than the head and thorax, and widened towards two-thirds their length, truncated at the apex, with both angles distinct; surface clothed with long black bristles, grey, with dusky specks lying over the setiferous punctures and arranged in lines. Body beneath and legs dull testaceous, clothed with grey pile; thighs not abruptly clubbed; tarsi very moderately elongated. Ovipositor short; apical dorsal plate of the abdomen broad at the apex, and produced into a point in the middle.

S. Paulo, Upper Amazons.

This species, in the length of the ovipositor of the female, forms a connecting link between *Sporetus* and *Trichonius*; the terminal dorsal plate, although short and broad, is pointed at the apex, and not rounded off as in *Trichonius*.

§ 2. Ovipositor of the ♀ long, projecting to the length of a line beyond the tips of the elytra; dorsal plate tapering to a sharp point.

2. *Sporetus seminalis*, n. sp.

S. fusco-niger, cinereo conspersus: thorace medio cinereo quadriguttato, lateribus utrinque bivittatis: elytris apice albis, oblique truncatis. Long. 3–4 lin. ♂ ♀..

Head dark brown, with a central line from vertex to epistome, another on each side from the upper inner margin of the eye, and the cheeks yellowish ashy. Antennæ black, basal joint red, bases of third, fourth, sixth, eighth, tenth, and eleventh joints white; setose both above and beneath. Thorax above dark brown, the disk having four ashy dots, namely, one each on the front and hind margins, and two, smaller and more rounded, placed transversely in the middle; the sides each with two oblique ashy lines, which sometimes meet at the hind angle. Elytra with the tips truncate, both angles distinct; surface densely clothed with short stiff hairs, finely punctured, dark brown, with a large number of ashy specks, of different sizes and shapes, the extreme apex having a distinct white spot on each side. Body beneath ashy; sides of breast striped with dark brown. Legs elongated; thighs abruptly clavate, black or pitchy red, ringed with grey; basal joint of hind tarsi longer than the three remaining taken together.

Pará and Ega; not uncommon: found also at Cayenne. At Ega a strongly marked variety occurred, which merits separate name and mention :—

Var. *agglomeratus*. The ashy spots of the elytra are partly collected into large cinereous patches, one on each side, placed transversely in the middle of the elytron, the space anterior and posterior to this large spot being nearly free from markings.

Long. 3–4 lin. ♂ ♀.

Ega; less common than the type*.

Genus Seriphus, nov. gen.

Body oblong-ovate, convex, setose. Forehead and muzzle short, as in the Leiopodinæ generally. Antennæ elongated,

* The following Rio-Janeiro species belongs also to section 2 of the genus *Sporetus*. It has a strong resemblance to *Probatius ludicrus*, and is confounded with it in some collections :—

S. *probatioides*. Oblongus sive ellipticus, purpureo-fuscus, sericeus, cinereo maculatus. Caput fuscum, sericeum. Antennæ longissimæ, breviter setosæ, purpureo-fuscæ, articulis (tribus basalibus, 7us et 9° exceptis) basi griseis. Thorax convexus; spinis lateralibus minutis mox pone medium sitis; dorso fusco angulis et margine posticis obscure cinereo maculatis. Elytra apice subsinuato-truncata, angulis obtusis, dorso nigro-setosa, punctata, purpureo-fusca, sericea, guttis sex basalibus, maculis duabus lateralibus (una magna ante medium, altera parva pone medium) et signaturis apicalibus cinereis. Corpus subtus griseum. Pedes subelongati, obscure castanei, griseo annulati, haud setosi; femoribus valde clavatis. Maris segmento apicali abdominis emarginato. Long. 4½ lin. ♂. Hab. Rio Janeiro, a Dom. Squires lecto.

hair-like, setose both above and beneath. Thorax convex; lateral spines tuberculiform, and placed behind the middle. Elytra free from tubercles and ridges, obtusely truncated. Legs moderate; thighs clavate; basal joint of hind tarsi about equal to the two following taken together.

♂ Apical ventral segment obtusely rounded; dorsal sharply truncated, with the angles distinct.

♀ unknown.

The species which constitutes this genus would probably be better placed in a section or subgenus of *Sporetus*. It differs greatly from the *Sporeti* in colour, being of a rich changeable silky-green hue.

Seriphus viridis, n. sp.

S. supra viridi-sericens, purpureo nitens, nigro setosus: thorace postice macula, elytris plagis tribus cinereo-tomentosis. Long. 3⅜ lin. ♂.

Head minutely punctured, black; vertex silky green. Antennæ black; base of the third joint and a broad ring on the fourth grey. Thorax shagreened silky green, the middle of the hind margin with a patch of ashy tomentum. Elytra briefly truncated at the tip; surface thickly punctured towards the base, and having besides many rows of setiferous punctures, running from base to apex; silky green, changing with the play of light into dullish purple; a rounded spot of ashy tomentum on the disk of each before the middle, and a similar common spot over the suture near the apex. Legs shining black. Body beneath black, clothed with scant ashy pile.

One example only of this peculiarly-coloured species occurred, namely at Ega, on the Upper Amazons.

Genus ŒDOPEZA, Serville.

Leiopus (§ *Œdopeza*), Serville, Ann. Soc. Ent. Fr. iv. p. 89.

This group was distinguished by Serville from *Leiopus* on account of the singular dilatation of the basal joint of the anterior tarsi, and the length of the basal joint of the hind tarsi, which "equals the three following taken together." The enlargement of the anterior tarsi, which is peculiar to the males, seems to be only a specific character, as several other species, agreeing with Serville's *Œdopeza* in shape of thorax and tarsi, style of coloration, and other minor features, do not present this peculiarity. The group seems to be distinguished from *Trypanidius*, to which it is otherwise closely related, by the great narrowness of the prosternum, the depressed mesosternum, and the length of the hind tarsi. The thorax is convex, and widens from the front to the tips of the lateral spines, which are conical and placed a little behind the middle. The elytra are somewhat uneven, with faint carinæ and centro-basal ridges; they are sparsely setose in some species, naked in others. The terminal ventral and dorsal plates in the ♂ are more or less emarginated; and the ovipositor of the ♀ with its sheath is elongated, the ventral plate being truncated, and the dorsal pointed.

1. Œdopeza pogonocheroides, Serv.

Leiopus (Œdopeza) Pogonocheroides, Serv. l. c. p. 88.

This species is sufficiently well known through the description of Serville. It is of a brown colour, tawny in some parts, and marked behind the middle of the elytra with a black angulated streak or spot, followed by an ashy club-shaped streak near the suture, which reaches to the apex; the faint dorsal carinæ are speckled with grey and black, and the antennæ and feet are spotted with brown and grey. The males are known by the enlarged basal joint of the fore tarsi. It varies much in the markings of the elytra, the subapical black spot and the apical ashy streak being both subject to become either enlarged in size or diminished so as to be scarcely visible.

The species has a wide range, and the varieties do not appear to be restricted to localities, specimens before me from Panama not differing from others taken on the banks of the Tapajos and the Upper Amazons. It is a common insect on felled trees in new clearings throughout the Amazons region.

2. Œdopeza leucostigma, n. sp.

Œ. oblongo-elongata, fulvo-brunnea, nigro alboque variegata: elytris obliquo sinuato-truncatis, ante medium macula suturali alba. Long. 7 lin. ♀.

Head tawny, vertex with two black spots close to the inner margin of the eyes. Antennæ reddish brown, second to fifth joints each with two whitish rings, the remaining joints (except the second) each with one pale ring. Thorax uneven, tawny varied with dusky, and marked with four short and crooked black vittæ. Elytra elongate-oblong, very long compared with the thorax, apex of each briefly sinuate-truncate; surface punctured, some of the punctures in rows, and these latter each covered by a black spot, and emitting a short bristle; the colour is tawny brown; beyond the middle is an undulated blackish fascia, and before the middle, on the suture, is a round white spot; besides which there are a few white specks on the sides of the faint dorsal carinæ, and an irregular thin white fascia near the apex. Body beneath clothed with fine silky iron-grey pile, the sides of the abdomen spotted with tawny ashy. Legs reddish; thighs prettily variegated with grey; tibiæ and tarsi ringed with grey. Ovipositor of the female short.

One example only, taken on a fallen bough in the forest at Ega. It seems to be closely allied to the South-Brazilian *Trypanidius litigiosus* of collections, which is also an *Œdopeza* *.

Genus Cosmotoma (Dej.), Blanchard.

Blanchard, Hist. des Insectes, ii. p. 155 (1845) (description very imperfect).
Syn. *Beltista*, Thoms. Classif. des Cérambycides, p. 16 (1860).

The present group differs from the neighbouring genera by a multitude of characters; but the length and shape of the basal joint of the antennæ, besides many other minor features, leave no doubt that its true position is in the Leiopodine group. The chief peculiarities reside in the ornamentation of the antennæ—a feature that reappears here and there throughout the family

* *Œdopeza litigiosa*. Oblongo-elongata, brunnea, nigro canoque variegata. Caput obscure fulvum, vertice nigro bimaculato. Antennæ brunneæ, articulis basi cinereis, primo cinereo annulato. Thorax supra inæqualis, brunneus, disco nigro quadrimaculatus, lateribus maculis nigris griseo cinctis. Elytra oblonga, postice attenuata, apice oblique truncata; dorso passim irregulariter punctata haud setosa, brunnea, maculis nigris sparsa quarum duabus majoribus pone medium sitis, et strigis canis varia præcipue apud medium et ante apicem. Corpus subtus tomento tenui griseo vestitum, abdominis lateribus fulvo-griseo maculatis. Pedes brunnei, griseo annulati. Maris segmento ventrali apicali semicirculariter emarginato, dorsali profunde inciso; tarsi antice simplices. Long. 7 lin. ♂. *Hab.* Rio Janeiro. Coll. Bakewell, Bates, &c.

The two following belong also to this genus:—

Œdopeza guttigera. Oblonga, convexa, setosa, brunnea, nigro canoque maculata. Caput cinereo-brunneum, vertice nigro quadrimaculato. Antennæ obscure ferrugineæ, articulis (duobus basalibus exceptis) basi griseis. Thorax supra subinæqualis, spinis lateralibus magnis acutis; dorso brunneo maculis obscurioribus; lateribus utrinque ante spinam macula nigro velutina cinereo marginata notatis. Elytra oblonga, obtuse truncata; dorso leniter carinata, punctata, setosa, setis e fasciculis setarum breviorum orientibus; brunnea, maculis liturisque nigris varia et guttulis canis (una majore utrinque prope apicem) ornata; lateribus medio griseis nigro punctatis. Corpus subtus griseum, sericeum. Pedes nigricantes, griseo annulati; articulo primo tarsorum posticorum elongato. Maris segmento ultimo abdominali emarginato. Long. 5 lin. ♂. *Hab.* Mexico. Coll. Pascoe.

Œdopeza apicalis. Oblonga, parum convexa, setosa, brunnea, sericea, atro-purpureo variegata. Caput fuliginosum. Antennæ robustæ, obscure ferrugineæ. Thorax brunneus, supra vitta latissima dorsali atro-purpurea sericea cinereo guttata; spinis lateralibus conicis. Elytra oblonga, transverse late sinuato-truncata, breviter fulvo setosa, passim punctata, brunnea plaga magna scutellari maculis lateralibus basalibus fasciaque pone medium atro-purpureis, apice utrinque macula triangulari laterali nigro velutina. Corpus subtus griseo sericeum. Pedes obscure ferruginei; articulo primo tarsorum posticorum elongato. Maris segmento ultimo abdominali sinuato-truncato. Long. 5 lin. ♂. *Hab.* Guatemala. Coll. Pascoe.

of Longicorns, in groups which have otherwise no resemblance—the fourth joint in *Cosmotoma* having a thick brush of hairs attached to its upper surface, and the second and third having thin pencils of hairs at their tips, besides being clothed with a few long hairs, like the remaining joints. The thorax has a thick conical protuberance behind the middle, in the place of the lateral spines, and its surface has two large obtuse tubercles. The elytra are clothed with long hairs instead of setæ, and the centro-basal ridges, which are very thick and large, are also crested with hairs, the rest of the surface being free from inequalities. The terminal abdominal segment is of normal size and obtuse in both sexes, the female not having an exserted ovipositor. The sterna, head, and muzzle are of the same shape and structure as in the majority of the Leiopodinæ; but the eyes are rather smaller and more pointed beneath than in many of the foregoing genera; they resemble, however, very closely the same organs in the restricted genus *Leiopus*.

1. *Cosmotoma rubella*, n. sp.

C. rufescens : thoracis lateribus obscurioribus : elytris tomento argenteo strigosis, postice nigro fasciatis. Long. 2¼–3¼ lin. (6 exempl.)

Head dull red. Antennæ red, the hairy clothing black. Thorax dull red, the sides behind tinged with dusky, and the under surface black. Elytra dull red, streaked with silvery tomentum; behind the middle is a broad black fascia, followed by a narrow silvery belt, the apex itself being dusky. Body beneath dusky, with patches of grey pile. Legs reddish testaceous.

A common insect on broken branches in the forest at Pará and on the banks of the Tapajos. I have received Cayenne specimens from Paris as *Cosmotoma venustulum* of Dejean's Catalogue; but, according to Chevrolat (Journal of Entomology, vol. i. p. 188), the *C. venustulum* of Dejean's Catalogue is the species described by M. Thomson as *Beltista adjuncta*, which, from the description given, cannot be the same as our *C. rubella*.

2. *Cosmotoma nigricollis*, n. sp.

C. rufescens: thorace nigro velutino : elytris tomento argenteo strigosis, postice nigro fasciatis. Long. 3¼–4 lin. (5 exempl.)

Head dull black. Antennæ red, the hairy clothing black. Thorax deep velvety black. Elytra dull red, streaked with silvery tomentum; behind the middle is a broad black fascia, followed by a narrow silvery belt, the apex itself being dusky. Body beneath dusky, with patches of grey pile. Legs reddish testaceous.

This form represents *C. rubella* on the Upper Amazons, being as common an insect at Ega as its sister form is at Pará. It is possible that it may be the species described by Thomson as *Bellista adjuncta*; but the following phrase in this author's diagnosis, "prothorax et elytra extremitate nigra, illo versus apicem maculis 2 albis nebulosis," is quite unsuited to our *C. nigricollis*, there being no white spots near the apex of the thorax. The locality of *Bellista adjuncta* is given as San Domingo; but M. Chevrolat (*l. c.* p. 188) states that this is an error, the species being from Cayenne, and identical with *Cosmotoma venustulum* of Dejean's Catalogue.

Genus STENOLIS, nov. gen.

Body elongate, slightly convex, free from setæ. Forehead short. Antennæ elongate, slender, furnished with short setæ. Thorax somewhat narrow; lateral spines existing as minute tubercles at a distance from the hind angles. Elytra smooth, truncated. Thighs clavate; basal joint of the hind tarsi about equal to the two following taken together. Apical abdominal segment in the males (the only sex known) somewhat elongated, rounded, and entire at the tip; the ventral plate with a longish pencil of hairs on each side.

The single species constituting this genus cannot be included in any of the allied genera, on account of its peculiarities in the form of body and shape of the terminal apical segment of the abdomen. In form it agrees pretty well with certain species of *Nyssodrys* (*N. guttula* and allies), but differs from them in the apical segment of its abdomen being entire. With *Lepturges* it has in common the entire apical segment; but the very different shape of the body and thorax forbids its being included in that group. The pencils of hairs at the tip of the abdomen may be only a specific character.

Stenolis undulata, n. sp.

S. elongata, gracilis, cano-grisea : elytris fascia undulata pone medium et maculis utrinque duabus lateralibus brunneis. Long. 3½ lin. ♂.

Forehead dark brown, inner margins of the eyes grey; vertex grey, with two blackish lines. Antennæ furnished beneath with short setæ, reddish, tips of joints dusky. Thorax gray. Elytra elongate-ovate, apex obliquely subsinuate-truncate, the angles slightly produced : surface even, finely punctured (except towards the apex), hoary grey, with a clear dark-brown zigzag fascia behind the middle, a curved line of the same colour near the scutellum, and two lateral spots—one, large, before the middle, and another, smaller, near the apex. Body beneath and legs clothed with hoary-grey pile.

One example, taken at Ega.

Genus Nyssodrys, nov. gen.

Body free from setæ, oblong-ovate or elongate. Forehead and muzzle short; eyes ample, their lower lobe subquadrate. Antennæ greatly elongated, sparingly furnished with setæ. Thorax even on the surface; lateral spines short, conical, placed near to or distant from the hind angles. Elytra free from tubercles, centro-basal ridges, and lateral carinæ, rarely having faint dorsal carinæ, truncated at the apex. Legs moderate; thighs clavate; tarsi with the basal joint scarcely longer than the two following taken together.

♂ Apical dorsal plate of the abdomen entire or sinuated at the tip, ventral notched.

♀ Ovipositor elongated beyond apex of elytra; apical dorsal plate pointed or obtuse, ventral truncated or (rarely) faintly notched at the tip.

1. *Nyssodrys sedata*, n. sp.

N. oblongo-ovata, convexiuscula : thoracis spinis lateralibus conicis, acutis, prope angulos posticos sitis; elytris sordide fulvis, punctis plagisque brunneis, ante medium plaga communi transversa obscure cæruleo-grisea, pone medium fascia obliqua cana : femoribus crassis. Long. 4¼ lin. ♂. (3 exempl.)

Head ashy brown. Antennæ dull reddish, with brown pile; tips of joints, from the third, black. Thorax widening from the front to the tips of the lateral spines, which are conical and placed near the hind angles, thence narrowed in an oblique line to the base; surface dark brown, with tawny marks, four small tawny spots being arranged in a quadrangle in the middle. Elytra subovate, convex, apex briefly and obliquely truncated; surface punctured, tawny brown, with a number of small spots and several larger patches dark brown; across the suture before the middle is a transverse bluish-grey patch, and behind the middle on each elytron is a broadish, oblique, hoary-white fascia, beginning on the lateral margin, but not reaching the suture. Body beneath and legs ashy brown; thighs thickly clubbed.

♂ Terminal dorsal plate of the abdomen obtuse; ventral deeply notched.

This was a rather common insect at Ega, on dead boughs in the forest.

2. *Nyssodrys lentiginosa*, n. sp.

N. oblongo-ovata, convexiuscula : thoracis spinis lateralibus conicis, prope angulos posticos sitis : elytris brunneis, disco cæruleo-griseis brunneo maculatis, utrinque plagis irregularibus fulvis plerisque subsuturalibus, una majore laterali : femoribus modice clavatis. Long. 3¼–4 lin. ♀. (3 exempl.)

Head tawny or ashy brown. Antennæ reddish; tips of joints, from the third, black. Thorax widened from the front to the tips of the lateral spines, which are conical and placed obliquely near the hind angles, thence narrowed in an oblique line to the base; surface dark brown, with scant ashy pile, and marked with four curved tawny spots arranged in a quadrangle on the disk, and embracing in their curves so many dark-brown spots. Elytra subovate, convex; apex briefly and obliquely truncated, outer angle of the truncature slightly produced; surface punctured towards the base, dark brown; disk and suture bluish grey, speckled with brown; parallel to the suture is a row of small tawny patches, and in the middle of the lateral margin a triangular patch of the same hue. Body beneath and legs clothed with fine ashy pile; thighs not very thickly clubbed.

♀ Ovipositor short, projecting very little beyond tips of elytra; dorsal plate plane, and rounded at the tip, ventral truncated.

Found both on the Upper and Lower Amazons, on dead branches*.

8. *Nyssodrys cinerascens*, n. sp.

N. oblongo-ovata, parum convexa, obscure brunnea, cinereo maculata et conspersa : thoracis spinis lateralibus conicis, prope angulos posticos sitis : elytris breviter oblique truncatis, angulis exteriori-bus productis. Long. 4–4½ lin. ♂ ♀. (6 exempl.)

Head black, with fine ashy pile. Antennæ reddish, tips of joints, from the third, black. Thorax widened from the front to the tips of the lateral spines, which are conical and placed near the hind angles, thence narrowed in a very oblique line to the base; disk dark brown, with several ashy marks, four of which form in the middle two subinterrupted vittæ. Elytra subovate, slightly convex; apex briefly and obliquely truncated, outer angle of the truncature distinctly produced; surface punctured towards the base, dark brown, with patches of ashy-grey on the disk, speckled with dark brown, and with a number of small patches of ashy grey lying parallel to the suture. Body beneath and legs black, clothed with fine ashy pile.

♂ Terminal dorsal plate obtusely sinuated at the tip, ventral rather deeply and semicircularly notched.

♀ Ovipositor projecting one-third of a line beyond the tips of the elytra; dorsal plate tapering to the tip, but rounded, ventral truncated.

Common at Pará.

4. *Nyssodrys corticalis*, n. sp.

N. elliptica, convexa, sordide fulva, brunneo vittata : thoracis spinis

---

* There is an example of this species in Mr. Bakewell's collection, ticketed "South America" (probably from Cayenne), which has the tawny marks very clear; but the four spots on the thorax are nearly straight, and almost form two tawny vittæ.

lateralibus conicis prope angulos posticos sitis: elytris oblique truncatis, angulis (praecipue suturalibus) obtusis. Long. 4¾ lin. ♀.

Head tawny, vertex with three dusky vittæ. Antennæ dull reddish, clothed with tawny-brown pile; tips of joints, from the third, dusky. Thorax widened from the front to the tips of the lateral spines, which are obtusely conical and placed near the hind angles, thence narrowed very obliquely to the base; surface dingy tawny; disk with three dark brown vittæ, the middle one intersected by the pale dorsal line. Elytra convex, narrowed curvilinearly from half the length to the apex, the latter very obliquely truncated; sutural angles of the truncature rounded off, external ones obtuse; surface punctured, except near the apex, tawny, streaked very irregularly with ashy and dark brown, the streaks short, longitudinal, and of unequal thickness, four, thicker than the rest, lying parallel to the suture. Body beneath and legs dingy light brown.

♀ Ovipositor short, scarcely projecting beyond the tips of the elytra; dorsal plate tapering and rounded at the tip.

Forests of the Tapajos. In markings this species much resembles *Leiopus contemptus* (Chevrolat, MS.) from Mexico, which is a *Nyssodrys* allied to *N. corticalis**.

5. *Nyssodrys spreta*, n. sp.

N. oblongo-ovata, parum convexa, sordide fulvo-brunnea, cinereo

---

* *Nyssodrys contempta.* Oblonga, convexiuscula, postice attenuata (♂). Caput cinereo-fulvum, vertice fusco trimaculato. Antennæ rufescentes, tomento cinereo vestitæ, articulis (duobus basalibus exceptis) apice nigris. Thorax usque ad apices spinarum lateralium dilatatus; spinis brevibus, conicis, prope angulos posticos sitis; dorso cinereo, fusco trivittato. Elytra apices versus paulo attenuata, subtransverse sinuato-truncata, passim punctata, utrinque leviter bicostata; sordide fulva griseo canoque varia, utrinque plagis elongatis tribus fuscis notata. Corpus subtus griseo vestitum. Pedes grisei, tibiis apice tarsisque supra nigricantibus. ♂ Segmento dorsali apicali obtuso, ventrali late emarginato. Long. 4½ lin. ♂. *Hab.* Mexico. Coll. Bates.

The following common South-Brazilian species also belongs to this part of the genus *Nyssodrys*:—

Nyssodrys lignaria. Oblongo-ovata, convexiuscula. Caput cinereum. Antennæ rufescentes, cinereo vestitæ, articulis (2 basalibus exceptis) apice nigris. Thorax breviusculus usque ad apices spinarum lateralium dilatatus, spinis magnis conicis acutis, ab angulis posticis paulo distantibus; dorso fusco-cinereo, medio maculis quatuor vel vittis duabus fulvis. Elytra postice paulo attenuata (♂) vel oblongo-ovata (♀), subtransverse breviter truncata, utrinque leviter bicostata, passim punctata; sordide grisea vel brunnea fulvo fuscoque maculata, utrinque plaga majore cana laterali. Corpus subtus cinereo vestitum. Pedes nigricantes. ♂ Segmento dorsali apicali truncato, ventrali late emarginato. ♀ Stylo brevissimo, segmento dorsali attenuato, apice rotundato, ventrali truncato. Long. 4–4½ lin. ♂♀. *Hab.* Rio Janeiro (D. Squires). Coll. Bakewell., Bates., &c.

maculata: thoracis spinis lateralibus conicis, subuncinatis, prope
angulos posticos sitis: elytris oblique truncatis. Long. 2¼
lin. ♂.

Head dingy brown. Antennæ dull reddish, furnished with
very short setæ above and beneath. Thorax moderately widened
from the front to the tips of the lateral spines, which are acutely
conical, placed obliquely, and separated from the base by a
sinuated space; surface dull ashy or tawny brown, with paler
marks forming two indistinct interrupted central vittæ. Elytra
not broader than the thorax, oblong-ovate, apex rather obliquely
truncated, angles distinct; surface slightly convex, punctured
(except near the apex) partly in rows, dingy tawny or ashy
brown, with paler greyish or ashy specks, four of which (larger
than the rest) form an interrupted flexuous fascia beyond the
middle. Body beneath and legs blackish, clothed with dingy-
brown pile. ♂ Terminal ventral segment faintly emarginated
at the apex.

One example, Santarem. The species is distinguished from
its nearest relatives by its narrower-oblong form and the sinuation
of the space between the spines and the base of the thorax.

6. *Nyssodrys binoculata*, n. sp.

N. parva, subovata, antice et postice attenuata; thoracis spinis lateralibus brevissimis: elytris convexis, valde transverse truncatis,
cinereis, fulvo fuscoque punctatis, apud medium utrinque macula
magna nigro velutina fulvo cincta. Long. 2½ lin. ♂.

Head dingy tawny. Antennæ red, apices of the joints (from
the third) blackish. Thorax convex, widened from the front to
the tips of the lateral spines, which are extremely short, thence
narrowed obliquely to the base; surface brown, with curved
fulvous spots. Elytra convex, narrowed to the apex, broadly
and transversely truncated; surface punctured towards the base,
grey, with numerous blackish specks; near the middle on each
side is a large round velvety-black spot, neatly margined with
fulvous, and touching the lateral margin; apex dusky brown,
with a central fulvous spot. Body beneath clothed with ashy-
yellow pile. Legs reddish.

♂ Terminal abdominal segment narrowed to the tip; both
dorsal and ventral plates slightly notched.

Santarem, on dead twigs. There is a closely allied species
found near Rio de Janeiro *.

* *Nyssodrys dioptica.* Subelongata, postice sensim attenuata (♀), convexa. Caput flavo-cinereum. Antennæ rufescentes, articulis (duobus
basalibus exceptis) apice obscurioribus. Thorax usque ad spinas laterales leniter ampliatus, deinde paulo attenuatus, spinis longis acutis,
ante basin sitis; dorso fusco, medio maculis quatuor fulvis. Elytra
elongata, postice sensim attenuata, apice oblique truncata, angulo

7. *Nyssodrys grisella*, n. sp.

N. oblonga, parum convexa : thorace griseo, brunneo trivittato, spinis lateralibus brevissimis acutis prope angulos posticos sitis : elytris brunneis, griseo maculatis, apice peroblique sinuato-truncatis. Long. 2¼ lin. ♂.

Head tawny grey, vertex with two dark brown spots. Antennæ pitchy red. Thorax slightly widened from the front to the tips of the lateral spines, the spines very small and conical and placed near the hind angles; surface tawny grey; disk with three dark brown vittæ. Elytra rather narrow, apex very obliquely sinuate-truncate, surface sparingly punctured, dark brown, with numerous tawny-grey spots and streaks, two, near the base on each side, more elongate than the rest. Body beneath and legs dark brown, clothed with grey pile; abdomen testaceous red.

♂ Apical ventral segment slightly emarginated.

Ega, on slender branches in the forest.

8. *Nyssodrys fulminans*, n. sp.

N. oblonga, parum convexa ; thorace griseo, nigro bivittato, spinis lateralibus brevissimis, paulo ante basin sitis : elytris nigricantibus fasciis quatuor griseis fortiter undulatis, apice sinuato-truncatis. Long. 3¼ lin. ♂ ♀.

Head dusky, with a grey line running from the top of the forehead to the occiput. Antennæ reddish, tips of joints dusky. Thorax scarcely widened from the front to the tips of the spines, which are extremely small and situated a short distance from the hind angles, the space between them and the base being very feebly narrowed; surface grey, with two broad and clear blackish dorsal vittæ. Scutellum grey. Elytra oblong, sinuate truncate, the outer angles of the truncature produced, feebly convex; surface scantily punctured, brownish black, with four thin grey zigzag fasciæ, the space between the first and second and between the third and fourth being darker grey; besides these lines, there is a grey ring on the margin touching the second fascia, and a dentated grey line continuing from the fourth fascia along the suture to the apex. Body beneath and legs dusky, clothed with silvery-grey pile.

♂ Terminal segment with both dorsal and ventral plates distinctly notched.

externo producto, convexa, passim crebre punctata, sordide grisea, fusco fulvoque conspersa, utrinque apud medium macula reniformi nigro velutina fulvo cincta marginem lateralem haud attingente. Corpus subtus obscure griseum. Pedes nigricantes. ♀ Stylo longiusculo, segmento dorsali attenuato, apice obtuso. Long. 3 lin. ♀. Hab. Rio Janeiro (D. Squires). Coll. Bakewell.

♀ Ovipositor projecting half a line beyond the elytra ; dorsal plate ending in a blunt point.

I took many examples of this elegantly marked species in the forest at Ega.

9. *Nyssodrys bispecularis*, White.

Letopus bispecularis, White, Cat. Long. Col. Brit. Mus. vol. ii. p. 384.

" *L.* fusco-cinereus ; thorace medio vitta lata nigro-fusca : elytris singulis macula magna subovata obliqua suturam non attingente pallidoque cincta, apice truncatis : antennis ferrugineis. Ega" (White, *l. c.*) Long. 2¼–3½ lin. ♂ ♀.

This pretty species is of oblong shape, slightly convex and depressed a little before the middle of the elytra. The thorax is but little widened from the front, and the lateral spines are scarcely perceptible at the point where the thorax is broadest— namely, a short distance from the hind angles. The apical ventral segment in the ♂ is broadly notched, and the ovipositor of the ♀ projects but little beyond the tips of the elytra.

Common at Ega, on broken boughs and trunks of fallen trees.

10. *Nyssodrys guttula*, n. sp.

N. oblonga, subdepressa, brunnea, cinereo guttata : thoracis spinis lateralibus brevibus conicis, ab angulis posticis distantibus : elytris subtransverse truncatis : antennis albo quadriannulatis. Long. 3¼ lin. ♂.

Head tawny, vertex black, with a central tawny line. Antennæ black, base of second, third, fourth, and fifth joints with a whitish ring. Thorax scarcely widened from the front to the tips of the spines, which are obtusely conical and situated at a distance from the hind angles ; ashy tawny ; surface dark brown, with three ashy spots in the middle, and two longer ones near each of the front angles. Elytra oblong, rather depressed, subtransversely and simply truncated, surface dark brown, sprinkled throughout with little spots and patches of a tawny-ashy hue. Body beneath and legs clothed with ashy-tawny pile.

♂ Apical dorsal and ventral plates very slightly emarginated.

Forests of the Tapajos.

11. *Nyssodrys incisa*, n. sp.

N. oblonga, subdepressa, olivaceo-brunnea, cinereo guttata : thoracis spinis lateralibus brevibus, conicis, ab angulis posticis distantibus : elytris sinuato-truncatis : antennis albo annulatis. Long. 3¼ lin. ♀.

Head blackish, vertex with a central ashy line. Antennæ black, base of second, third, fourth, and fifth joints with a whitish ring. Thorax scarcely widened from the front to the tips of the spines, which are obtusely conical and situated at a distance from the hind angles ; dingy ashy, sides each with two short

N

blackish stripes; disk dark olivaceous brown, with three small
ashy spots. Elytra oblong, rather depressed, apex sinuate-
truncate, with the angles prominent; surface punctured towards
the base, dark olivaceous brown, silky, with a number of dingy
ashy specks and cross streaks, some of which unite to form a
fascia just before the apex, leaving a clear space before and after
it of the ground-colour of the elytra. Body beneath and legs
clothed with silky-grey pile; base of thighs testaceous.
  ♀ Ovipositor slender, projecting about half a line beyond the
apices of the elytra; dorsal plate tapering, obtusely pointed.
Taken at Ega.

12. *Nyssodrys anceps*, n. sp.

N. oblonga, subdepressa, olivaceo-brunnea : thoracis spinis lateralibus
  brevibus, conicis, ab angulis posticis paulo distantibus : elytris
  sinuato-truncatis, obscure cinereo-brunneis, dimidiis basalibus cæruleo-griseis brunneo punctatis, apice fulvo-cinereo plagiatis :
  antennis albo annulatis. Long. 3⅔ lin. ♀.

  Head dusky, vertex with a pale line. Antennæ rusty red, tips
of joints dusky, bases of second to fifth joints pallid. Thorax
scarcely widened from the front to the tips of the spines, which
are conical and situated at a distance from the hind angles;
sides ashy, streaked with dark brown; disk dark brown, with a
dingy ashy dorsal line, and a speck of the same colour on each
side of it. Elytra oblong, rather depressed, apex sinuate-truncate, with the angles prominent; surface punctured towards the
base, dark olivaceous brown, basal half bluish grey, with brown
specks and ashy-tawny patches, apex with a larger ashy-tawny
patch, indented with dark brown. Body beneath clothed with
dingy-ashy pile. Legs ferruginous, apex of tibiæ dusky.
  ♀ Ovipositor projecting a little beyond the tips of the elytra.
Santarem, on dead trees.

13. *Nyssodrys stillata*, n. sp.

N. oblonga, parum convexa, olivaceo-nigra, sericea : thorace supra
  vittis tribus, elytris guttis numerosis distinctis, cinereis : fœminæ
  stylo elongato, fistuloso. Long. 4 lin. ♀.

  Head olivaceous black, a yellowish-ashy line from the forehead to the occiput. Antennæ slender, nearly three times the
length of the body (♀), blackish, base of second to fifth joints
pallid. Thorax scarcely widened from the front; spines very
short and obtusely conical, space between them and the hind
angles indented; surface olive black, with three dorsal and (on
each side) one lateral vitta yellowish ashy. Elytra oblong-
oval, very slightly convex, apex obliquely sinuate-truncate; surface (except at the apex) punctured, olive-black, sprinkled with
smallish yellow-ashy spots, two near the apex transverse and

larger than the rest; apices themselves margined with ashy. Body beneath clothed with yellowish-ashy pile; legs ringed with ashy and black.

♀ Ovipositor greatly elongated and tubular, projecting one and a half line beyond the tips of the elytra; dorsal plate narrowed to the tip and pointed, ventral truncated.

Ega, on trunks of felled trees.

14. *Nyssodrys vitticollis*, n. sp.

N. oblonga, robusta, subconvexa, atro-fusca, sericea: thorace vittis quinque cinereis: elytris cinereo multiguttatis, apice transverse sinuato-truncatis: antennis albo annulatis. Long. 4½ lin. ♂.

Head black; margins of the eyes and a central vitta, from the middle of the forehead to the middle of the vertex, ashy. Antennæ stout, scarcely twice the length of the body (♂), black, base of third joint pallid, bases of fourth, fifth, and six joints with a whitish ring. Thorax slightly widened from the front to the tips of the lateral spines, which are prominent and conical, the space between them and the hind angles being indented; surface blackish brown, with three clear ashy vittæ on the disk and one on each side. Elytra oblong-oval, apex transversely subsinuate-truncate, angles prominent; surface punctured, except towards the apex, deep silky brownish black, sprinkled with a large number of ashy spots. Body beneath clothed with yellowish-grey pile. Legs stout, ringed with grey and black.

♂ Apical ventral segment rather deeply and broadly notched.

S. Paulo, Upper Amazons.

15. *Nyssodrys caudata*, n. sp.

N. oblongo-elongata, subdepressa, brunnea: elytris utrinque fasciis tribus fortiter angulatis atro-brunneis (prima interrupta) suturam haud attingentibus cinereo cinctis, apice oblique sinuato-truncatis. Long. 3¼–5½. ♂♀.

Head ashy brown. Antennæ rusty red, tips of joints darker. Thorax slightly widened from the front to the tips of the spines, which are short and conical, the space between them and the hind angles being indented or slightly sinuated; surface light brown, sometimes with greyish marks. Elytra elongate, slightly narrowed towards the tips, scarcely convex, apex obliquely sinuate-truncate (less obliquely and angles more prominent in the ♀ than in the ♂); surface with the basal half punctured, light brown, each elytron with three zigzag or irregular transverse spots or fasciæ of a dark-brown hue encircled with ashy; the first of these, near the base, consists, on each elytron, of two (sometimes three or even four) separated spots; the second is a zigzag belt, broad on the margin, but narrow on the disk, and terminating before reaching the suture; the third is an oblique spot near the apex. The second and third fasciæ vary much in form—both, in some examples, being reduced to spots which do not touch the lateral margin. In well-developed individuals, the space between the second and third fasciæ is of an ashy hue. Body beneath and legs clothed with ashy pile; the legs sometimes reddish.

♂ Apical ventral segment triangularly notched at the apex.

♀ Ovipositor elongated, projecting 1½ line beyond the tips of the elytra; dorsal plate pointed, ventral truncated.

This is an extremely common species in the Amazonian forests, on fallen trees. The varieties do not seem to be confined to particular localities, as I found extreme forms (as to the development of the markings of the elytra) living together at Ega. It is found also at Cayenne, and exists in French collections under the name of *Leiopus caudatus*, Lacordaire, MS.

16. *Nyssodrys signifera*, n. sp.

N. oblonga-elongata, parum convexa, brunnea: elytris fasciis duabus plagam magnam cineream includentibus maculaque parva subapicali nigro-brunneis, apice subtransverse truncatis: ♀ segmento ventrali ultimo breviter emarginato. Long. 4½–5½ lin. ♂ ♀. (6 exempl.)

Head ashy brown. Antennæ rusty red, tips of joints dusky. Thorax very slightly widened to the tips of the spines, the latter short, but prominent, and placed nearer the middle than the hind angles, the space between them and the base being moderately narrowed; surface silky brown. Elytra oblong, apex scarcely obliquely truncated in the ♀, more obliquely in the ♂; surface thickly punctured, except over the apical third (which is very smooth), light brown, with two dark-brown fasciæ on each elytron, the basal one oblique, the second angulated at the middle, neither touching the suture, the space between them being of a light ashy colour; near the apex is a small rounded dark-brown spot; the disk on the apical portion is tinged with ashy; the sides are silky brown. Body beneath and legs clothed with tawny-ashy pile.

♂ Apical ventral segment broadly emarginated.

♀ Ovipositor projecting 1½ line beyond the tips of the elytra; dorsal plate obtusely pointed, ventral briefly emarginated at the apex.

This elegant species is found throughout the Amazons region, from Pará to Ega. I have seen it, in Parisian collections, under the name *Leiopus hieroglyphicus* (Buquet), Dej. Cat.; but as the following species also exists in the same collections under this name, and no diagnosis has been published to guide us in deciding to which of the two it should be applied, I am obliged to pass it over without further notice.

17. *Nyssodrys propinqua*, n. sp.

N. oblonga, subdepressa, brunnea: elytris fasciis duabus obscurioribus plagam magnam cineream includentibus maculaque parva subapicali, apice oblique sinuato-truncatis, angulis prominentibus: ♀ segmento ventrali ultimo integro, truncato. Long. 3–4 lin. ♂ ♀. (7 exempl.)

Head ashy brown. Antennæ rusty red, tips of joints dusky. Thorax widened from the front to the tips of the spines, which are short and conical, and placed nearer the middle than the hind angles of the thorax, the space between them and the base being moderately narrowed; surface silky brown. Elytra oblong, apex obliquely and rather strongly sinuate-truncate in both sexes, angles prominent; surface thickly punctured over the basal half, dingy brown, with two dentated fasciæ of a darker shade on each elytron,—the basal one anteriorly blending with the ground-colour of the wing-case, the second more distinct and very broad on the lateral margin, neither touching the suture, the space between them being of a light ashy colour; near the apex is a small rounded dark-brown spot, encircled with ashy, which in many examples extends laterally towards the margin. Body beneath and legs clothed with tawny-ashy pile.

♂ Apical ventral segment broadly emarginated.

♀ Ovipositor rather short, extending only ¾ of a line beyond the tips of the elytra; dorsal plate acute, ventral truncated and entire at the tip.

The present species is almost identical in markings with the preceding, but it differs by its much smaller size, dingier colours, more sinuated truncature of the elytra, and by the apical ventral segment of the female being entire instead of notched at the apex. It is a generally distributed species in the Amazons region, but occurs much more commonly than *N. signifera*. Specimens from Cayenne, under the name of *Leiopus hieroglyphicus*, have been sent to me from Paris.

18. *Nyssodrys simulata*, n. sp.

N. oblonga, subdepressa, brunnea: elytris utrinque plaga canescente fusco maculata ante medium, macula dentata laterali alteraque parva subapicali fuscia, apice subtransverse sinuato-truncatis. Long. 4½ lin. ♂ ♀.

Head ashy brown. Antennæ rusty red, tips of joints dusky. Thorax widened from the front to the tips of the spines, which are short and conical and placed nearer the middle than the hind angles; surface light brown. Elytra oblong, apex in both sexes scarcely obliquely sinuate-truncate; surface faintly punctured towards the base, light brown, each elytron before the middle with a large hoary-white patch sprinkled with clear dark-brown spots, not corresponding with the punctures; behind this white patch there is the usual angulated silky-brown lateral spot, besides a subapical smaller spot on the disk. Body beneath and legs clothed with lightish-brown pile.

♂ Apical dorsal plate feebly, ventral deeply notched.

♀ Ovipositor projecting very little beyond the tips of the elytra; dorsal plate obtusely pointed, ventral truncated.

Ega, Upper Amazons. It has also been found in the interior of French Guiana by M. Bar. The species is readily distinguishable from the two preceding by the white patch of the elytra being sprinkled with distinct dark-brown spots.

19. *Nyssodrys afflicta*, n. sp.

N. oblonga, parum convexa, brunnea: elytris utrinque plaga angulata pone basin guttulisque numerosis griseis, apice oblique sinuato-truncatis. Long. 4½ lin. ♂ ♀.

Head ashy brown. Antennæ rusty red, tips of joints dusky. Thorax widened slightly from the front to the tips of the lateral spines, which are small and acutely conical and placed about as near the middle as the hind angles; surface light brown, with paler specks. Scutellum dark brown, with a central line ashy. Elytra oblong, apex in both sexes obliquely sinuate-truncate; surface thickly punctured near the base, light brown, each with an angulated greyish patch on the side before the middle, and sprinkled with short greyish streaks or spots. Body beneath and legs clothed with lightish-brown pile.

♂ Apical dorsal plate broadly emarginated, ventral deeply notched.

♀ Ovipositor projecting a line and a half beyond the tips of the elytra; dorsal plate pointed, ventral slightly notched.

A common insect on branches of fallen trees in the forest, both on the Upper and Lower Amazons, and at Pará.

20. *Nyssodrys deleta*, n. sp.

N. oblonga, parum convexa, brunnea, sericea; scutello, plaga laterali elytrorum sæpe obsoleta maculaque subapicali cinereis: elytris apice peroblique sinuato-truncatis, angulis externis fortiter productis. Long. 3–5 lin. ♂ ♀.

Head ashy brown. Antennæ rusty red, tips of joints darker. Thorax slightly widened from the front to the tips of the lateral spines, which are small and acute and placed near to the hind angles, the space between them and the base scarcely narrowed; surface light brown. Scutellum ashy. Elytra elongate-oblong, very obliquely sinuate-truncate, external angles of the truncature strongly produced, almost mucronate; surface punctured near the base, light brown, each with a large faint ashy patch on the side before the middle, margined and spotted with dark brown, and an ashy crescent near the apex enclosing a brown dot; the patch obsolete in many examples, and the apical half of the elytron having sometimes three or four ashy specks. Body beneath and legs clothed with dingy brown pile.

♂ Apical dorsal plate broadly emarginated, ventral sharply notched.

♀ Ovipositor projecting the length of a line beyond the tips of the elytra; dorsal plate pointed, ventral truncated.

This is an equally common species with *N. caudata*, being found at all stations throughout the Amazons region, on dead branches. I have seen it in some collections under the name of *Leiopus deletus*.

21. *Nyssodrys rodens*, n. sp.

N. oblonga, subdepressa, postice sensim attenuata (♂ ♀), nigro-brunnea : thorace vittis quatuor vel sex, elytris lineolis confluentibus suturaque cinereo-brunneis; his apice oblique sinuato-truncatis, angulis exterioribus valde productis. Long. 3½–4½ lin. ♂ ♀.

Head light brown, occiput blackish. Antennæ rusty red, tips of joints dusky. Thorax widened curvilinearly to the tips of the spines, which are short and acute, and placed nearer the middle than the hind angles; surface blackish brown, with four (sometimes six) light-brown vittæ (the alternate ones sometimes grey), besides a thin dorsal line, which is often absent. Scutellum light brown. Elytra gradually narrowed from base to apex, the latter obliquely sinuate-truncate, outer angles of the truncature strongly produced, almost spiniform; surface blackish brown, varied with several light-brown (partially grey) streaks of unequal length and very irregular in position, but always with an angulated one near the apex; suture greyish. Body beneath and legs clothed with light-brown pile.

♂ Apical dorsal and ventral segments both notched.

♀ Ovipositor short, projecting to the length of scarcely half a line beyond the tips of the elytra; dorsal plate narrow and pointed.

Found throughout the Amazons region, on slender branches and twigs; beaten once out of a mango-tree.

22. *Nyssodrys lineolata*, n. sp.

N. oblonga, robusta, subdepressa, nigro-brunnea : thorace vittis septem plus minusve indistinctis, elytris fasciis duabus lineolarum maculisque subapicalibus cinereo-brunneis : his apice sinuato-truncatis, angulis exterioribus modice productis. Long. 4¾–5¼ lin. ♂♀.

Head ashy brown. Antennæ rusty red, spotless. Thorax widened from the front to the tips of the spines, which are conical and placed a short distance from the hind angles; surface dark brown, with seven more or less incomplete light-brown or ashy vittæ. Scutellum ashy. Elytra tapering from base to apex (♂), or more oblong-ovate (♀), apex obliquely sinuate-truncate, outer angles produced; surface shining dark brown, with two broad fasciæ (interrupted at the suture) composed of a number of short ashy longitudinal lines; a few specks near the base and apex and a short line along the outer point of the apex also of an ashy colour. Body beneath and legs clothed with ashy-brown pile.

♂ Apical dorsal plate scarcely emarginated, ventral notched.

♀ Ovipositor very short and broad; dorsal plate broad and obtuse at the tip.

Ega; rare.

23. *Nyssodrys promeces*, n. sp.

N. angustata, parum convexa, nigro-brunnea: thorace elytrisque vittis tribus fulvis, his oblique truncatis, angulis suturalibus obtusis. Long. 3¾ lin. ♂.

Head ashy brown. Antennæ four times the length of the body, scantily furnished with short setæ, black. Thorax scarcely widened to the tips of the spines, which are conical and placed nearer the middle than the hind angles; surface blackish brown, with three tawny vittæ. Elytra elongate, narrow, obliquely and obtusely truncated at the apex; surface punctured, except near the apex, blackish brown; each elytron with three tawny vittæ terminating before reaching the apex, the sutural and central ones having a shorter faint grey streak between them; the apical part has two angular fulvous spots. Body beneath and legs clothed with silky grey pile; sides of sternum and abdomen with a fulvous line.

♂ Apical dorsal plate truncated, ventral broadly notched.

This curious species approximates in length of antennæ, shape, and colours to the Hippopsine group of Lamiaires; but all its essential features show that it is a true Acanthocinite of the Leiopodine section, the basal joint of the antennæ having a waved outline beneath, the thorax and head having the shape usual in the Leiopodinæ, and the sternums the same outline. Its habits are those of a *Hippopsis*, clinging, like the species of this and the neighbouring genera, to slender dead twigs; consequently the claw-joints of the tarsi (especially of the middle legs) are longer than is usual in the Acanthocinitæ, and have some analogy to those of the subtribe Onciderites to which the

Hippopsinæ belong; but the claws are not thickened and sub-parallel, and the claw-joints of the fore tarsi not elongated—characters which further distinguish the present species from the Hippopsinæ. I do not think the slight elongation of the middle and posterior claw-joints warrants the establishment of a new genus for this species.

24. *Nyssodrys ptericopta*, n. sp.

N. elongata, postice sensim attenuata, fuliginosa, griseo obscure lineata: elytris pone medium cinereo biguttatis, apice oblique valde truncatis. Long. 3¾ lin. ♂.

Head tawny yellow, forehead with two brown spots. Antennæ rusty red, basal joint darker. Thorax widened from the front to the tips of the short conical spines, thence narrowed in a sinuated line to the base; surface sooty brown, with seven very indistinct greyish lines. Elytra narrowed from base to apex, the latter obliquely and broadly truncated, angles obtuse; surface punctured towards the base, sooty brown, with several very indistinct greyish lines, and on the disk of each elytron a rounded ashy spot a little after the middle. Body beneath and legs clothed with greyish pile.

♂ Dorsal and ventral apical plates equally notched.
Banks of the Tapajos.

25. *Nyssodrys ramea*, n. sp.

N. oblonga, convexiuscula, grisea, fusco plagiata: thoracis spinis lateralibus tuberculiformibus, mox pone medium sitis: elytris apice truncatis, angulis distinctis. Long. 4¼–5¼ lin. ♂.

Head dingy brown, vertex grey. Antennæ dusky; bases of fourth to seventh joints pale. Thorax widened from the front to the tips of the lateral spines, which are short and conical and placed soon after the middle; surface greyish, with confluent blackish patches. Elytra scarcely obliquely truncated at the apex, angles distinct; basal portion covered with large punctures; surface greyish, with blackish-brown spots and patches, some of them confluent and forming near the middle a zigzag fascia interrupted at the suture. Body beneath and legs clothed with grey pile.

♂ Apical dorsal segment faintly, ventral broadly emarginated.
Ega, rare. It has been since found also in the interior of French Guiana by M. Bar.

26. *Nyssodrys excelsa*, n. sp.

N. oblonga, subconvexa, brunnea, sericea: thoracis spinis lateralibus ab angulis posticis distantibus: elytris griseis, macula magna communi basali alteraque laterali utrinque pone medium brunneis, apice oblique truncatis, angulis distinctis. Long. 5½ lin. ♀.

Head ashy brown. Antennæ clothed with ashy-brown pile, apices of the joints darker. Thorax widened from the front to the tips of the lateral spines, which are conical and distinct and placed nearer the middle than the hind angles. Elytra obliquely truncated at the apex, angles distinct; surface, except near the apex, covered with large punctures, greyish; a large patch in the middle of the base and an oblique lateral spot or belt on each side behind the middle light brown; the apical part has also a faint brownish cloud. Body beneath and legs clothed with greyish pile.

♀ Ovipositor projecting to the length of a line beyond the tips of the elytra; dorsal plate narrow, pointed, ventral truncated.

Ega, rare.

27. *Nyssodrys alboplagiata*, White.

Leiopus alboplagiatus, White, Cat. Long. Coll. Brit. Mus. ii. p. 381.

" *L.* pallide fulvo-ochraceus, sericeus: elytris plaga magna laterali alba, elytris punctatis: metathoracis lateribus albis. Ega." Long. 6½ lin. ♂ ♀.

This fine species is similar in shape to the two preceding, the lateral spines of the thorax being conical, short, and nearer the middle than the hind angles. The elytra are rather obliquely truncated, with both angles slightly prominent.

♂ Both dorsal and ventral plates of the terminal segment notched.

· ♂ Ovipositor projecting to the length of nearly two lines from the tips of the elytra.

Ega, closely adhering to slender branches of dead trees.

Genus HYLETTUS, nov. gen.

Body elongate-oblong, more or less depressed, free from setæ. Head, as in all the allied genera, much narrower than the thorax, with the antennæ approximated at the base; muzzle short and obtuse; lower lobe of the eyes subquadrate. Antennæ greatly elongated, sparingly furnished beneath with short bristles. Thorax uneven on the surface; lateral tubercles prominent and placed near the middle of the sides. Elytra without smooth lateral keels proceeding from the shoulders. Sterna simple. Terminal abdominal segment in the males with both dorsal and ventral plates notched or emarginated. Ovipositor of the female elongated, tubular; dorsal plate pointed, ventral truncated. Legs moderate; thighs clavate, thickly so in the males; basal joint of the posterior tarsi as long as, or longer than, the two following taken together. Fore and middle tarsi in the male dilated and fringed with hairs.

The chief character which distinguishes this group from *Nyssodrys* is the dilatation and ciliation of the anterior and (in less degree) of the intermediate tarsi in the males. Some of the larger species of *Nyssodrys* have the male anterior tarsi much broader than those of the hind legs, but in none of them are they furnished with the marginal fringe of hairs. The *Hyletti* are somewhat larger insects than the *Nyssodryes*, and the shape of their thorax is somewhat different, the lateral spines being in the form of large or distinct tubercles, and placed near the middle of the sides. The genus approaches *Acanthocinus* and *Graphisurus* (groups characteristic of North America and Europe) nearer than any we have yet passed in review.

Hylettus cœnobita, Erichs.

Leiopus cœnobita, Erichson, Consp. Ins. Col. Peruana, p. 145.

" *L.* fuscus, dense cinereo-tomentosus, supra flavo irroratus, scutello nigro cincto : elytris puncto infra scutellum maculaque transversa atro-tomentosis, flavo cinctis, apice emarginatis, spina brevi terminatis. Long. 5½–8 lin." ♂ ♀.

The examples which served Erichson for his description were obtained by Von Tschudi in Eastern Peru, in the same forest region where, further east, at Ega, I met with it in abundance. The elytra are sinuate-truncate at the apex, and it is only in the male that the outer angle of the truncature is produced into a spine; in the female both angles are acute. The thoracic tubercles are rather small, but stand out distinct from the sides of the thorax.

The *Ædilis griseofasciatus* of Serville (Ann. Soc. Ent. Fr. iv. p. 83), a common South-Brazilian insect, belongs to the present genus, and there are doubtless many other tropical American species yet to be added to it*.

Genus PALAME, nov. gen.

Body oblong, narrow; elytra clothed with short setæ. Head

---

* *Hylettus decorticans*, n. sp. Oblongus, subdepressus, griseo-fulvus, brunneo variegatus. Caput griseum. Antennæ rufescentes, articulis apice obscurioribus. Thorax inæqualis, griseo-fulvus, tuberculis lateralibus magnis conicis, mox pone medium sitis. Elytra oblonga, postice sensim attenuata (♂), apice oblique truncata, supra passim punctata, punctis basalibus granulis elevatis adjunctis, fulvo-grisea, utrinque maculis lateralibus tribus fuscis, prima elongata, obliqua, pone basin suturam fere attingente, secunda latiore pone medium, tertiaque interrupta prope apicem. Corpus subtus griseum. Pedes rufescentes, tibiis apice tarsisque nigricantibus. Long. 6 lin. ♂. *Hab*. Venezuela. Coll. Bakewell., Bates.

not much narrower than thorax or elytra. Antennæ moderate in length, setose both above and beneath. Thorax with lateral spines extremely small and placed near the hind angles. Elytra free from ridges and lateral keels. Terminal abdominal segment with the ventral plate in the males sharply notched; ovipositor in the female not prolonged, the apical segment being only a little longer than that of the male, with the ventral plate convex and truncated, and the angles of the truncature produced. Legs stout; fore and middle tarsi dilated and fringed with hairs in the male; coxæ and under surface of body also densely hairy in the same sex.

In many points (for example, the setose elytra and antennæ, shape of thorax, and style of coloration) the curious insect forming this genus shows a near degree of relationship with the *Sporeti*, especially with *S. seminalis*. It exhibits, however, an almost equally close approximation to the *Colobotheæ*, showing that, notwithstanding the great amount of apparent difference between the elongate *Colobotheæ* and short flattened *Leinpi*, from which the *Sporeti* differ little, the two extremes are in reality closely bound together by connecting links, and, notwithstanding the almost endless multiplication of specific forms, have not diverged widely from a common plan of structure. The genus *Palame* is readily distinguishable from all allied genera by the hairy coxæ and sterna of the male, and the absence of ovipositor in the female.

Palame crassimanus, n. sp.

P. oblonga, subcylindrica, nigro-olivacea, sericea: thorace vittis quinque cinereis: elytris late subsinuato-truncatis, plagis cinereis nigro maculatis. Long. 3¾ lin. ♂ ♀.

Head black; forehead with three ashy lines, and outer orbits of the eyes ashy. Antennæ black, bases of joints paler, those of the fourth to the sixth joints ashy. Thorax convex, above silky black, with five ashy vittæ, the two lateral ones on each side, in some examples, being interrupted, and in others confluent. Elytra slightly narrowed from base to apex; apex broadly subsinuate-truncate, angles of the truncature obtuse, surface silky olive black, and with large ashy patches speckled with black; sides speckled with ashy tomentum: besides the setiferous punctures over the whole elytra, the basal part has a number of simple punctures. Body beneath ashy. Legs black, with ashy pile; tarsi and under surface of body naked in the female; in the male the fore and middle tarsi are dilated and fringed with hairs, and have fulvous brush-like palms, the coxæ and middle of the breast and abdomen being thickly clothed with brownish hairs.

Generally distributed throughout the Amazons region, on both sides of the river. It is found on slender branches of fallen trees in the forest. The terminal segment of the abdomen of the female is scarcely visible beyond the tips of the elytra.

Genus Toronæus, nov. gen.

Body oblong, somewhat convex. Head and thorax of nearly equal breadth, and much narrower than the elytra. Antennæ greatly elongated; joints long and slender, sparingly furnished with setæ both above and beneath. Thorax with a slight protuberance on the sides a little behind the middle, in place of the lateral spines. Elytra without setæ, and free from ridges and lateral carinæ. Terminal segment of the abdomen in the males with both dorsal and ventral plates more or less notched at the tip. Ovipositor of the females greatly elongated, and generally exserted beyond its sheath, tubular; dorsal plate of the terminal abdominal segment (constituting the sheath) slender and pointed, ventral deeply cleft at the apex. Legs moderately slender; thighs clavate; tarsi undilated and simple in both sexes; basal joint of posterior tarsi as long as, or longer than, the three succeeding taken together.

This genus is distinguished from all the preceding by the cleft or deeply notched apex of the terminal ventral segment in the females; in this it agrees with *Graphisurus* of Kirby[*], which, again, is connected by intermediate species with *Acanthocinus*, a group containing the well-known *A. ædilis*, or carpenter-beetle, an inhabitant of the wooded parts of our own island. Thus all the numerous genera of Acanthocinitæ are closely linked together; for species of *Nyssodrys* (e. g. *N. signifera*) exhibit to a slight extent the character of a cleft apex of the terminal ventral segment, and this genus leads on without any sharp line of demarcation to *Leiopus*,—showing that the European genera *Leiopus* and *Acanthocinus*, which appear to us so far asunder, are connected together by insensible gradations of form. The typical species of *Toronæus* (namely those which have no thoracic spines) are easily distinguishable from *Graphisurus*; but if the bounds of the genus be extended a little, so as to embrace a few

---

[*] This genus comprehends the following North-American species:—

1. *G. fasciatus*, De Geer, Mém. v. p. 114, t. 14. f. 7.
   ——, Kirby, Fauna Boreali-Americana, Ins. p. 169.
   ?= *Lamia mixta*, Fabr. E. S. Suppl. 144. 26.
2. *G. obsoletus*, Oliv. Col. iv. p. 130, t. 13. f. 90.
   = *Astynomus lævicollis*, Dj. Cat.
3. *G. pusillus*, Kirby, Fauna Bor.-Americana, p. 169.

Acanthocinus atomarius (F.), of Europe, is also probably a *Graphisurus*.

134

closely allied species which have small thoracic spines*, the only difference between the two genera will be one of general form, the *Graphisuri* being much flattened, with comparatively short antennal joints, whilst the *Toronæi* have convex shapes and very slender antennæ.

1. *Toronæus figuratus*, n. sp.

T. oblongus, convexiusculus, nigro-castaneus, capite thoraceque vitta centrali ochracea: elytris litura humerali, macula magna communi ante medium antice et postice per suturam excurrente, fasciaque lata inflecta prope apicem cinereo-ochraceis. Long. 4–5 lin. ♂ ♀.

Head dark brown, sides of forehead and cheeks each with a yellowish streak, vertex with a broad central yellowish stripe. Antennæ slender, twice the length of the body in both sexes, reddish, tips of joints dusky, and bases of third to sixth joints whitish. Thorax not much broader than the head, and with a slight protuberance on each side about the middle, but no trace of spine or tubercle; above dark chestnut-colour, silky, sides and a central vitta continuous with that of the head yellowish. Scutellum ochreous. Elytra in both sexes very slightly narrowed to three-fourths of their length, then abruptly narrowed in a curved line to the apex, which is subsinuate-truncate; surface punctured, except near the apex, dark brownish chestnut ornamented with marks of a yellowish-ashy hue; there is a small spot on each side of the scutellum, an angulated streak under each shoulder, and a large common spot a little before the middle extending along the suture both towards the base and apex, and connected with an angulated streak which touches the side on each elytron; this patch has a small blackish speck in its middle over the suture: besides these marks, the apex has on each side a flexuous streak enclosing a tooth-shaped spot of the ground-colour of the elytron. Body beneath hoary white. Legs reddish, with ashy pile; apex of thighs, tibiæ, and tarsi black.

♂ Terminal abdominal segment feebly emarginated at the apex.

♀ Ovipositor greatly elongated and exserted beyond its sheath, apical dorsal plate of its sheath pointed, ventral deeply cleft.

I met with this elegantly marked insect only at Obydos, on the Guiana side of the Lower Amazons, where it was abundant, in March 1859. It has been found also in the interior of Cayenne by M. Bar, and exists in French collections under the names of *Eutrypanus figuratus* and *E. elegans*, the former of which I have adopted.

---

* Such as *Eutrypanus tessellatus*, White, Cat. p. 372 (= *E. variegatus*, Dej. Cat.), and others, not found in the Amazons region.

2. *Toronæus suavis*, n. sp.

T. oblongus, convexiusculus, nigro-castaneus, capite thoraceque vitta centrali ochracea : elytris litura humerali, fascia obliqua pone medium, linea arcuata laterali prope apicem, suturaque postice cinereo-ochraceis. Long. 3¼–5¼ lin. ♂ ♀.

Head dark brown, sides of forehead and cheeks each with a yellowish streak, vertex with a broad central yellowish stripe. Antennæ slender, reddish, tips of joints dusky, bases of third to sixth joints pale. Thorax not much broader than the head, and with a slight protuberance on each side about the middle, but no trace of spine or tubercle ; above dark chestnut, silky, sides whitish, the middle traversed by a yellowish stripe continuous with that of the head. Scutellum ochreous. Elytra in both sexes gradually narrowed from base to apex, the latter sinuate-truncate; surface punctured, except towards the apex, dark brownish chestnut ornamented with yellowish-ashy marks ; there is a small spot on each side of the scutellum, a patch beneath and a curved line above the shoulder, an oblique stripe beginning about the middle of each side, and extending to the suture, connected with a lighter streak on the disk, and, lastly, a distinct arcuated yellowish line on each side near the apex ; the suture near the base and apex is also bordered with yellowish ashy. Body beneath hoary white, breast and base of abdomen on each side with dark oblique stripes: abdomen sometimes reddish. Legs reddish ; apical halves of tibiæ and tarsi black ; basal joints of tarsi ashy.

♂ Terminal abdominal segment with dorsal and ventral plates rather deeply notched.

♀ Ovipositor greatly elongated and exserted beyond its sheath, apical dorsal plate pointed, ventral deeply and narrowly cleft.

This pretty species, which differs from *T. figuratus* by the more tapering shape of its elytra, and by the markings on the surface of the wing-cases, was met with at various places on the southern side of the Lower Amazons, and on the banks of the Tapajos, but never in abundance.

3. *Toronæus perforator*, n. sp.

T. oblongus, convexiusculus, fuscus, nigro fulvo canoque variegatus : elytris apice cinereo marginatis et fasciatis ; fœminæ stylo elongatissimo. Long. 3¼–5¼ lin. ♂ ♀.

Head velvety black, cheeks ashy, vertex with a short yellow line. Antennæ slender, more than twice the length of the body in both sexes, reddish testaceous, all the joints except the first and second with a pale ring at their bases. Thorax very little broader than the head, the sides in the middle with a slight protuberance, surface dark brown, with blackish spots on the disk and fulvous spots on the sides, a curved ashy streak below the lateral protuberance. Scutellum black. Elytra oblong, not narrowed until near the apex, at which point they are suddenly narrowed to the tip, which is obliquely truncated; surface thickly punctured, except near the apex, dark purplish brown, sides with greyish marks, and disk spotted with black, sometimes varied also with obscure greyish and fulvous streaks and spots, a more distinct but short oblique pale line existing, in all examples, on each elytron a little before the middle near the suture; the apical margin, both sutural and external, has a neat ashy border, which, being joined to a præapical fascia of the same hue, encloses a transverse blackish spot. Body beneath clothed with silky grey pile. Legs more or less reddish, with ashy and black rings.

♂ Terminal abdominal segment with dorsal plate semicircularly notched at the tip, ventral with a shallower notch.

♀ Ovipositor greatly elongated, the sheath extending more than two lines beyond the tips of the elytra; dorsal plate finely pointed, ventral cleft at the tip.

A widely distributed insect in the Amazons region, being found on the banks of the Tapajos and near Ega on the Upper Amazons. The species has also been met with by M. Bar in the interior of French Guiana. Cayenne examples agree precisely with those found at Ega; but those brought from the Tapajos are much lighter in colour, and have many tawny spots on the upper surface of the thorax and elytra, which are wanting in those of other localities.

4. *Toronæus terebrans*, n. sp.

T. oblongus, convexiusculus, fuscus : thorace antice maculis quatuor fulvis in serie transversa dispositis : elytris nigro griseoque nebulosis, medio macula communi cinerea, ante apicem linea transversa fulva. Long. 4 lin. ♂.

Head velvety black, cheeks ashy, vertex with a short ashy line. Antennæ reddish testaceous, bases of joints (except the basal two) pallid, apices dusky. Thorax very little broader than the head, the sides in the middle with a distinct conical protuberance ; surface blackish, sides streaked with ashy, fore part with a transverse row of four distinct tawny spots, an obscure oblique line of the same hue extending from the base towards the disk on each side. Elytra oblong, not narrowed until near the apex, at which point they are suddenly narrowed to the tip, the latter obliquely truncated ; surface thickly punctured, except near the apex, purplish brown, varied throughout with pale bluish grey and patches of a black colour, apical part clear brown (including the margins), but crossed by a thin yellowish line from lateral margin to suture. Body beneath silky ashy. Legs reddish, ringed with grey and black.

♂ Terminal abdominal segment with dorsal plate semicircularly notched at the tip, ventral with a shallower notch.

Found only at S. Paulo, Upper Amazons.

5. *Toronæus virens*, n. sp.

T. oblongus, convexiusculus, fusco sericeus, viridi micans: elytris plaga magna ante medium cinerea, apicibus canis utrinque macula transversa fusca. Long. 3½–4½ lin. ♂ ♀.

Head sooty black, cheeks yellowish ashy. Antennæ reddish, bases of third to sixth joints pallid. Thorax very little broader than the head, the sides in the middle with a conical protuberance, surface dark brown, becoming green in certain lights; disk speckled with tawny ashy, sides ashy, with a brown streak. Elytra oblong, obliquely truncated, surface punctured, except at the apex; dark brown, with a large patch before the middle, and the apical region ashy, the apical spot enclosing a transverse curved blackish streak: the whole surface has a silky green lustre in certain lights. Body beneath ashy. Legs blackish, ringed with ashy.

♂ Terminal abdominal segment with both dorsal and ventral plates deeply notched.

♀ Ovipositor projecting one line and a half beyond the tips of the elytra; ventral plate deeply notched.

A common insect on branches of fallen trees in the forest, both on the Upper and Lower Amazons.

Genus CALLIPERO, nov. gen.

Body elongate, narrow; head and thorax of nearly equal width, and narrower than the elytra. Muzzle short, lower lobe of the eyes short, and narrower below than above. Thorax with a slight protuberance behind the middle, but free from lateral spines or tubercles. Elytra without lateral keels, clothed with short setæ. Sterna simple. Antennæ moderately elongated; third to seventh joints thickened (the seventh thicker than the rest), and densely clothed on their under surface with short setæ, besides the usual longer bristles which exist on all the joints (except the first) both above and beneath. Ovipositor of the female not exserted; terminal abdominal segment elongated and conical, with the dorsal plate pointed, and ventral truncated. Legs moderately elongated; thighs clavate; basal joint of posterior tarsi as long as the three following taken together.

This genus differs from all the genera of Acanthocinitæ known

P

to me by the shape and clothing of the third to the seventh
joints of the antennae. In shape of body and style of coloration
the species composing it might easily be mistaken for Cerambycideous insects of the genus *Rhopalophora*.

Callipero bella, n. sp.

C. elongata, capite thoraceque chalybeis, azureo vittatis : elytris
purpureis, sutura azurea, maculis duabus basalibus aurantiacis :
corpore subtus azureo. Long. 5 lin. ♀.

Head steel-blue, forehead dusky, cheeks grey, a pale blue
vitta extending from the middle of the front to the occiput.
Antennæ black. Thorax steel-blue, a narrow central vitta, and
on each side a broad lateral one, pale blue. Elytra elongated,
broader than the thorax, tapering to the apex, and broadly truncated ; surface in the middle with three faint, smoothed, raised
lines, thickly punctured towards the base, and covered with finer
punctures, each emitting a longish, erect, black bristle ; dark
blue, changing to purple, suture and apical margin bordered
with light cobalt-blue ; base of each elytron with a large orange-
coloured spot. Body beneath pale blue. Legs black, with grey
pile.

I met with one example only of this most charming species,
at S. Paulo on the Upper Amazons, where it was found sunning
itself on a leaf on the banks of one of the brooks which run
through the virgin forest.

Genus COBELURA, Erichson.

Erichson, Conspectus Ins. Coleop. Peruan, p. 149.

The founder of this genus likened it to *Colobothea*, mention-
ing as the only characters which distinguish it the depressed
body and tumid mesosternum. *Cobelura*, however, differs from
all the genera of the group Colobotheinæ in wanting the acute
prominent shoulders and sharp lateral carinæ of the elytra which
are characteristic of the group. The genus is more nearly allied
to *Nyssodrys* and *Hylettus*, differing from both chiefly in the
elongate-elliptical shape of the body (which assimilates the species to the *Colobotheæ*), unarmed sides of the thorax, tumid
mesosternum, and small size of the lower lobe of the eyes. The
only species described by Erichson is the *C. lorigera*, inhabiting
the forest region of Eastern Peru, which differs greatly from the
following in colours and markings.

Cobelura prolixa, n. sp.

C. elongata, subdepressa, postice paulo attenuata, olivaceo-grisea :
thorace vitta lata mediana fusca, nigro marginata : elytris maculis

irregularibus discoidalibus alternaque laterali majore triangulari ante apicem fuscis, leviter tricostatis. Long. 7½ lin. ♂.

Head clothed with tawny pile. Antennæ reddish, bases of the joints pallid or ashy, apices dusky. Thorax much broader than the head, and much narrower than the elytra, convex and rounded on the sides, the broadest part being the middle; surface olivaceous or tawny ashy, the middle occupied by a broad dusky vitta bordered by black lines; there is also a dusky vitta on each side below the lateral dilatation. Elytra elongated and rather depressed; shoulders prominent, but obtuse; apex obliquely sinuate-truncate, with both angles of the truncature produced (the external one most so), sides destitute of carinæ; surface of each with three smooth costæ, the innermost only strongly pronounced, covered with minute punctures, each bearing a short bristle; dull greenish ashy, with small dark-brown specks and a larger triangular dark-brown spot on the sides near the apex. Body beneath obscure tawny; middle of breast and abdomen, and terminal segment of the latter, blackish. Legs greenish tawny; tibiæ and tarsi ringed with black. Mesosternum with a very large rounded tubercle.

♂ Terminal abdominal segment with both ventral and dorsal plates deeply notched.

I met with a few examples only of this species in the dry woods near Santarem, at the mouth of the Tapajos.

Genus XYLERGATES, nov. gen.

Body oblong, robust. Antennæ stout, moderately elongated, sparingly furnished with short bristles beneath. Thorax tubercular on the disk; lateral tubercles large and placed near the middle of the sides. Elytra much broader than the thorax, their deflexed sides broad and vertical, but not separated from the dorsal surface by smooth keels; surface costate and roughened by small tubercles surmounted by short bristles; apices truncated. Sterna narrow. Terminal abdominal segment in the males with dorsal and ventral plates notched. Ovipositor of the females moderately elongated, conico-tubular; dorsal plate obtuse, ventral truncated. Legs stout; thighs thickly clavate; fore and middle tarsi dilated in the males; first joint of the hind tarsi about equal to the two following taken together.

The robust forms and tubercular thoraces of the species composing this genus give them a strong general resemblance to the Acanthoderes; they are distinguished, however, by the elongate gradually thickened basal joint of the antennæ, the closure of the anterior acetabular sutures, the ovipositor of the females, and other characters. The genus is very closely related to

Eurypanus, no constant mark of difference existing other than the absence of smooth lateral keels proceeding from the shoulders of the elytra. From *Acanthocinus* it is distinguished by the high vertical sides of the wing-cases, the tuberculose surface of the body, and the dilated anterior and middle tarsi of the males.

Xylergates lacteus, n. sp.

X. oblongus, supra planiusculus, postice sensim attenuatus, brunneo sericeus : elytris strigis curvatis lacteis plagas griseas includentibus, apice sinuato-truncatis, angulis exterioribus productis. Long. 6½–7 lin. ♂ ♀.

Head tawny brown. Antennæ ringed with grey and black. Thorax with large obtusely conical lateral tubercles near the middle of the sides, and with two obtuse tubercles on the fore part of the disk, besides three other smaller ones on the posterior part; surface purplish brown, silky, sides below the tubercles ashy. Elytra broad and straight at the base, thence gradually narrowed to the apex, which is somewhat broadly sinuate-truncate, the external angles of the truncature produced; deflexed sides (towards the base) thickly granulate-punctate; surface with numerous small punctures towards the base, and with four or five interrupted rows of acute blackish tubercles surmounted by short bristles, the middle ones lying along the faint dorsal carinæ; the colour is silky purplish brown, with (on each elytron) a curved milk-white streak from the shoulders to near the apex bending towards the suture, and two obliquely transverse similar streaks near the apex, all enclosing patches of a light-grey colour and shorter milk-white streaks, the anterior curved lateral lines being connected across the suture by a thin straight line of the same hue. Body beneath tawny ashy. Legs grey, with dusky rings.

♂ Middle of breast and coxæ thickly clothed with brown pubescence. Terminal abdominal segment with ventral and dorsal plates deeply notched, the angles of the ventral notch acute, of the dorsal obtuse. Fore and middle tarsi dilated and fringed with hairs.

♀ Breast, coxæ, and tarsi simple and naked. Ovipositor projecting the length of a line beyond the tips of the elytra; dorsal plate broad and obtuse at the tip.

This elegant and rare species occurred only at Ega and S. Paulo, Upper Amazons. It has since been found also in the interior of French Guiana by M. Bar *.

* The following common South-Brazilian insect belongs to the genus *Xylergates* :—

Xylergates asper, n. sp. Oblongus, supra convexiusculus, postice rotundatim attenuatus, cinereo-fulvus, sericeo-brunneo plagiatus. Caput

Group *Colobotheinæ.*

Genus EURYPANUS (Dej. Cat.), Thomson.

Thomson, Classif. des Cérambyc. p. 13.

Char. emend. Body oblong or subelongate, above somewhat plane. Thorax with stout lateral spines or tubercles placed near the middle of the sides, above tubercled or convex. Elytra much broader than the thorax, their deflexed sides broad and vertical, and separated from the dorsal surface by a sharp keel proceeding from the shoulder; surface furnished with setæ, apices truncated. Prosternum narrow; mesosternum broad, nearly square. Terminal abdominal segment in the males more or less notched at the tip. Ovipositor of the females short, projecting but slightly beyond the tips of the elytra, and subconical in form. Legs stout; thighs strongly clavate; fore and middle tarsi of the males slightly dilated; basal joint of posterior tarsi longer than the two following taken together.

The species selected by M. Thomson as the type of this genus is the *E. nitidus* of White (Cat. Long. Col. Brit. Mus. p. 371, pl. 9. fig. 4), which he has redescribed in the 'Classification des Cérambycides' under the name of *E. Venezuelensis*. A considerable number of species will be found to associate with *E. nitidus*, the principal generic feature of which (omitted in M. Thomson's definition) is the sharp lateral keels proceeding from the shoulders of the elytra. This distinguishes the *Eurypani* well from *Xylergates*, to which some of the species (e. g. *E. ellipticus* of Germar) are otherwise closely related. There is not, however, any positive character whereby to distinguish *Eurypanus* from *Colobothea*; for some species, by their elongated shapes, might almost be mistaken for *Colobothea*, and the aberrant forms of the latter genus have lateral thoracic tubercles and fore tarsi in the males not differing from those of the intermediate legs, as in the *Eurypani*. The best distinguishing character is probably this:—in *Eurypanus* the lateral outlines of the head and thorax are not continuous, and therefore the

sordido fulvo-cinereum. Antennæ robustæ, breviusculæ, cinereæ, articulis apice fuliginosis. Thorax supra inæqualis, trituberculatus, fulvo-cinereus, disco plaga obscura brunnea, tuberculis lateralibus magnis acutis. Elytra oblonga, postice (♂ ♀) rotundato-attenuata, breviter oblique truncata, supra punctata, utrinque quadricostata, costis ante apicem abbreviatis, tuberculo nigra hispido gerentibus; fulvo-cinerea, plaga indistincta scutellari alterisque duabus apud medium lateralibus angulatis sericeo-brunneis. Corpus subtus fulvo-cinereum. Pedes cinerei, fusco annulati. Maris pectore nudo, segmento ultimo abdominali fortiter inciso, tarsis anticis intermediisque dilatatis, nec ciliatis. Fœminæ stylo modice elongato, segmento ultimo dorsali subacuto. Hab. in Brasilia meridionali.

fore part of the body has not that conical form which gives so peculiar a facies to the *Colobothea*. A less trenchant point of difference is presented by the elytra, which in the great majority of the *Colobothea* are nearly straight to the apex, but in *Eutrypanus* are curvilinearly attenuated before the apex.

1. *Eutrypanus nobilis*, n. sp.

E. oblongus, robustus, brunneus: thoracis lateribus late ochraceo vittatis: elytris maculis trilobis duabus communibus ochraceis, una apud medium suturali, altera majore subapicali: spinis thoracicis acutis, retrorsum spectantibus. Long. 7 lin. *♂*.

Head dusky, with scant tawny pile. Antennæ twice the length of the body (*♂*), brown, tips of all the joints blackish, bases pallid. Thorax widened from the front to the tips of the thoracic spines, which are large, acute, and oblique, and placed behind the middle of the sides ; surface convex, slightly uneven, dark brown, with a broad ochreous vitta on each side margined with black. Elytra broad at the base, gradually narrowed to near the apex, thence more abruptly narrowed ; apex transversely sinuate-truncate, both angles equally and moderately produced ; lateral carina extending beyond the middle of the elytra, acute, but not smooth ; whole surface thickly punctured, punctures setiferous, colour dark brown mixed with tawny ; over the suture near the middle is a trilobed ochreous spot, and near the apex over the suture is a much broader but similar spot, the two connected by an ochreous sutural line. Body beneath tawny ashy. Legs moderately long ; thighs abruptly and strongly clavate, dusky, with ashy pile ; two basal joints of the tarsi grey.

♂ Terminal abdominal segment with dorsal and ventral plates very slightly emarginated. Fore and middle tarsi broader than those of the hind legs.

Obydos, on the Guiana side of the Lower Amazons ; rare.

2. *Eutrypanus assula*, n. sp.

E. oblongus, brunneus : thorace nigro vittato, elytris nigro cinereoque strigosis : spinis thoracicis brevibus, conicis, pone medium sitis : elytris breviter oblique truncatis. Long. 4½ lin. *♀*.

Head brown, vertex with two black spots. Antennæ dull reddish, bases of joints greyish. Thorax with small and conical lateral tubercles placed a little behind the middle, disk uneven, brown, the middle part with two black vittæ, the sides above the tubercles each with two short black lines, below the tubercle a broad black streak. Elytra moderately broad and convex, curvilinearly narrowed from near the base to the apex, the latter briefly and obliquely truncated ; lateral carinæ moderately acute and smooth, and reaching beyond the middle of the elytra ;

surface and sides scantily punctured towards the base, brown, with many black and ill-defined longitudinal streaks, besides a broad indistinct ashy streak beginning at the shoulder, bending towards the suture, and then continuing, parallel to the suture, to the apex ; the mode of coloration gives to the insect a striking resemblance to a small chip of bark. Body beneath dusky, with scant ashy pile. Legs reddish, ringed with ashy.

♀ Ovipositor projecting very slightly beyond the tips of the elytra ; dorsal plate obtusely rounded at the tip, ventral truncated.

Banks of the Cupari, a branch of the river Tapajos.

8. *Eutrypanus incertus*, n. sp.

E. elongatus, subangustatus, fulvo-griseus, nigro vittatus et maculatus : spinis thoracicis parvis, conicis, pone medium sitis : elytris postice attenuatis, apice breviter truncatis, nec dentatis. Long. 4½–6 lin. ♂ .

Head blackish, orbits of eyes fulvous. Antennæ black or dull red, third to sixth joints ringed at the base with grey. Thorax not much broader than the head ; lateral tubercles small, placed a little behind the middle ; disk slightly uneven, ashy tawny, with six black vittæ, the two outermost of which are below the lateral tubercles. Elytra elongate, gradually narrowed to near the apex, thence more abruptly narrowed, apex briefly and obtusely truncated ; lateral carinæ sharp and smooth, surface faintly punctured towards the base, and covered besides with minute setiferous punctures, clothed with tawny pile, much spotted and patched with black, the apical region on each elytron being occupied by a large clear black spot margined with ashy. Body beneath ashy tawny. Legs blackish, with scant tawny clothing ; tibiæ ringed with ashy ; tarsi with the two basal joints grey.

♂ Coxæ and breast densely hairy, as also (in well-developed examples) the middle of the abdomen. Terminal abdominal segment with ventral plate sharply notched, dorsal moderately so. Fore and middle tarsi dilated and fringed with hairs.

Also found on the banks of the Cupari. M. Bar has since met with it in the interior of French Guiana. The species, although having an elongated form of body like the *Colobothea*, does not offer the peculiar facies of that genus, owing to the different shape of the apex of the elytra.

Genus CARTERICA, Pascoe.

Pascoe, Trans. Ent. Soc. Lond. iv. (1858) p. 250.

With this genus commence the more elongated and narrow forms which distinguish the typical Colobotheinæ. The elytra

are nearly parallelogrammic, especially in the male sex; in
the females slightly dilated a little before the apex. The head
is somewhat narrow, and the bases of the antennæ rise from
distinct antenniferous tubercles. The antennæ are greatly
elongated, and, from the third joint, very slender—the basal
joint being longer than the third, and thickened from the base
to near the apex. The prothorax is rather short, much narrower at the base than the elytra, and its widest part is at some
distance from the base, where it forms, on each side, an obtuse
prominence. The humeral angles of the elytra are prominent,
and a distinct, but not polished, ridge proceeds from them towards the apex; the surface of the elytra is ribbed, and the apex
is truncate, with the outer angle alone prominent and dentiform.
The sternums are narrow and plane. The abdomen is slender
and tapering, and the terminal segment elongated, especially in
the female. In *C. cinctipennis* the ventral plate of the female is
subtubular and truncated at the apex, the dorsal obtusely
rounded: in the male the dorsal plate is notched at the apex;
in *C. cincticornis* the apical segment is shorter and obtuse at the
apex. The legs are slender, the basal joint of the tarsi much
elongated: the fore tarsi are simple in both sexes.

1. *Carterica cinctipennis*, Pascoe.

Carterica cinctipennis, Pasc. Tr. Ent. Soc. Lond. iv. 1858, p. 250.
C. oolobotheoides, Thomson, Classif. des Cérambycides, p. 19 (1860), sec.
   Chevrolat, Journ. Ent. i. 188.

C. ochraceo-fulva, vertice vittis duabus, thorace vitta lata mediana
   alteraque angustiore laterali usque ad oculos extensa, pectoris
   lateribus, femoribus tarsisque apice, tibiis et antennis nigris, his
   articulis intermediis basi piceis: elytris nigris, utrinque tricostatis
   macula humerali margineque fulvis, pone medium fascia testacea
   ad suturam interrupta. Long. 4½–6 lin. ♂ ♀.

Mr. Pascoe described this as a new species, believing it, after
careful examination, to be distinct from the *S. mucronata* of
Olivier, a species closely resembling it; but Prof. Gerstaecker, in
the Berlin 'Bericht' for 1858 (p. 117), believes the two to be
the same, "the description of Olivier being much more indica-
tive than his figure." It is a generally distributed insect
throughout the Amazons region. I did not find it on timber,
but on the leaves of trees in the forest.

2. *Carterica cincticornis*, n. sp.

C. minor, modice elongata, depressa, setosa, nigra; capite (occipite
   excepto), vitta lata laterali thoracis, macula parva humerali femorumque basi fulvo-testaceis; antennarum articulo quarto late testaceo annulato, primo infra ciliato; elytris pone medium paulo

ampliatis, apices versus leviter attenuatis, supra grosse punctatis, bicostatis. Long. 2–3 lin. ♂.

Head short, forehead convex, tawny testaceous; antenniferous tubercles and two broad stripes behind them, united on the occiput, black. Antennæ twice the length of the body, black, the fourth joint, with the exception of the apex, pale testaceous; clothed with short setæ, the basal joint furnished beneath with a fringe of long hairs. Thorax scarcely convex, lateral prominences placed at a short distance from the base; black, with a silky fulvous vitta on each side. Elytra depressed, shoulders obtuse, lateral carina proceeding thence prominent, but not visible from above, slightly dilated from the middle to near the apex, then more suddenly attenuated, apex sinuate-truncate with the sutural angle rounded and external angle produced into a stout tooth; surface clothed with erect brown setæ, coarsely punctured, except near the apex, and traversed by two faintly elevated costæ, both of which disappear before reaching the apex. Prosternum reduced to a very narrow thread; mesosternum also extremely narrow. Abdomen blackish, clothed with grey pile. Legs moderately slender, basal joint of the posterior tarsi a little longer than the remaining joints taken together; black; coxæ and basal halves of the thighs tawny testaceous.

Ega, rare. I met with two examples only of this pretty little species: its habits are probably very similar to those of *C. cincti-pennis*, it being found only on the leaves of trees in the shades of the forest. The depressed body, somewhat dilated elytra, and fringed basal joint of the antennæ are so many points of approximation to the genus *Sparna* of Thomson (Systema Cerambycidarum, Liége, 1864, p. 30), the species of which resemble the dilated forms of the family Lycidæ.

Genus COLOBOTHEA, Serville.

Serville, Ann. Soc. Ent. Fr. 1835, p. 69.

The typical forms of this genus are well known to all who occupy themselves with the study of exotic Coleoptera. They are known by their elongate, narrow, and compressed form of body—the vertical, deflexed sides of the clytra being separated from the dorsal surface by an elevated line, which proceeds from the ridge formed by the shoulders, and disappears before reaching the apex. The elongated basal joint of the antennæ has the same outline as in the great body of the Acanthocinitæ previously described. The anterior coxæ are somewhat globular, and the acetabular suture is quite closed; both pro- and meso-sterna are plane, the former being very narrow and the latter subquadrate, narrowed behind. The apical segment of the abdomen is not prolonged into an ovipositor in the female; it varies so much in form in the two sexes, especially as to the outline of the apices of the ventral and dorsal plates, that it affords no constant characters for the formation of groups within the genus. The males are larger and more robust than the females, the anterior legs also being longer and stouter, and having dilated and fringed tarsi. In these typical forms the body is somewhat depressed above, with a very gradual and slight slope posteriorly; with this the elytra are narrowed nearly in a uniform degree from base to apex, and the thorax is widest at its hind angles, with a gradual attenuation from its base to its apex.

These characters, however, do not hold together so as to form a well-defined genus. Some species, which in all other respects are true *Colobotheœ*, recede from the typical forms in the shape of the thorax. Thus *C. Schmidtii* has a thorax approximating to that of some members of the Leiopodine group, having a lateral tubercle towards the hind angles; and *C. lineola* presents a thorax of nearly the same form as *Œdopeza*, *Trypanidius*, and the allied genera. The dilatation of the male tarsi also fades away from species to species, and some of these aberrant forms have the elytra less depressed and more narrowed near their apices than in the more typical *Colobotheæ*. Notwithstanding this diversity, I have failed in my attempts to divide the genus. One of the aberrant forms constitutes the genus *Priscilla* of Thomson (Systema Ceramb. p. 80). It is much less elongate and more convex than the true *Colobothea*, and the shoulders of the elytra form a larger and more elevated ridge; I have not ventured, however, to separate it from the rest whilst many other species equally entitled to form distinct genera remain in the genus.

§ I. Fore tarsi not more dilated in the male than in the female. Thorax narrowed at the base, and tumid or tuberculated behind the middle on each side.

1. *Colobothea lignicolor.*

C. modice elongata, brunnea cinereo nigroque variegata, corticis fragmentum simulans; elytris apices versus subito attenuatis, apicibus minus late sinuato-truncatis utrinque bispinosis, dorso costatis. Long. 6 lin. ♀.

Head clothed with tawny-brown pile. Antennæ twice the length of the body (♀), brown, bases of the joints pale ashy, basal joint ringed with ashy. Thorax widened from the front to the lateral tubercles, which are short and acute, then strongly attenuated and incurved to the base; surface convex, varied with light and dark brown, and with two black vittæ each interrupted in the middle; side, below the tubercle, black, shiny. Elytra short

for this genus, broad at the base, gradually attenuated to near
the apex, thence suddenly attenuated, making the truncated apex
narrow; sutural spine short, external one elongated; shoulders
advanced and rounded, lateral carina strongly pronounced and
polished, deflexed sides coarsely punctured and with a smooth
carina; surface longitudinally convex, setose, and punctured;
two short, rugose, slightly elevated carinæ near the base, and
one longer and smoother along the disk; the colour is brown
varied with ashy, near the scutellum is a dull blackish patch, and
behind the middle is an oblique black streak; the anterior part
of the disk is ashy, and there is a triangular ashy spot near the
apex. Body beneath black, with grey pile; a row of ashy spots
on each side the abdomen. Legs shining pitchy red, spotted
with grey.

♀ Terminal ventral segment broadly and triangularly excised,
angles prolonged into acute spines. Dorsal segment broadly
truncated.

On a bough of a dead tree, forest, Ega. There is a closely
allied and similarly coloured species found at Cayenne*.

2. *Colobothea velutina*, n. sp.

C. elongata, parallela, convexa, antice et postice declivia, thorace velu-
tino-nigro vitta laterali fulva; elytris griseis fulvo nigroque maculatis, regione scutellari, maculis lateralibus duabus undulatis
plagaque quadrata apicali purpureo-nigris carneo-fulvo cinctis.
Long. 6–7 lin. ♂ ♀.

Head brown, a fine central line on the vertex and a broad
stripe down each cheek tawny ashy. Antennæ one-third longer
than the body, black (dark red towards the base), the fourth
joint with a grey, the sixth with a white ring. Thorax with a
distinct tubercle standing out from each side at a short distance
from the base, scarcely narrowed behind the tubercle, hind angles
slightly prominent; surface velvety purplish black, with a tawny-

* *Colobothea ligneola.* Parum elongata, angustata, brunnea, cinereo
nigroque varia, corticis fragmento simillima. Caput piceo-fuscum.
Antennæ piceæ, articulis basi cinereis. Thorax cinereo-fuscus, vittis
duobus nigris, lateribus nigris politis; convexus, prope basin sinuato-
attenuatus, tuberculis lateralibus obtusis. Elytra brevia angustata,
apices versus citius attenuata, apicibus sinuato-truncatis angulis suturalibus distinctis exterioribus productis; supra grosse punctata, prope
basin et disco breviter costata, brunneo cinereoque varia, vitta brevi
suturali maculæque discoidoli (lineola cinerea divisa) saturatioribus.
Corpus subtus nigrum politum, cinereo varium, abdomine lateribus
cinereo maculato: pedibus piceo-rufis, griseo maculatis. Feminæ segmento ultimo ventrali attenuato, apice sinuato-truncato hispinoso;
dorsali apice rotundato, medio unidentato. Long. 3½ lin. ♀. *Hab.*
in Cayenna, a Dom. Bar lecto.

ashy stripe on each side having a blackish line in its middle, sides below this with a shining black stripe. Elytra moderately elongate, and scarcely tapering from their base to near their apex, whence they are distinctly narrowed to the apex, the latter broadly truncated, the sutural angle of the truncature scarcely distinct, outer angle produced into a longish and acute tooth; the surface is convex, setose, and moderately punctured, partly in rows; the colour is grey sprinkled with blackish spots, and ornamented with large purplish-black patches—namely, one semicircular, over the scutellum, a second angular, on the side near the base, a third of zigzag outline, beyond the middle, and a fourth quadrate, close to the apex; all these spots are margined with pinkish tawny, but the apex is narrowly edged with grey. Body beneath tawny; abdomen grey in the middle and spotted with black on the sides, the apical segment shining black with two basal greyish spots. Legs black, with grey and tawny-grey rings; fore tarsi simple in both sexes, but the legs of the male are visibly stouter than those of the female.

♂ ♀. Terminal ventral segment sinuate-truncate, angles produced into short and not very acute spines: dorsal segment obtuse. The whole segment is much longer in the female than in the male.

Common on felled trees in the forest throughout the Amazons region. Also taken at Cayenne. An allied but quite distinct species is found in Venezuela*.

8. *Colobothea decemmaculata*, n. sp.

C. elongata, angustata, postice flexuoso-attenuata, carneo-cinerea maculis oblongis lateralibus nigro-velutinis laete ornata: thorace utrinque paulo ante basin tumido, deinde paulo constricto; pedibus rufis, griseo annulatis. Long. 5–6½ lin. ♂ ♀.

Head reddish, cheeks and vertex each with a pinkish-ashy

---

* *Colobothea maculicollis* (Chevrol. MS. sec. Dom. Deyrolle). Elongata, parallela, modice convexa. Caput sordide cinereum. Antennæ vix corpore longiores (♀?), fuscæ, articulis 4to 6to 8vo et 9o cinereo annulatis. Thorax paulo ante basin tuberculo majore conico armatus, sordide cinereus (lateribus inclusis), medio dorsi macula oblonga velutino-purpurea ornato. Scutellum velutino-purpureum, medio macula parva cinerea. Elytra imprimis paulo, apices versus citius attenuata, flexuoso-truncata, angulis suturalibus nullis, exterioribus spinosis; supra punctata, grisea, fusco maculata, utrinque maculis majoribus velutino-purpureis tribus ornata, prima parva laterali ante medium, secunda magna triloba pone medium, tertia obliqua valde angulata ante apicem, totis carneo-fulvo partim marginatis. Corpus subtus fulvo tomentosum, abdomine nigro lateribus fulvo maculatis. Pedes nigri, fulvo annulati. Fœmina (?) segmento ultimo ventrali truncato angulis vix productis, dorsali obtuso. Long. 5½ lin. *Hab.* Venezuela. Coll. Bates.

stripe. Antennæ one-fourth longer than the body, dark red, becoming blacker towards the apices and greyish at the bases of the joints. Thorax widened from the front to a short distance from the base, where it is tumid on each side and after that constricted; surface pinkish ashy (the tomentum very compact), each side occupied by a broad velvety-black vitta, below this is a pinkish-ashy stripe succeeded by another black one. Scutellum pinkish tawny. Elytra elongated and tapering from base to apex, but appearing to have a flexuous outline, from the great prominence, after the middle, of a raised line which runs along the deflexed sides very near to the extreme margin; apex obliquely sinuate-truncate, sutural angles prominent, external angles produced into a long spine; the lateral carina proceeding from the prominent shoulders runs in a strongly flexuous course to near the apex; surface punctured, pinkish grey, and ornamented, on each elytron, with four rich velvety-black spots, namely, one triangular in the middle of the base, a second long and oblique, stretching from under the shoulder to the disk of the elytron, a third, angulated, behind the middle, and a fifth, oblique, near the apex, all margined with pinkish tomentum. Body beneath blackish, clothed with grey pile; breast red. Legs red, ringed with grey; fore tarsi simple in both sexes.

♂ Terminal ventral segment with a broad triangular excision, angles acute; dorsal segment obtuse, narrowly notched in the middle.

♀ Terminal ventral segment with a deep semioval excision, angles acute but not produced; dorsal segment with a broad notch in the middle.

This extremely beautiful species was rare. I met with it only at Obydos and on the banks of the Tapajos. It is found in Cayenne, and I have adopted the MS. name under which it exists in some collections in Paris.

4. *Colobothea flavomaculata*, n. sp.

C. parva, angustata, postice attenuata, purpureo-nigra; capite lineis tribus, thorace lineolis transversis lateralibus alteraque dorsali, elytris maculis sex apiceque sulphureis. Long. 3½–4 lin. ♂ ♀.

Head black, front with three sulphur-yellow lines, the middle one extending to the occiput; cheeks with a yellow line behind the eyes. Antennæ twice the length of the body in both sexes, pitchy black, bases of the fourth and sixth joints with pale grey rings. Thorax rather small, tumid on the sides in the middle, constricted near the base; purplish black, the sides each with three transverse sulphur-coloured lines, one along the front margin extending to the upper surface, and two shorter, near the hind margin; there is also a short line above, in the middle of the fore margin, and a round spot in the middle, near the hind margin. Scutellum purplish. Elytra prominent at the shoulders, then gradually attenuated to near the apex, afterwards more quickly narrowed, apex truncated in a straight line, sutural angle simple, external angle produced into a short and acute tooth; surface clothed with strong erect bristles, each proceeding from a puncture, dark purplish, with a silky gloss; a small oblong spot on each side near the scutellum, and two larger, rounded, on the disk (one before, the other after, the middle) and a transverse spot at the apex sulphur-yellow. Body beneath blackish, clothed with grey pile and with an oblique stripe on each side of the breast, and a row of linear spots on each side of the abdomen, densely ashy tomentose. Legs pitchy red, ringed with ashy; fore tarsi simple in both sexes.

♂ Terminal abdominal segment moderately short, depressed, slightly narrowed towards the apex; both dorsal and ventral plates truncated and slightly emarginated.

♀ Terminal abdominal segment greatly elongated, tubular; ventral plate simply truncated, dorsal lanceolate, longer than the ventral.

This very beautiful little species occurred sparingly at Ega, on slender branches of trees in the forest*.

5. *Colobothea luctuosa*, Pascoe.

Colobothea luctuosa, Pascoe, Trans. Ent. Soc. Lond. v., n. 4. i. 42.

" *C. nigra*; capite vittis tribus, prothorace quinque, elytris singulis duabus, una humerali altera medio-suturali, ochraceis, his fascia apicali, macula tertia terminali, antennarum articulis quarto sexto-que basi, albis. Long. 4½ lin. Pará." *Pascoe, loc. cit.* ♂ ♀.

This elegant species is readily distinguishable from all others that I have seen, by its peculiar colouring. The thorax is tumid on the sides behind the middle, and constricted between that point and the base. The elytra are gradually attenuated from

---

* To this section of the genus belongs the following:—

Colobothea biguttata, n. sp. Parum elongata, convexa, postice a medio elytrorum declivis, grisea. Caput obscure griseum. Antennæ piceæ, articulis basi griseis. Thorax lateribus longe ante basin tuberculatis, deinde angustatus, griseus, dorso fulvo quadrimaculatus. Elytra basi lata, humeris paulo obliquis, imprimis sensim, apices versus citius angustato, apicibus truncatis, angulis suturalibus obtusis exterioribus breviter spinosis; supra punctata, haud setosa, grisea, maculis rotundatis fulvis et altera pone medium majore et discoidali nigra ornata. Pedes rufescentes, tibiis tarsisque nigricantibus, his articulo primo griseo. Maris (?) segmento ultimo abdominali simplici obtuso. Long. 4½ lin. ♂ (?). *Hab.* in Brasilia, a Dom. Jekel sub nomine *C. biguttata* Dej. missa.

base to apex, and the latter is rather obliquely truncated, with the external angle alone produced into a spine. The scutellum is black, with a yellow spot at its tip. The apex of the elytra is ashy white, and there is also a white dot on the disk not far from the apex. Besides the yellow line on each extending from the base to the middle and the sutural streak, there is also a yellow dot near the suture, a short distance behind the scutellum. The white ring at the base of the sixth antennal joint is obsolete in the female. The body beneath is clothed with grey pile, and has an ochreous-ashy streak of denser pile on each side. The terminal antennal joints are much longer in the male than in the female, and there is but little sexual difference in the form of the terminal abdominal segment.

I met with the species at Ega on the Upper Amazons, and not at Pará, as erroneously recorded by Mr. Pascoe.

6. *Colobothea dioptica,* n. sp.

C. brevis, lata, convexa, brunnea, supra nullomodo setosa, thorace prope basin utrinque tuberculo acuto, deinde subiter angustato; elytris pone medium macula rotundata atro-velutina flavo cincta. Long. 4½ lin. ♂ ♀.

Head black, vertex grey. Antennæ pitchy, bases of the middle joints slightly grey. Thorax widened from the front to near the hind margins, and each side forming at that point an acute prominence, after which it is suddenly narrowed to the base; surface brown, varied with indistinct lighter brown marks. Scutellum dark brown, with a central tawny-ashy spot. Elytra short, broad, and convex, shoulders forming a short and very prominent ridge, the lateral carina proceeding thence being scarcely elevated, and disappearing before the middle of the elytron; apex truncated in a slightly flexuous line, sutural angles rounded off, external angles produced into a short and broad tooth: surface free from setæ, brown, speckled with light tawny brown, and each elytron having, behind the middle, a large round velvety-black spot encircled with yellow. Body beneath black, clothed with grey pile; sides of abdomen spotted with grey; terminal segment shining black. Legs black or reddish, ringed with grey. Fore tarsi simple in both sexes.

♂. Terminal ventral segment deeply notched; dorsal broad and obtuse.

♀. Terminal ventral segment simply truncated; dorsal tapering and obtuse.

On slender dead twigs in the forest; Pará and banks of the Tapajos. Rare.

This species seems to resemble much in colours and shape

Priscilla hypsiomoïdes, Thoms. (Systema Ceramb. p. 81); but the character he gives, "brunneo-setosa," does not at all suit, as our insect is one of the few *Colobothea*-forms which are destitute of setæ on the surface of the body.

§ II. Fore tarsi dilated and ciliated in the male.

a. Thorax tumid on each side behind the middle, or furnished with a tubercle: narrowed at the base.

7. *Colobothea pictilis*, n. sp.

C. elongata, postice modice angustata, grisea; thorace pone medium acute tuberculato, vitta latiore dorsali altera laterali lineolisque duabus utrinque intermediis nigris; elytris apice utrinque biden- tatis, fulvo maculatis, fasciis duabus interruptis nigris. Long. 3¼–4½ lin. ♂ ♀.

Head dusky grey, vertex with two ashy lines, diverging on the occiput. Antennæ black, bases of the joints grey. Thorax widest a little behind the middle, where a conical projection is formed on each side, behind constricted: surface grey, with a central vitta (unequal in width) and a lateral stripe, below the tubercles, black; there are also on each side of the upper surface two fine black lines, sometimes partially united. Scutellum black, with a central ashy spot. Elytra moderately elongated and attenuated, apex sinuate-truncate, sutural angles produced into a short tooth, external ones into an elongate spine; surface grey, sprinkled with tawny patchy spots; each elytron has besides two short angulated lateral fasciæ of a black colour, and more or less distinct indications of a third near the apex. Body beneath clothed with ashy tomentum; abdomen of the female spotted with black. Legs black, ringed with grey: fore tarsi moderately dilated and fringed in both sexes.

♂. Apical ventral segment greatly distorted, its surface forming an angular elevation with an elevated ridge on each side: the concavity thus formed shining black; the dorsal segment is notched in the middle. The middle segments of the abdomen are greatly contracted in the middle.

♀. Apical ventral segment with its terminal angles produced into long spines; dorsal segment broadly notched.

Pará, on branches of dead trees; rare.

8. *Colobothea pulchella*, n. sp.

C. parva, postice sensim attenuata, carneo-grisea; thorace pone me- dium prominulo angulato, deinde constricto, vitta dorsali (medio constricta) altera laterali lineolisque duabus utrinque intermediis nigris; elytris utrinque apice bidentatis, carneo maculatis, humeris, fasciis duabus interruptis lituraque subapicali nigris. Long. 4 lin. ♂.

Head pinkish grey, occiput with two stripes and a posterior spot black. Antennæ pitchy black; bases of joints grey, those of sixth, eighth, and tenth joints whiter. Thorax widened behind the middle, and forming there an acute prominence, constricted behind. Surface pinkish grey, with a dorsal vitta (constricted behind the middle), a lateral stripe, and two fine lines on each side black. Scutellum black, with a minute grey spot at the base. Elytra gradually narrowed from base to apex, sinuate-truncate, with the sutural angles produced into a short, the external into a long tooth: surface sparingly clothed with fine setæ, punctured, grey, sprinkled with pinkish patchy spots, a short stripe under each shoulder (continuous with the lateral thoracic stripe), a very short streak proceeding from each angle of the scutellum, a spot on the disk near the base, a short fascia behind the middle, and a curved letter near the apex black. Body beneath grey. Legs grey, ringed with black: fore tarsi in the ♂ moderately dilated and fringed.

♂. Terminal ventral segment strongly elevated towards the tip, the elevation surmounted by a curved ridge, leaving a smooth concave space within; second segment strongly contracted in the middle.

Banks of the Tapajos; one example. This and the preceding species are closely related to a Venezuelan species of much larger size *.

9. *Colobothea obtusa*, n. sp.

C. modice elongata, postice attenuata, fusco-nigra, cinereo maculata; thorace brevi, pone medium parum tumidulo, deinde leviter con-

---

* *Colobothea lineola* (Chevrol. MS. sec. Dom. Deyrolle).—Elongata, postice sensim attenuata, grisea, fulvo nigroque variegata. Caput nigricans. Antennæ nigricantes articulis basi griseis. Thorax usque ad medium dilatatus, lateribus pone medium valde acute tuberculatis, deinde basin versus sinuato-attenuatus; supra griseus fulvo variegatus, medio vitta postice dilatata, altera laterali lineolisque duabus intermediis nigris. Scutellum postice angustatum, nigrum, macula grisea. Elytra gradatim attenuata, truncata, angulis suturalibus simplicibus, exterioribus valde productis, humeris prominulis parum obliquis; supra breviter setosa, punctata, grisea, punctis nigris maculisque carneo-griseis variegata, utrinque maculis angulatis tribus quarum una pone medium major. Corpus subtus cinereum: abdominis medio et lateribus nigro maculatis; segmento apicali nigro, basi cinereo maculato. Pedes cinerei, nigro annulati: tarsis anticis maris valde dilatatis et ciliatis. Maris segmentum ultimum ventrale simplex, late irregulariter truncatum, angulis haud productis; dorsale obtusum: fœminæ segmentum ultimum ventrale angulis dentatis; dorsale magis attenuatum, apice breviter emarginatum. Long. 6¼-8. ♂ ♀. *Hab.* Venezuela.

Q

stricto; elytris apice sinuato-truncatis angulis haud productis. Long. 4½ lin. ♂ ♀.

Head brown, vertex with one, forehead with two ashy stripes. Antennæ clothed with stiff setæ, pitchy black, base of the fifth joint with a white ring; in the male the base of the fourth joint is also ashy. Thorax short and rather rounded on the sides, slightly tumid not far from the base, and then gradually narrowed to the base: surface black, centre with two short lines in front and a longer line behind (pointing between the two short ones) ashy; the sides have each two ashy lines, besides the ashy stripe lying over the fore coxæ. Elytra moderately prominent at the shoulders, apex somewhat narrow and truncated in a slightly incurved line, with the angles not at all prominent; surface punctured and clothed with fine setæ, black, varied with a large number of ashy spots of an oblong or short linear form; apex white. Body beneath greyish. Legs pitchy, clothed with grey pile; femora with a grey central ring; tarsi grey, two terminal joints black: fore tarsi of the male strongly dilated and ciliated.

♂. Terminal ventral segment broadly truncated, angles produced into long spines; dorsal notched.

♀. Terminal ventral segment narrowed towards the tip, angles produced into spines; dorsal also narrowed, notched at the apex.

Ega, on branches of dead trees. There is a handsome species in collections from Mexico, which much resembles *C. obtusa* in the form of the thorax *.

10. *Colobothea humerosa*, n. sp.

C. elongata, variegata, thorace tuberculis acutis lateralibus retrorsum spectantibus mox ante basin sitis; elytris humeris antice dilatatis, griseis, carneo maculatis, utrinque fasciis macularibus tribus nigris; pedibus carneo nigro griseoque variis. Long. 4½–6 lin. ♂ ♀.

Head blackish, forehead with three indistinct yellowish lines, vertex with two similar lines diverging on the occiput, cheeks striped with ashy ochreous. Antennæ pitchy, bases of the fourth

---

* *Colobothea leucophæa* (Chevrol. MS. sec. Dom. Deyrolle). Latiuscula, depressa, nigra, cinereo variegata. Caput nigrum, fronte cinereo obscure lineata, vertice vittis duabus cinereis postice divergentibus. Antennæ piceæ, articulis basi cinereis. Thorax brevis, lateribus rotundatis, ante basin tumidis, deinde constrictis; dorso nigro, medio vitta lata (lineola nigra includente) lateribusque maculis tribus cinereis. Scutellum triangulare, nigrum, cinereo marginatum. Elytra latiuscula, depressa, postice modice attenuata, truncata, angulis exterioribus spinosis; nigra, maculis cinereis confluentibus conspersa, relictis fasciis interruptis angulatis duabus nigris, una ante, altera pone medium. Corpus subtus dense cinereo tomentosum. Pedes cinerei, tibiis piceis cinereo annulatis, tarsis cinereis apice nigris. Fœminæ segmentum ultimum abdominale attenuatum; lamina ventrali longe bispinosa. Long: 6¼ lin. ♀. *Hab.* in Mexico.

and sixth joints grey, middle of the terminal joints grey. Thorax at first sight appearing to be gradually narrowed from base to apex; but the base itself is narrowed, and each side has an acute projection, pointed backwards, and nearly touching the humeral callus; surface with thirteen stripes alternately black and tawny, the central (black) stripe with a grey line down its middle. Scutellum black, streaked with grey. Elytra with the shoulders not advanced laterally but vertically and forward, so that the humeral ridge fits into the narrow space between the lateral tubercle and the base of the thorax; the surface is setose, punctured, and grey, with numerous pinkish marks which are chiefly collected round the black fasciæ; the latter are three in number—one, short, before the middle, the second, oblique and angular, behind the middle, and the third, quadrate, at the apex; the extreme apex is bordered with grey or pinkish, and is truncated, with the sutural angle simple, the external produced into a long spine. Body beneath greyish, sides with a stripe of fulvous tomentum, abdomen with the sides spotted in the middle. Legs grey, femora with a pinkish spot on their upper surface; tibiæ ringed with grey and black; tarsi black, with the two basal joints grey; fore tarsi of the male moderately dilated and fringed.

♂. Terminal ventral segment broadly emarginated, angles acute; dorsal narrowed to the tip, broadly notched.

♀. Terminal ventral segment broadly emarginated, with a pencil of stiff hairs proceeding from each angle; dorsal truncated.

Branches of dead trees, forest, Pará. In the colours of the elytra this species resembles *C. velutina*.

To this section of the genus belong also *C. pœcila*, Germar (Ins. Nov. p. 488), *C. subcincta*, Castelnau (Anim. Artic. ii. p. 491), *C. strigosa*, Mannerheim *, and *C. vidua* † (Chevrol. MS.);

* *C. strigosa* (Mannh. sec. Dom. Deyrolle). Elongata, postice paulo attenuata. Caput et antennæ rufescentia, his articulis basi pallidioribus. Thorax ante basin tuberculo conico instructus, deinde parum angustatus, dorso brunneo vittis quinque lateribusque nigris. Scutellum nigrum, macula grisea. Elytra brunnea, cinereo fulvoque varia, macula obliqua angulata ante medium, fascia valde undulata pone medium liturisque angulatis prope apicem nigris; apicibus sinuato-truncatis, angulis suturalibus paulo, exterioribus valde productis. Corpus subtus ochraceum, abdomine nigro maculato. Pedes rufescentes, cinereo annulati. Long. 7 lin. ♀. *Hab.* in Brazil.

† *C. vidua.* Minor, nigra, cinereo maculata. Caput nigrum, vertice lineolis duabus divergentibus cinereis. Antennæ nigræ, articulis basi griseis. Thorax angustior, lateribus pone medium tuberculo parvo armatis; niger, dorso vittis duabus maculam includentibus lineolaque laterali cinereis. Elytra nigra, cinereo maxim maculata, apice truncata, angulis exterioribus productis. Corpus subtus cinereum. Pedes nigri, cinereo annulati. Long. 4½ lin. ♀. *Hab.* in Mexico.

the last mentioned from Mexico, the other two from Rio Janeiro. *C. Schmidtii* * of French collections (Brazil) from the very slight, if any, narrowing of the thorax near the base, seems to stand on the confines of subsections *a* and *b*.

b. Thorax widest at the basal angles, gradually narrowed thence to the apex.

11. *Colobothea pimplea*, n. sp.

C. minus elongata et attenuata, cinereo- vel griseo-fulva; capite fusco, vertice lineis duabus divergentibus cinereo-fulvis; thorace vittis septem fuscis, quarum una mediana latiore; elytris griseis, fusco irroratis, cinereo-fulvo maculatis, fasciis tribus (apud suturam interruptis) fuscis, apice sinuato-truncatis, angulis interioribus prominulis, exterioribus spiniformibus. Long. 4½–5 lin. ♂ ♀.

Head dingy brown, forehead streaked with tawny, vertex with two fine tawny lines diverging on the occiput. Antennæ black or reddish, fourth, sixth, eighth, and tenth joints with a whitish ring. Thorax depressed at the base, ashy or tawny, with a broad central vitta, and, on each side, three narrower vittæ, purplish brown. Elytra moderately elongated, apex sinuate-truncate, with exterior angles produced into spines, and sutural angles dentiform; surface grey, minutely speckled with dusky and sprinkled with larger tawny (most often rounded) spots: each side has three transverse-quadrate purplish-brown spots or fasciæ, which do not reach the suture, the apex edged with tawny. Body beneath clothed with tawny-ashy pile; abdomen spotted on the sides with black. Legs reddish, spotted with ashy.

Terminal abdominal segment rather more tapering in the female than in the male; dorsal plate notched in both sexes; ventral plate terminating in spines in the male, angles simply acute in the female. Fore tarsi in the male moderately dilated, fringed with long hairs.

Branches of felled trees: Pará, Obydos, and banks of the Tapajos. Also found at Cayenne. I have seen it, in French collections, under the name of *C. sexlineata* (Reiche, MS.)—a name which I have not adopted, as the thorax has seven lines, and not six.

* *C. Schmidtii.* Elongata, postice attenuata. Caput obscurum, vertice cinereo macula trigona nigra. Antennæ nigræ, articulis basi griseis. Thorax ante basin tuberculo parvo, deinde vix angustatus; dorso griseus, vittis quinque nigris, mediana latiore lineolam griseam includente. Elytra grisea, nigro punctata, maculis confluentibus ochraceo-cinereis conspersa, maculis majoribus vel fasciis tribus nigris, una (interdum obsoleta) ante medium, altera majore angulata pone medium, tertiaque prope apicem; apicibus sinuato-truncatis, angulis suturalibus prominulis, exterioribus productis. Pedes picei, cinereo annulati. Fœminæ segmentum ultimum dorsale attenuatum bifidum; maris emarginatum. Long. 7½ lin. ♂ ♀. *Hab.* in Rio Janeiro, a Rev. Hamlet Clark lecta.

12. *Colobothea destituta*, n. sp.

C. minus elongata, obscure grisea ; capite nigro, vertice lineis duabus
divergentibus griseis ; thorace vittis septem nigris, quarum una
mediana latiore ; elytris obscure griseis, nigro confertim irroratis,
cinereo-griseo maculatis, fasciis interruptis tribus vel duabus (plus
minusve obsoletis) nigris, apice sinuato-truncatis, angulis interioribus prominulis, exterioribus spiniformibus. Long. 4½–6 lin. ♂ ♀.

Head blackish, forehead streaked with ashy-grey, vertex with
two fine ashy lines diverging on the occiput. Antennæ black,
fourth, sixth, eighth, and tenth joints with a whitish ring. Thorax dull grey, with seven black vittæ, central one twice as thick as
any of the rest. Elytra moderately elongated and tapering, apex
sinuate-truncate, sutural angles dentiform, external spiniform ;
surface obscure grey, thickly irrorated with blackish, and having
a few larger ashy, mostly rounded spots ; each with two (and sometimes an indication of a third) transverse quadrate black spots,
not distinctly limited. Body beneath ashy ; abdomen spotted
on the sides with black. Legs black, spotted with grey.

♂. Terminal ventral plate sinuate-truncate, angles acute ;
dorsal plate narrower, sinuate-truncate.

♀. Terminal abdominal segment elongated and tapering ;
both plates sinuate-truncate, not spinose.

On branches of dead trees, Pará.

13. *Colobothea seminalis*, n. sp.

C. minus elongata, fusco-nigra ; capitis vertice lineis duabus divergentibus ; thorace vittis sex, elytris maculis parvis partim confluentibus, cinereo-fulvis, his apice cano marginatis sinuato-truncatis, angulis exterioribus spinosis. Long. 3¾–5 lin. ♀.

Head blackish, forehead streaked with ashy-tawny, vertex
with two ashy-tawny lines diverging on the occiput. Antennæ
black or reddish, fourth, sixth, eighth, and tenth joints with a
whitish ring at their bases. Thorax black, with three ashy-
tawny longitudinal lines on each side ; the sternum and the
sides above the coxæ also tawny-ashy : on the surface near the
base are two distinct punctures, besides the row along the
hind margin. Elytra moderately elongated and tapering, deep
brownish black, covered with little oblong tawny-ashy spots,
which are collected together irregularly in some places, leaving
small spaces of the ground-colour ; the apex has a hoary spot.
Body beneath tawny-ashy ; abdomen spotted (as in the allied
species) with black. Legs blackish, spotted with grey.

♀. Terminal ventral plate simply sinuate-truncate, angles not
produced ; dorsal plate narrower, emarginate at apex.

Branches of dead trees, Pará.

14. *Colobothea paulina*, n. sp.

C. robustior, modice elongata, fusco-nigra ; capitis vertice lineis duabus divergentibus ; thorace vittis sex, elytris maculis parvis oblongis discretis, cinereo-fulvis, apice cano marginatis, sinuato-truncatis, angulis exterioribus spinosis. Long. 4–6 lin. ♀.

Head dusky, streaked with ashy-tawny, vertex with two ashy-tawny lines diverging on the occiput. Thorax black, with three tawny-ashy lines on each side ; the sides above the coxæ and the sternum of the same colour ; surface wanting the two punctures near the base which are distinctive of *C. seminalis*. Elytra brownish black, sprinkled with a number of small oblong tawny-ashy spots, which are so arranged as to leave black undefined spaces in the situations where lateral spots or fasciæ are usually situated in the allied species ; apex edged with hoary white. Body beneath ashy-tawny ; abdomen spotted with black on the sides. Legs black, spotted with grey.

♀. Terminal ventral plate broadly emarginated at the apex, and with a tooth in the middle of the emargination ; angles produced into spines. Dorsal plate tapering, obtuse.

Upper Amazons, at S. Paulo, on branches of dead trees. The species has also been found in the interior of French Guiana by M. Bar. A closely allied form is found in Venezuela ; but it differs greatly in the shape of the terminal abdominal segment in the female*.

15. *Colobothea varica*, n. sp.

C. modice elongata et attenuata, fusca ; thorace dorso vittis duabus cinereo-fulvis, postice divaricatis, lateribus cinereo-fulvis vitta nigra ; elytris maculis parvis cinereo-fulvis, partim discretis, partim subconfluentibus. Long. 4½–5 lin. ♂ ♀.

Head dusky, forehead streaked with tawny-ashy ; two diverging lines of the same colour on the vertex. Antennæ black, reddish towards the base ; fourth, sixth, eighth, and tenth joints with whitish rings. Thorax black on the surface, with two thickish tawny vittæ diverging behind ; sides and under surface ashy, each with a black stripe. Elytra moderately tapering, apex truncate, exterior angles spinose ; surface dark brownish, sprinkled with

---

* *Colobothea mosaica* (Deyrolle, MS.). Modice elongata, nigra, griseo maculata. Caput nigrum, orbita oculorum griseo marginata. Antennæ fuscæ, griseo tomentosæ, articulis basi pallidioribus. Thorax griseus, vittis octo nigris, dorso punctis duobus prope basin. Elytra modice attenuata, apice truncata, angulis exterioribus spiniformibus, nigra, maculis quadratis griseis in seriebus sex vel septem ordinatis, pone medium et prope apicem interruptis. Corpus subtus cinereo-fulvo tomentosum, nigro maculatum. Pedes picei, cinereo tomentosi. Fœminæ segmento terminali abdominis attenuato, apice fisso. Long. 6 lin. *Hab.* in Venezuela. Coll. Bates.

small tawny spots, sometimes arranged in rows over the basal half, but agglomerated more or less beyond the middle, leaving clear spaces; in other examples more irregular, apex edged with whitish. Body beneath somewhat uniformly clothed with tawny-ashy tomentum (except, as usual, the terminal segment). Legs reddish, spotted with ashy.

♂ ♀. Terminal abdominal segment tapering; dorsal plate rounded at tip; ventral broadly truncate, with angles produced into short and broad spines, and middle of the truncation slightly advanced or festooned. The fore tarsi in the ♂ are simple.

Branches of dead trees, Ega; abundant.

16. *Colobothea propinqua,* n. sp.

C. modice elongata et attenuata, fusca; thorace cinereo-fulvo, vittis septem fusco-nigris, una mediana et tribus utrinque lateralibus quarum duabus postice conjunctis tertiaque inferiore tenuissima; elytris maculis parvis cinereo-fulvis in seriebus interruptis ordinatis. Long. 5 lin. ♀.

Head blackish, streaked with greyish; vertex with two divergent grey lines; antennæ black, reddish towards the base, fourth, sixth, eighth, and tenth joints ringed with white. Thorax ashy, with seven black vittæ—namely, one in the middle, broader, two on each side converging and blending before reaching the base, and one below them very slender. Elytra truncate at apex, with exterior angles spiniform; surface dark brown, covered with distinct ashy-tawny spots, arranged partly in rows, but interrupted by oblique clear spaces near the base, at the middle, and near the apex; apex with an ashy spot on each elytron, much enlarged towards the suture. Body beneath clothed with tawny-ashy tomentum; abdomen spotted with black. Legs reddish, spotted with black and ashy.

♀. Terminal ventral segment strongly tapering; ventral plate truncate, angles produced into lengthy spines; dorsal plate rounded at apex.

S. Paulo, Upper Amazons. It is very closely allied to *C. varica,* but differs in the thoracic markings from all the numerous specimens which I have examined of that species.

17. *Colobothea nævia,* n. sp.

C. elongata, nigra; thorace vittis quatuor tenuibus, elytris maculis parvis, rotundatis, dispersis, cinereis; corpore subtus vitta laterali fulvo-cinerea tomentosa infra nigro marginata. Long. 4½–6 lin. ♂ ♀.

Head black, streaked with tawny, vertex with two diverging lines of the same colour. Thorax black, with four tawny-ashy lines, the two dorsal ones not at all divergent. Elytra somewhat more elongated than in the preceding species; apex truncate, outer angles spinose; surface black, shining, and sprinkled with rounded tawny-ashy scattered spots, which sometimes leave a clear space behind the middle and near the apex; apex margined with whitish. Body beneath greyish; each side with a broad stripe of dense tawny tomentum extending from the front margin of the prothorax to the tip of the abdomen, interrupted on each segment of the latter by a black spot. Legs ashy, spotted with black.

♂. Terminal ventral segment elongated, flattened, tapering, very much longer than the dorsal, and deeply notched at the apex. Anterior tarsi not dilated, but fringed with long fine hairs.

♀. Terminal ventral segment tapering, apex sinuate-truncate, angles spinose; dorsal segment obtuse at apex, slightly notched in the middle.

On branches of dead trees, Ega. An abundant species.

18. *Colobothea juncea*, n. sp.

C. gracilior, angustata, fusca; thorace cinereo-fulvo, vittis septem fuscis; elytris pone humeros sensim, apices versus citius attenuatis, maculis parvis cinereo-fulvis plerumque confluentibus. Long. 4½ lin. ♀.

Head blackish, streaked with ashy-tawny, vertex with two diverging tawny lines. Thorax ashy-tawny, with seven blackish vittæ. Elytra slender, tapering gradually from the shoulders to near the apical spines, thence more quickly narrowed; apex truncate, outer angles spinose; surface sprinkled with ashy-tawny specks, agglomerated here and there into irregular larger spots, and leaving a clear space near the apex; apex broadly edged with white. Body beneath reddish, clothed with ashy tomentum, which is denser on the sides of the breast, and more scanty along the middle of the abdomen. Legs reddish, spotted with grey and black.

♀. Terminal abdominal segment tapering and narrow; ventral plate sinuate-truncate, angles not prominent; dorsal plate with a shallow angular emargination.

Pará.

19. *Colobothea securifera*, n. sp.

C. modice elongata, postice attenuata, fusca; thorace vittis sex cinereo-fulvis; elytris maculis parvis cinereo-fulvis conspersis, singulis spatio magno discoidali fusco maculam majorem cinereo-fulvam includente: maris segmento terminali ventrali angulis in lobos securiformes productis. Long. 4½ lin. ♂ ♀. (7 exempl.)

Head reddish brown, streaked with tawny, and with two divergent tawny lines on the crown. Antennæ reddish, bases

of alternate joints from the sixth ashy. Thorax chestnut-brown
or dark brown, with six vittæ and the under surface ashy-tawny.
Elytra rather short, tapering gradually and rather strongly from
base to apex; apex sinuate-truncate, external angles spinose;
surface dark castaneous brown or blackish brown, the basal and
apical parts dusted with irregular-sized tawny-ashy specks,
leaving a broad clear middle space, in the centre of which (on
each elytron) is a large irregular tawny-ashy spot; apex with a
triangular broadish ashy spot. Body beneath clothed with
tawny-ashy pile. Legs reddish; tarsi and tibiæ spotted with
ashy and black.

♂. Terminal ventral segment short and broad, each apical
angle produced into a long, deflexed, horny, hatchet-shaped lobe;
dorsal segment narrowed and emarginated at the apex: fore
tarsi moderately dilated, not fringed; first joint not broader
than the second.

♀. Terminal ventral segment tridentated, middle tooth shorter
and broader than the outer ones; dorsal segment narrow and
obtuse.

Pará and Lower Amazons.

20. *Colobothea sejuncta*, n. sp.

C. modice elongata, postice attenuata, fusca; thorace vittis sex
cinereo-fulvis; elytris maculis parvis cinereo-fulvis conspersis,
singulis spatio magno discoidali fusco maculam majorem cinereo-
fulvam includente: maris segmento terminali ventrali obtuso,
inermi, angulis penicillatis. Long. 4¼ lin. ♂ ♀. (3 exempl.)

Head reddish brown, streaked with tawny, and with two divergent tawny lines on the crown. Antennæ reddish, bases of
alternate joints from the fourth or sixth ashy. Thorax chestnut-
brown or darker, with six vittæ and the under surface ashy-
tawny. Elytra rather short and slender, gradually and rather
strongly tapering from base to apex; apex sinuate-truncate,
both sutural and external angles spinose, the sutural shorter;
surface dark castaneous brown, the basal and apical parts sprinkled with irregular-sized tawny-ashy specks, leaving a broad
clear space on the disk of each, in the centre of which is a larger
irregular tawny-ashy spot; apex with an ashy margin of regular
width. Body beneath clothed with tawny-ashy pile. Legs
reddish; tibiæ and tarsi spotted with ashy and black.

♂. Terminal abdominal segment rather elongate, thickened
before the apex; the ventral plate with obtuse angles, from each
of which proceeds a line of thick bristles; dorsal plate simple at
the apex, and closely applied to the sloping front margin of the
ventral. Fore tarsi with the first joint greatly dilated.

♀ . Terminal abdominal segment strongly tapering and notched at the apex.

Ega, Upper Amazons.

The very great and striking difference in the accessory genital organs between these two closely allied species (*Colobothea securifera* and *C. sejuncta*) merits a few words of especial mention. When I was separating my specimens of *Colobothea* into species, I placed together all the individuals belonging to these two as one and the same, and could not find anything in their form or markings to warrant their being treated as anything more than mere local varieties, even after I had given them a second examination. A species has so often proved to exist under distinct local forms on the Upper and Lower Amazons, that I concluded this was simply another example of the rule. When I came, however, to separate the sexes previous to describing the species, I discovered the remarkable difference of structure described above, and then noticed the two or three other small points of difference in the general shape and tips of the elytra which I have noted in the descriptions. A pair of elongated horny processes, which I suppose to be the sheath of the penis, project from between the terminal abdominal segments in two out of the three males I possess; in the third they appear to be withdrawn into the abdomen. It is a remarkable circumstance, that in many families of Insects which have accessory sexual parts easy of examination, it is found that these differ very considerably in structure in closely allied species. It has been remarked that they offer some of the best characters to distinguish species, and they have been made use of to separate species which scarcely offered any other distinguishable characters. Mr. Baly has also discovered that the horny penis concealed in the male abdomen of Phytophagous Coleoptera differs in form in closely allied species; and he has shown me a long series of specimens mounted for examination under the microscope, belonging chiefly to the genera *Chrysomela* and *Eumolpus*, which offer a most instructive study, since by their means some forms before considered as varieties turn out to be distinct species. This class of facts seems to me of great significance, as throwing light on the segregation of varieties and their passage into true species. For if we admit that the only sound difference between allied varieties and allied species is that the former intermarry, and the latter do not, then the abrupt and great diversities of structure in those organs most directly involved in the matter must be considered as affording an explanation why many varieties do not intercross with the parent stock, and therefore remain as independent forms or species. The difference in the accessory male organs of our two allied species or local forms of

Colobothea is so great that no one who examines them can believe both to be adapted to the corresponding organs of the females of each form. At the same time I have no doubt that, were it not for the great difference between these organs in our two forms, no entomologist would doubt their being mere local varieties of one and the same stock. Scores of other local varieties occur in the same countries, presenting all the successive steps of segregation, from the most partial variation to the full-formed local race.

Thus we have only to admit that species disseminate themselves over wide areas, and adjust themselves to the diversities of local conditions, or, in other words, segregate local varieties, to open the way towards an explanation of the way in which the world has become peopled by its myriads of species. The inevitable law of Natural Selection which governs the general process of the adjustment of the local races to new conditions will explain the changes of conditions of life in time; and the laws of variation, diversified in details as are the species themselves, will explain the rest.

21. *Colobothea bisignata*, n. sp.

C. modice elongata, fusca; thorace vittis novem cinereis; elytris maculis parvis subconfluentibus cinereis, relicto spatio medio fusco maculam magnam albam includente. Long. 5 lin. ♂ ♀.

Head rusty brown, streaked with ashy, vertex with two ashy lines divergent towards the occiput. Antennæ rusty brown, tips of joints blackish, bases of alternate joints whitish. Thorax with nine ashy longitudinal lines, the central one the slenderest, the second (from the central one) not reaching the hind margin, and the two lateral ones on each side very oblique. Elytra moderately short and tapering, apex sinuate-truncate, external angle produced into a long tooth; dark purplish brown, sprinkled near the base and apex with ashy dots, which unite here and there in irregular strigæ; the central space clear, and having in the middle of each elytron a large round white spot; there is also a small white spot on the suture near the scutellum. Body beneath clothed with dingy-ashy pile; abdomen spotted with black. Legs purplish brown, ringed with ashy.

♂ ♀. Terminal abdominal segment similar in form in the two sexes, longer and tapering in the female; the ventral plate in both truncated, with angles simply acute; the dorsal plate distinctly notched in the middle of its apex in the female, obtuse in the male. Tarsi simple in the male.

A common insect on dead branches, &c., at Ega.

22. *Colobothea latevittata*, n. sp.

C. elongata, angustior; thorace sordide fulvo-cinereo, vitta lata mediana alteraque tenui laterali fuscis; elytris postice attenuatis, humeris valde obliquis, fuscis, maculis fulvo- vel sordide cinereis plagiatim conspersis. Long. 4–5 lin. ♂ ♀.

Head blackish, streaked with tawny, vertex with a single tawny line extending to the occiput. Antennæ long and slender, dusky brownish at the tips and ashy at the bases of the joints. Thorax with a small acute prominence on each side near the base; clothed with dingy tawny or ashy tomentum, leaving a broad stripe in the middle and a slender line on each side of the disk dark brown. Elytra rather slender and tapering; shoulders very oblique, apex somewhat narrow and sinuate-truncate, with sutural angle slightly prominent, external dentiform; surface brown, covered with dingy ashy or tawny spots, which unite together in patches, leaving irregular brown spaces. Body beneath dingy tawny; abdomen not spotted with black on the sides. Legs rusty brown, varied with black, and ringed with ashy.

♂. Anterior tarsi dilated and fringed. Terminal abdominal segment short, truncated; ventral plate emarginated at the apex.

♀. Terminal abdominal segment greatly elongated and subtubular; dorsal plate slender, obtuse; ventral truncated, angles not produced.

Var. *Obydensis*. A female example in my collection from Obydos, on the Guiana side of the Lower Amazons, differs from the typical form in having a distinct quadrate silky-brown spot on each elytron close to the apex, and also a distinct broad dusky stripe along the episterna of the prothorax.

Taken at Carepí, near Pará; found also, but sparingly, at Santarem and at Ega.

23. *Colobothea styligera*, n. sp.

C. elongata, thorace fulvo-cinereo, vitta lata mediana alteraque tenui laterali velutino-nigris, lateribus infra cinereo-fusco late vittatis; elytris postice attenuatis, cinereo-fulvo dense confluenter maculatis, relictis utrinque macula rotundata pone medium alteraque quadrata apicali nigro-fuscis. Long. 6 lin. ♂ ♀.

Head blackish, streaked with tawny; vertex with a single tawny line. Antennæ blackish, bases of fourth to sixth joints white. Thorax with the posterior angles extending laterally towards the shoulders of the elytra, above tawny fulvous, with a broad velvety-black central vitta, and a narrow lateral line of the same colour. The episterna have a broad ashy-brown stripe. Elytra tapering, shoulders less oblique, densely clothed with confluent ashy-tawny spots, leaving a rounded discoidal spot on each behind the middle, and a quadrate one close to the apex, dark brown; apex sinuate-truncate, sutural angle slightly prominent, external dentiform. Body beneath ashy, varied with tawny patches; abdomen spotted with dusky. Legs tawny-ashy, spotted with black.

♂. Anterior tarsi dilated and fringed. Terminal abdominal segment not reaching the tip of the elytra, narrowed towards the apex; dorsal plate obtuse, ventral truncate-emarginate.

♀. Terminal abdominal segment tubular, prolonged considerably beyond the apex of the elytra; dorsal plate lanceolate, ventral truncated, angles not produced.

Ega.

24. *Colobothea grallatrix*, n. sp.

C. elongata, postice valde regulariter attenuata, nigra; thorace vittis sex cinereo-fulvis, lineolaque mediana cinerea; elytris cinereo fulvoque dense confluenter irroratis, maculis tribus utrinque discoidalibus plagaque magna apicali nigris; maris pedibus longissimis validis. Long. ♂ 6½, ♀ 4½ lin.

Head blackish, streaked with tawny, vertex with two tawny-ashy slightly divergent lines. Antennæ robust, black, bases of alternate joints ringed with whitish. Thorax black, and having on each side three tawny-ashy vittæ, and a thin grey line down the middle of the black central streak. Elytra with shoulders moderately prominent laterally and vertically, thence regularly tapering to the apex, which is truncated, with the sutural angle moderately produced, and the apical angle spiniform; the surface is thickly covered with confluent spots, partly grey and partly fulvous, which leave, on the disk of each, three spots (one before the middle, and two, placed obliquely, after the middle) and a large square black apical spot of a fine black colour; apex margined with ashy. Body beneath grey; sides, from the front of the prothorax to the penultimate ventral segment, occupied by a broad ochreous-tawny stripe; sides of abdomen spotted with black. Legs ashy, spotted with black.

♂. Legs greatly elongated, and stouter than in the female; anterior tarsi broadly dilated and hirsute. Terminal abdominal segment short; apex both of the dorsal and ventral plates emarginated.

♀. Terminal abdominal segment narrow and moderately elongated, obtuse, angles not produced.

Ega and S. Paulo; rare.

25. *Colobothea olivencia*, n. sp.

C. elongata, postice regulariter attenuata, nigra; thorace vittis quatuor cinereo-fulvis; elytris confluenter fulvo-cinereo maculatis, plaga quadrata apicali nigra. Long. 5½–6½ lin. ♂ ♀.

Head blackish, streaked with tawny, vertex with two divergent tawny-ashy lines. Antennæ black, fourth, sixth, and tenth joints white at the base. Thorax black; disk with two tawny-ashy vittæ continuous with the lines on the crown of the head, and, like them, divergent posteriorly; besides these, there is a narrower vitta on each side at the extreme edge of the pronotum, and scarcely visible from above. Elytra with a prominent black tubercle at the apex of the prominent shoulders, gradually tapering, apex truncate, sutural angles scarcely prominent, external spiniform; surface sprinkled with tawny (and a few grey) spots, which are confluent, but do not leave very distinct black spaces; close to the apex, on each, is a large square black patch, the apex itself being edged with whitish. Body beneath ashy-tawny; breast, and abdomen on the sides, streaked or spotted with black. Legs varied with ashy and black.

♂. Larger and more robust than the female, both in body and limbs; anterior tarsi dilated and fringed. Terminal abdominal segment short, apex of both the dorsal and ventral plates emarginated.

♀. Terminal abdominal segment elongated, and projecting beyond the apex of the elytra, but not tubular, and somewhat flattened, with the apex both of the dorsal and ventral plates truncated and notched in the middle.

S. Paulo, Upper Amazons; rare.

26. *Colobothea pura*, n. sp.

C. elongata, postice attenuata, nigra; thorace vittis quatuor elytrisque maculis confluentibus cinereo-fulvis, his spatio apicali nigro; antennis robustissimis (♂), nigris, articulo sexto annulo lato albo, 8vo et 10mo basi cinereis. Long. 5½–6½ lin. ♂.

Head black, with ashy lines, vertex with two divergent tawny-ashy lines. Antennæ (♂) extremely stout, gradually tapering to the apex, deep black; the joints from the base to the sixth spotless; the sixth has a white ring occupying two-thirds of the length of the joint; the base of the eighth joint is grey (on one side only), and the tenth joint has an ashy ring. Thorax deep black, the central part with two ashy-tawny vittæ not continuous with the lines of the crown, and parallel; the sides near the episternum have also each a tawny-ashy line. Elytra tapering in straight lines to the apex, the latter truncated, with sutural angle not produced, external dentiform; surface thinly and irregularly sprinkled with punctures, each of which has a very short strong bristle, and being surmounted by a granule; olivaceous-black, sprinkled with ashy or tawny-ashy specks, everywhere confluent and forming a marbled pattern, but leaving a black space near the apex; apex itself edged with ashy. Body beneath grey, varied with tawny, and having, in fresh examples, a tawny-ochreous lateral vitta from the front edge of the prothorax to the last segment of the abdomen; abdomen thinly clothed with grey pile, sides spotted with black. Legs black, thinly clothed with grey pile; knees and tips of tibiæ and tarsi black.

♂. Legs elongated and robust; anterior tarsi diluted and fringed. Terminal ventral segment semicircularly sinuated at the apex, with the angles acute and produced; dorsal plate broad, obtuse, faintly emarginated.

Obydos, Lower Amazons.

27. *Colobothea carneola*, n. sp.

C. elongata, postice modice attenuata, nigra; thorace vittis quatuor, elytrisque maculis numerosis discretis, carneo-fulvis, his spatio apicali nigro; antennis (♂) normalibus, articulis 4to, 6to, 8vo, 10mo albo annulatis. Long. 5¾ lin. ♂.

Very closely allied to *C. pura*, but differs in the degree of robustness and coloration of the antennæ, and in the spots on the elytra being nearly all quite separate and inclining in colour towards pinkish red. The elytra are sprinkled with punctures, as in *C. pura*, but they are not so conspicuous, nor surmounted by elevated points; the bristles are more numerous towards the apex, although the punctures from which they arise are not conspicuous. Body beneath and legs as in *C. pura*.

♂. Terminal ventral segment semicircularly sinuated at the apex, with the angles acute; dorsal obtusely truncated. Anterior tarsi dilated and fringed.

Obydos.

28. *Colobothea forcipata*, n. sp.

C. gracilis, postice valde attenuata, nigra, vertice thoraceque vitta lata communi cinerea; elytris cinereo nebulosis, relictis plagis lateralibus et vitta lata apicali nigris; antennis nigris, articulo 6to albo annulato; maris segmento ventrali terminali forcipato. Long. 4-6½ lin. ♂.

Head black, forehead spotless, vertex with a broadish ashy line, which continues along the middle of the thorax, enlarging posteriorly, the rest of the surface of the thorax deep black. Antennæ black, sixth joint alone marked with a white ring. Elytra gradually attenuated from base to apex, the latter sinuate-truncate, sutural angle prominent, external spiniform; surface punctured, setose, and marked with an ashy cloud extending from the scutellum to near the apex, and emitting several irregular branches; the apical part is crossed by a broad black vitta, the apex itself being edged, as usual, with white. Body beneath ashy; sides of thorax and abdomen with a broad yellowish vitta. Legs ashy, spotted with black.

♂. Terminal ventral segment with each side produced into a long, compressed, incurved, horny lobe, the apex of which is obliquely truncated; dorsal plate obtusely rounded at apex.

Ega, rare.

29. *Colobothea navigera*, n. sp.

C. modice elongata, postice regulariter attenuata, nigricans, sericea, vertice thoraceque supra lineis duabus, elytris maculis paucis discretis, cinereis; his truncatis, angulis externis spinosis. Long. 4½–7½ lin. ♂ ♀.

Head black, forehead with three ashy lines, cheeks with a spot of the same colour, and vertex marked with two ashy lines diverging on the occiput. Antennæ greatly elongated and robust, black, sixth joint ringed with white, tenth joint with an exterior white line (♂), in the ♀ the eighth and eleventh joints also streaked with white. Thorax blackish, clothed with an olivaceous silky pile, the upper surface with two tawny-ashy, slender, nearly parallel lines; sides each with a single similar line, besides a broader streak above the coxæ. Elytra broad at the base, with prominent and not markedly oblique shoulders, regularly attenuated thence to the apex, which is truncated and has the external angles produced into spines; the surface has a few fine punctures surmounted by acute granulations towards the base, and beset with short black bristles; the colour is blackish, clothed with silky olivaceous pile, and ornamented with a small number of scattered and distinct, rounded, tawny-ashy spots, the extreme apex having an ashy-white border decreasing in width from the suture to the external angle. Body beneath black, thinly clothed with ashy tomentum; the sides of the breast have a tawny-ashy streak in continuation of the one on the prothorax, and the sides of the abdomen are spotted with the same colour. The legs are blackish, ringed with grey.

♂. Terminal ventral segment narrowed to the apex, truncated, with the angles produced into stout spines; dorsal segment obtuse. Legs stout; anterior tarsi moderately dilated and fringed. In the smaller males the legs are not perceptibly thicker than in the females.

♀. Terminal abdominal segment projecting considerably beyond the apex of the elytra, broad; dorsal segment notched, ventral truncated, angles not produced.

A common insect at Ega and S. Paulo, Upper Amazons.

80. *Colobothea lucaria*, n. sp.

C. modice elongata, nigra, vertice lineis duabus divergentibus, thorace lineis tenuibus quatuor, elytris maculis paucis hic illic congregatis, griseis; his apice cano marginatis, oblique truncatis, angulis externis spinosis. Long. 5 lin. ♂.

Head black, forehead with three obscure grey lines, vertex with two divergent lines of similar colour, and the posterior part of the orbits also grey. Antennæ black, base of fourth, eighth, and tenth joints grey on one side, sixth joint with a whitish ring. Thorax black, with a silky olivaceous gloss, upper surface with two slender parallel grey lines, each side also with a similar line visible in part when the insect is regarded from above; there is also a grey line above the coxæ. Elytra prominent, and scarcely oblique at the shoulders, thence gradually attenuated to the apex, which is on each side obliquely truncated, i. e. the sutural portion is more advanced than the lateral angles, which are produced into spines; the surface is finely punctate-granulate, and of the same colour as the thorax; the grey spots are nearly all of equal size and distinct; but they are collected partly into groups, and here and there confluent; the grey apical margin is of equal width from the sutural to the external angle. Body beneath thinly clothed with grey pile; sides of breast not striped with thicker tomentum. Legs black, ringed with grey.

♂. Terminal ventral segment truncated, angles produced into short spines; dorsal segment rounded. Anterior tarsi moderately dilated and fringed.

S. Paulo, Upper Amazons. Very closely related to *C. navigera*, differing only in the oblique truncature and somewhat different arrangement of spots of the elytra.

81. *Colobothea crassa*, n. sp.

C. major, robusta, nigra, tomento olivaceo-griseo vestita, vertice thoraceque dorso lineis duabus divergentibus, elytris maculis numerosis, minimis, discretis, fulvo-griseis, apice cano marginatis. Long. 8–10 lin. ♂ ♀.

Differs from *C. navigera* in being of much larger size, in the spots of the elytra being very much smaller and more numerous, and in the dorsal lines of the thorax being posteriorly divergent. In shape and in colour the two species offer no tangible point of difference. As in *C. navigera*, there are only two thoracic lines visible from above, although there is a lateral line on each side and a broader streak above the coxæ (yellower in colour and extending to the abdomen); the form of the terminal abdominal segment in both sexes offers also no difference in the two species. *C. crassa* is still more closely allied to a Cayenne species, *C. lineatocollis** (Dej. Cat.), which is similar to it in size and other respects, and differs chiefly in the multitudinous grey specks of the elytra being confluent and forming irregular marbled lines. *C. Osculatii* of Guérin (Cat. des Ins. Col. recueillis par Gaetano Osculati, no. 261) appears to be another allied form similar in size and colours to *C. crassa* and *C. lineatocollis*; but the description given of the thoracic markings ("quatre fines lignes longitudinales blanches") leaves us in doubt whether there are not four lines on the upper surface, which would remove the species from the neighbourhood of the two mentioned; for, if the lateral lines are to be included, the description ought to mention six instead of four. The distinctive character of *C. crassa* is the minute and equal size, great number, and equidistant position of the grey specks of the elytra.

Common in the neighbourhood of Pará. *C. Osculatii* is probably a native of the banks of the Napo, where M. Osculati formed his collection.

82. *Colobothea ordinata*, n. sp.

C. elongata, postice attenuata, olivaceo-nigra, vertice postice bilineato; thorace supra lineis quatuor crassiusculis vittaque lata supracoxali fulvo-cinereis; elytris maculis numerosis subquadratis fulvo-cinereis in seriebus subordinatis; thorace ante basin utrinque breviter tuberculato. Long. 7½ lin. ♂.

Head black, forehead with three slender lines, vertex with two divergent lines, and cheeks with a broad streak, tawny ashy; there is also a tawny-ashy streak behind each eye. Antennæ stout, black, sixth joint with a narrow white ring, the bases of the fourth, eighth, tenth, and eleventh joints with an ashy streak on one side (♂). Thorax slightly constricted at the base, and with a small tubercle on each side; surface black, with four rather thick tawny-ashy lines; there is also a broad tawny-ashy

* *Colobothea lineatocollis* (Dej. Cat. sec. Dom. Chevrolat). Elongata, antice et postice attenuata, nigra, obscure olivaceo-grisea, sericea, griseo lineata et maculata. Caput nigrum, griseo lineatum, vertice lineis griseis duabus postice divergentibus, genis griseo plagiatis. Antennæ validæ, nigræ, articulo sexto albo annulato. Thorax lineis tenuibus duabus dorsalibus subparallelis, alteris duabus lateralibus, vittaque utrinque supracoxali, griseis. Elytra postice modice attenuata, humeris parum obliquis, apicibus truncatis, angulis externis dentiformibus, supra sparse punctata maculis minutis griseis confluentibus, reliquo spatio subapicali immaculato, ipso apice albo marginato. Corpus subtus nigrum, griseo sparse tomentosum, abdomine maculato. Pedes nigri, griseo annulati. Maris segmento dorsali terminali truncato, angulis prominulis; ventrali profunde emarginato, angulis spinosis. Fœminæ segmento ultimo dorsali apice lato; ventrali profunde emarginato, angulis productis. *Hab.* in Cayenna.

vitta above the coxa on each side. Elytra with prominent and rather acute shoulders, thence gradually attenuated to the apex, which latter is truncated, the external angles produced each into a longish spine; surface olivaceous black, marked with a large number of well-separated and squarish tawny-ashy spots, mostly arranged in rows, and leaving a distinct belt beyond the middle and another near the apex unspotted; apex itself edged with whitish. Body beneath ochraceous ashy. Legs greyish, varied with black.

♂. Terminal ventral segment broadly truncated, angles produced; dorsal segment obtuse, entire.

Ega; rare.

33. *Colobothea subtessellata*, n. sp.

C. elongata, postice attenuata, olivaceo-nigra, vertice postice bilineato, thorace lineis duabus dorsalibus crassiusculis alteraque laterali et vitta supracoxali cinereo-ochraceis; elytris maculis numerosis cinereo-ochraceis in seriebus subordinatis, spatio lato apicali immaculato; thorace absque tuberculis. Long. 8½ lin. ♀.

Head black; forehead with three slender lines, vertex with two divergent lines, and cheeks with a broad streak tawny ashy. Antennæ stout, black, sixth joint with a broad white ring, tenth joint with an ashy streak on one side (♀). Thorax not constricted at the base, broadest at its basal angles, and free from tubercles; surface black, with two moderately thick tawny lines, sides each with a similar line, not visible from above, and a broad tawny vitta above the coxa. Elytra moderately broad at the shoulders, and narrowed thence to the apex, the latter truncated, with the outer angles spinose; surface olivaceous black, marked with a large number of tawny spots, which are in some examples arranged in rows, and in others more or less confused; there is a broad immaculate space at the apex, and the apex itself is broadly margined with white. Body beneath black, thinly clothed with ashy pile, and having a broad, distinct, ochreous lateral vitta. Legs blackish, ringed with grey.

♀. Terminal abdominal segment elongated and tapering; dorsal plate broadly notched; ventral truncated, angles acute.

Banks of River Tapajos; rare.

34. *Colobothea octolineata*, n. sp.

C. valde elongata, postice attenuata, olivaceo-nigra, vertice linea unica, genis utrinque lineis duabus cinereis; thorace lineis tenuibus cinereis octo, quarum quatuor dorsalibus; elytris humeris prominentibus, maculis cinereis discretis irregulariter dispersis. Long. 7½–11 lin. ♂ ♀.

Head black, forehead with two greyish lines, vertex with a single narrow line, and cheeks on each side with two oblique greyish lines. Antennæ black, sixth joint thickened, with a ring of dense white hairs in both sexes. Thorax marked with eight slender, greyish or tawny lines, of which four are on the upper surface and two on each side, including the supracoxal streak, which in this species is slender, like the other lines. Elytra greatly elongated; shoulders very prominent, thence gradually narrowing to the apex, the latter truncated, with outer angles spinose; surface olivaceous black, marked with a moderate number of larger and smaller spots, widely separated from each other, but very irregularly dispersed; apex edged with whitish. Body beneath black, marked with ashy or tawny streaks and spots. Legs black, ringed with tawny and grey.

♂. Terminal abdominal segment narrowed from the base; apex of both dorsal and ventral plates emarginate-truncate. Anterior tarsi very broadly dilated and fringed.

♀. Terminal abdominal segment elongate and tapering; apex of both dorsal and ventral plates emarginate-truncate.

Pará, also Ega, Upper Amazons; common.

85. *Colobothea contaminata*, Serville.

Colobothea contaminata, Serv. Encycl. Méth. x. p. 337.

C. valde elongata, angustata, postice vix attenuata, olivaceo-nigra, vertice linea unica cinereo-fulva, thorace lineis quatuor, quarum externa utrinque usque ad oculum extensa et vitta supracoxali supra genas continuata; elytris maculis cinereo-fulvis passim confluentibus, vel cinereo-fulvis nigro irregulariter maculatis, fascia lata subapicali nigra; antennis utroque sexu nigris, articulo sexto annulo incrassato albo. Segmento ultimo abdominali maris attenuato, apice emarginato; fœminæ angustato, lamina dorsali obtusa, ventrali angulis productis; maris tarsis anticis valde dilatatis. Long. 6½–10 lin. ♂ ♀.

Generally distributed and common throughout the Amazons region; also found at Cayenne.

86. *Colobothea geminata*, n. sp.

C. elongata, postice vix attenuata, olivaceo-nigra, vertice linea unica, thorace lineis duabus antice et postice conjunctis; elytris maculis numerosis in lineis curvatis confluentibus fulvo-griseis. Long. 7½–8½ lin. ♂ ♀.

Head black, forehead with three tawny-grey lines, and vertex with a single line; cheeks with a tawny-ashy stripe. Antennæ black, sixth joint with a broad white ring. Thorax black, sides each with two tawny-ashy stripes joined together near the front and posterior margins, and continuous with the cheek-stripe. Elytra elongated, of very nearly the same width from base to apex in both sexes; external angle of the truncature spinose;

surface blackish olivaceous, sprinkled with a large number of
tawny-ashy spots, which are mostly confluent, and tend to form
a pattern consisting of three irregular pale rings, on each elytron, enclosing a blackish space; apex edged with tawny whitish.
Body beneath ashy, but tawny towards the sides; abdomen
spotted with black. Legs ashy, ringed with black.

♂. Terminal abdominal segment short; ventral plate emarginate-truncate, angles produced; dorsal plate obtuse and
notched in the middle. Anterior tarsi not dilated.

♀. Terminal abdominal segment tapering; dorsal plate
notched in the middle; ventral truncate, angles not produced.

Guiana side of the Lower Amazons and banks of the Tapajos;
also found at Cayenne.

87. *Colobothea concreta*, n. sp.

C. valde elongata, angustata, olivaceo-nigra, vertice linea unica, thorace vittis quatuor (quarum duabus externis usque ad oculos extensis) fulvo-cinereis; elytris basi thorace vix latioribus, apice
truncatis, angulis externis spinosis, maculis cinereo-fulvis confluentibus dense vestitis, apice macula magna nigra. Long. 6–9
lin. ♂ ♀.

Head black, forehead streaked with tawny ashy, vertex with a
single line; occiput on each side with a short line continuous with
the external thoracic stripe, cheeks with a transverse stripe continuous with the supracoxal vitta. Antennæ black, sixth joint with
a broad white ring. Thorax black, surface with four rather thick
tawny-ashy lines, sides having only the supracoxal vitta. Elytra
elongated, scarcely tapering; shoulders very oblique, and not at
all prominent; apex truncate, external angles spinose; surface
very thickly clothed with tawny or tawny-ashy spots, mostly
confluent, but leaving a broad unspotted space at the apex, the
latter margined with tawny white. Body beneath ashy, sides
streaked with tawny; abdomen spotted with black. Legs black,
ringed with tawny and grey.

♂. Terminal abdominal segment narrowed from the base;
dorsal plate deeply notched; ventral plate semicircularly emarginated, with angles much produced. Anterior tarsi widely
dilated and fringed.

♀. Terminal abdominal segment elongate and much narrowed; dorsal plate very obtuse, ventral truncated, angles
slightly prominent.

Pará, and banks of the Tapajos.

38. *Colobothea bilineata*, n. sp.

C. valde elongata, postice vix attenuata, nigra, vertice linea unica,
thorace lineis duabus usque ad oculos extensis, griseis; elytris

griseis, nigro dense maculatis, apice macula magna nigra. Long. 7–10¼ lin. ♂.

Head black, forehead streaked with ashy, vertex with a single line, occiput on each side with a short line continuous with the thoracic stripe; cheeks crossed by an ashy streak continuous with the supracoxal vitta. Antennæ black, sixth joint with a broad white ring. Thorax black, surface with only two ashy stripes, each continuous to the hind margin of the eye. Elytra elongate and scarcely tapering, very little broader at the base than the thorax, but shoulders prominent and conical; apex sinuate-truncate, the sutural angles being prominent and acute, the outer angles spinose; surface grey, thickly spotted with black; some of the spots confluent, and a large spot at the apex spotless; apex itself edged with white. Body beneath thinly clothed with grey; abdomen spotted with black. Legs black, ringed with grey.

♂. Terminal abdominal segment with the ventral plate semi-circularly emarginated, angles acute; dorsal plate triangularly emarginated. Anterior tarsi dilated and fringed.

Ega and S. Paulo, Upper Amazons; rare.

89. *Colobothea lunulata*, Lucas.

Colobothea lunulata, Lucas, Voyage de Castelnau, Entomologie, p. 190, pl. 13. f. 5 (1857).
—— *Fryi*, Pascoe, Trans. Ent. Soc. vol. i. 41 (1861).

C. elongato-elliptica, nigra; vertice, thorace et elytris albo bivittatis, vittis longe ante apicem elytrorum convergentibus et annulo albo utrinque connexis. Long. 7½–9½ lin. ♂ ♀.

This very distinct and handsome species was one of the commonest of its genus at Ega, on the trunks of fallen trees in the forest. The shoulders are extremely oblique and scarcely prominent, so that the insect has the form of an elongated ellipse truncated at the elytral end. The terminal abdominal segment in the male has both the dorsal and ventral plates truncated; in the female it is elongated, and the angles of the ventral plate are produced. The anterior male tarsi are widely dilated and fringed*.

* The following species of *Colobothea* have not yet been described:—
  Colobothea hebraica (Chevrolat, MS.). Modice elongata, postice attenuata, fusco-nigra, griseo maculata. Caput nigrum, fronte griseo trilineata, occipite maculis duabus, genis vitta lata, griseis. Antennæ nigræ, articulis basi griseis. Thorax basi paulo angustatus, dorso linea abbreviata, disco utrimque maculis parvis, lateribus vitta latiuscula cinereo-griseis. Elytra apud humeros lata, deinde usque ad apices attenuata, truncaturæ angulis externis spinosis, supra fusco-nigra maculis cinereo-griseis (partim subagglomeratis) adspersa, apice haud pallide marginato. Corpus subtus griseum, lateribus cinereis.

Subtribe Lamiitæ.

Genus Tæniotes, Serv.

Serville, Ann. Soc. Ent. Fr. iv.

This well-known and handsome genus is the only one belonging to the typical Lamiaires found in the Amazonian forests, the allied genus *Ptychodes*, common in other parts of Tropical America, being absent from the low-lying Equatorial region. The other Tropical American representants of this subtribe, so rich in forms in the Old World (namely, *Plectodera*, *Hammoderus*, and *Deliathis*), seem to be confined to the northern portion of the zone—Central America, Mexico, and thence extending into the Southern States of North America.

1. *Tæniotes decoratus*, Castelnau.

Tæniotes decoratus, Casteln., Animaux articulés, ii. p. 479.

T. nigro-velutinus, capite fascia utrinque infra oculos, vitta laterali alteraque coronali per thoracem et scutellum continuata, maculisque rotundis elytrorum utrinque circa 13 læte flavis; corpore subtus vitta flava laterali: maris pedibus anticis vix elongatis, tarsis haud pilosis. Long. 13 lin. ♂ ♀.

I met with this fine species only in the neighbourhood of

---

nigro maculatis. Pedes nigri, cinereo annulati. Fœminæ segmentum ultimum abdominale attenuatum; lamina dorsali apice rotundata, ventrali truncata, angulis productis. Long. 5–7 lin. ♀. *Hab.* in Mexico.

Colobothea fasciata. Modice elongata, postice valde attenuata, tomento brunneo fulvo-maculato vestita; elytris fascia lata nigro-velutina. Caput nigrum, fulvo-brunneo vestitum, vertice linea unica fulva. Antennæ breviores, nigræ, breviter setosæ, articulis basi griseis. Thorax fusco-niger, dorso vittis duabus fulvo-brunneis. Elytra apud humeros lata, deinde valde attenuata, apice sinuato-truncata, angulis externis longe spinosis, supra brunnea obscure fulvo maculata, pone medium fascia nigro-velutina apud dorsum dilatata, apices versus nigro liturata. Corpus subtus rufescens, medio nigricans. Pedes nigri. Maris segmentum ultimum ventrale subtumidum, apice obtuse truncatum; tarsi antici haud dilatati. Fœminæ segmentum ultimum paulo elongatum, valde attenuatum, lamina ventrali sinuato-truncata, haud spinosa. Long. 4–6. ♂ ♀. *Hab.* in Rio Janeiro.

Colobothea lateralis. Elongata, postice valde attenuata; corpore supra cinereo-ochraceo, rufo variegato, lateribus nigris. Caput nigrum, fronte fulvescente, vertice et maculis quatuor occipitalibus cinereis fulvo maculatis. Antennæ griseæ, articulis apice nigris. Thorax antice angustatus, dorso cinereo-ochraceus, rufo maculatus, lateribus nigris. Elytra apud humeros lata, deinde attenuata, apice truncata, angulis externis spinosis, supra cinereo-ochracea, rufo maculata, lateribus irregulariter nigris, nigredine ramos tres dentatos in discum emittente, his rufo marginatis. Corpus subtus cinereum, medio nigrum, segmentis primo et ultimo abdominalibus nigris. Fœminæ segmentum ultimum attenuatum, lamina ventrali truncata, dorsali medio emarginata. Long. 7½ lin. ♀. *Hab.* in Brasilia.

Pará, on felled trees in broad roads through the forest. The terminal ventral segment in both sexes is broadly truncated, with a distinct spine at each angle. M. Guérin-Méneville (Icon. Règne Animal, p. 243) believes this species to be the same as the *T. subocellatus* of Olivier (Ent. no. 67. pp. 69, 80, pl. 2. f. 12 *a*, *b*), and that the latter is founded on a worn or immature individual.

2. *Tæniotes D'Orbignyi*, Guérin.

Tæniotes D'Orbignyi, Guérin-Méneville, Icon. Règne Animal, p. 444.

T. nigro-velutinus, capite fascia utrinque infra oculos, vitta laterali, alteraque coronali per thoracem et scutellum continuata, vittaque elytrorum utrinque medio interrupta et maculiformi læte flavis ; corpore subtus vitta flava laterali : maris pedibus anticis vix elongatis, tarsis haud pilosis. Long. 8–13 lin. ♂ ♀.

This species, originally discovered in the wooded plains of Bolivia by M. D'Orbigny, was common on the Upper Amazons at Ega. The yellow (partially macular) stripe of the elytra varies a little in the degree in which it is broken up into spots ; but it never forms a double row of distinct round spots from base to apex, as shown in *T. decoratus*, and can scarcely be considered a local form of the same stock.

3. *Tæniotes Amazonum*, Thomson.

Tæniotes Amazonum, Thoms. Archives Entomologiques, i. p. 172.

T. niger, capite linea curvata frontali, vitta utrinque laterali, altera coronali per thoracem scutellum et elytros continuata (hic dentata) pallide flavis ; thorace utrinque linea tenuissima grisea ; elytris maculis parvis numerosis, quarum duabus vel tribus discoidalibus majoribus, flavis ; corpore subtus vitta flava laterali : maris pedibus anticis valde elongatis, tibiis curvatis, tarsis haud pilosis. Long. 9–16 lin. ♂ ♀.

A common insect in the forest at Ega, on the Upper Amazons. It is probably a local form of *T. scalaris*, Fabr., but differs much from the description given by that author. The terminal ventral plate is formed as in *T. decoratus*.

4. *Tæniotes farinosus*, Linnæus.

Cerambyx farinosus, Linn. Syst. Nat. ii. 626. 24 ; Oliv. Ent. lxvii. p. 50, f. 46 *a*.
—— *pulverulentus*, Oliv. Ent. lxvii. p. 50, f. 46 *b*.

T. niger, griseo vestitus ; capite thoraceque lineis tenuibus tribus, elytris maculis numerosis parvis, flavo-griseis, his apice acutis ; corpore subtus flavo maculato : maris pedibus anticis valde elongatis, tibiis curvatis, tarsis hirsutis. Long. 13 lin. ♂.

This species was a rare one in the Amazons region, and found only in the dry forests of the Tapajos. The spines of the terminal ventral segment are more elongated than in the other species.

Subtribe Oncideritæ.

Group *Onciderina.*

Genus HYPSELOMUS, Perty.

Perty, Delectus Anim. Articul. Brasil. p. 95 (1830–34).
Syn. *Hypsioma,* Serv. Ann. Soc. Ent. Fr. p. 38 (1835).

This genus is distinguished from its allies by its short sub-trigonal form of body, with projecting and often acute shoulders of the elytra. The claw-joint of the tarsi is not so much elongated as in *Oncideres,* or even *Clytemnestra.* It is very closely allied to the latter genus, but is distinguishable at once by the abrupt clavate form of the basal joint of the antennæ and the curved shape of the third. The males of most species have a short, slender, curved joint at the tip of the eleventh joint of the antennæ, which is sometimes visible (but much smaller) in the female.

1. *Hypselomus basalis,* Thomson.

Hypsioma basalis, Thomson, Classif. des Cérambyc. p. 117.

H. modice elongatus, brunneus; capite, thorace et elytrorum parte antica rufescenti-ochraceis; summa fronte acute bituberculata; antennis nigris, basi rufescenti-ochraceis, articulis cæteris basi rufescentibus; elytris basi utrinque vix elevatis, nigro tuberculatis humeris, apice nigris; abdomine lateribus rufo vittatis; pedibus nigricantibus, tibiis compressis, posticis (♂) apice dilatatis. Long. 6–9 lin. ♂ ♀.

A common insect throughout the Amazons region, being found, like the rest of the species, on dead branches, closely adhering to them, and gnawing the bark and wood all round, until the bough is sometimes severed. The face and parts of the mouth are much elongated and directed a little backwards between the anterior haunches, so that when the legs are extended, grasping a branch, the jaws are in a good position to gnaw effectually. The supplementary joint of the antennæ is very conspicuous in the males of this species.

2. *Hypselomus picticornis,* n. sp.

H. suboblongus, brunneus, elytris fascia obliqua indistincta pallidiore; antennis brunneis, articulo 2do toto et cæteris basi rufescentibus; elytris basi haud tuberculatis, humeris oblique conicis modice productis. Long. 7 lin. ♀.

Head brown, forehead near base of antennæ with two very small conical tubercles. Antennæ about the length of the body, setose beneath; basal joint strongly and abruptly clavate, third much bent, dark brown; second joint, basal half of third, and bases of each remaining joint pallid-reddish. Thorax scarcely uneven on the surface, uniform dingy brown. Elytra oblong trigonal; shoulders moderately prominent, and thence gradually narrowed to the apex, which is broadly rounded; surface convex; centrobasal ridges not at all prominent, and quite destitute of tubercles, the basal half of the elytra being simply punctured. Body beneath rufescent tawny, centre of abdomen black; legs brown, claw-joints of tarsi with their basal halves pale reddish.

Ega; rare.

3. *Hypselomus Amazonicus*, Thomson.

Hypseloma Amazonica, Thomson, Classif. des Céramb. p. 119.

H. convexus, brunneus; elytris humeris conicis, subuncinatis, pone medium fascia irregulari pallidiore, deinde ad apices pallide marmorntis; antennis articulis basi rufescentibus: maris tibiis posticis apice valde dilatato-compressis. Long. 9 lin. ♂ ♀.

Closely allied to *H. picticornis*, but larger and darker, with the elytra behind the middle much more variegated with pale ashy brown, and the conical protuberances of the shoulders strongly curved anteriorly. The antennæ are coloured as in *H. picticornis*, the second and basal half of the third, with bases of the remaining joints being pale reddish. The underside of the body is tawny brown, with the centre of the abdomen black. The elytra are smoothly and strongly convex from base to apex, without any trace of centrobasal ridge or tubercles.

Ega, Upper Amazons.

4. *Hypselomus dimidiatus*, n. sp.

H. modice convexus, fuscus, fulvo irroratus; elytris apud medium ochraceo fasciatis, deinde usque ad apices pallide ochraceo-brunneis fusco striatis et maculatis; thorace supra quinquetuberculato, lateribus acute tuberculatis. Long. 6–7 lin. ♂ ♀.

Head dingy brown. Antennæ dull brown, base of each joint, from the third, pallid-reddish. Thorax uneven, disk on each side with two prominent tubercles, and dorsal line elevated behind into a ridge, sides each with an acute tubercle; dingy brown. Elytra with very prominent shoulders, the anterior side of the subconical projection oblique; centrobasal ridges slightly elevated, but not tuberculated; dark brown, sprinkled with fulvous; behind the middle a pale oblique belt or broad triangular spot darker in the middle, thence to the apex light brown with darker lines and spots. Body beneath tawny brown, middle of abdomen black. Legs black, apex of thighs fulvous, claw-joint red, apex black. Posterior tibiæ in the male dilated at the apex; supplementary antennal joint in the same sex very short or wanting.

Ega. Rather variable in the colour of the posterior part of

the elytra, the pale belt being sometimes extended into a large triangular patch, and sometimes blended with the pale-brown shade of the apical half of the wing-cases. The species seems to be very closely allied to *H. subfasciata*, Thomson (Classif. des Céramb. p. 118).

5. *Hypselomus rodens*, n. sp.

H. oblongus, nigro-fuscus, carneo-fulvo strigatus; thorace supra haud tuberculato; elytris humeris apice truncatis, postice uncinatis, pone medium fascia obliqua pallida. Long. 6 lin. ♀.

Head dingy black, crown sprinkled with reddish tawny. Antennæ black, sprinkled with tawny; base of each joint, from the fourth, pale. Thorax convex above, and free from tubercles, sides with an inconspicuous tubercle. Elytra oblong, shoulders prominent, but the apex of the cone largely truncated, with the posterior edge of the truncature projecting; surface coarsely punctured, blackish, streaked with reddish tawny, behind the middle tawny streaked with black, the tawny part separated from the anterior darker portion by a pale-ochreous fascia. Body beneath tawny, middle of abdomen black. Legs tawny, sprinkled with black, base of claw-joint reddish.

Parä.

6. *Hypselomus pagamus*, Pascoe.

H. sordide fuscus, nigro obscure irroratus; thorace dorso tuberoso, lateribus tuberculo acuto; elytris humeris subconicis, antice curvato angulatis, cristis centrobasalibus prominulis, obtusis. Long. 7–8 lin. ♂♀.

Head dingy brown. Antennæ blackish brown, bases of the joints (from the fourth) pallid. Thorax with prominent dorsal ridge and, on each side, two well-marked tubercles, sides each with a small acute tubercle; colour dingy tawny brown, speckled with dusky. Elytra with projecting shoulders, the projection somewhat conical, but anterior slope curved or angulated, the apex formed by a thick black tubercle; centro-basal ridges pronounced, but not crested with tubercles; surface dingy tawny brown, speckled or irregularly marked with dusky. Body beneath dingy brown; abdomen black in the middle. Legs blackish, speckled with tawny; base of claw-joint reddish: posterior tibiæ in the male dilated at apex. Supplementary antennal joint of male wanting.

Ega and S. Paulo, Upper Amazons.

7. *Hypselomus seniculus*, n. sp.

H. parvus, fuscus griseo vestitus, summa fronte acute bituberculata; elytris grosse punctatis, humeris modice productis, obtuse truncatis, truncaturæ angulo postico acuto; maris articulo 12ᵐᵒ antennarum longiusculo, curvato. Long. 4½ lin. ♂.

Head clothed with thick tawny-grey pubescence, vertex spotted with brown; inner side of each antenniferous tubercle (♂) produced into an acute tooth. Antennæ towards the base grey, spotted with dark brown; apices of third to eleventh joints dusky, bases of joints from the fourth testaceous. Thorax convex, unarmed, grey, coarsely punctured (especially on the sides) and spotted with dark brown. Elytra moderately broad at the shoulders, the latter not conically produced, but obtusely truncated, with the posterior end of the truncature acute; surface thinly clothed with grey pile, and coarsely punctured, simply convex. Body beneath and legs clothed with tawny-grey pile, spotted with blackish, base of claw-joint testaceous; apical half of posterior tibiæ strongly dilated (♂).

Ega.

8. *Hypselomus crassipes*, n. sp.

H. robustus, brunneus; thorace lateribus pallidis; elytris utrinque macula oblonga transversa cretacea; pedibus crassis, nigris, tibiis posticis maris trigonis. Long. 8½ lin. ♂.

Head coarsely wrinkled, black; antenniferous tubercles produced on the inner side into a stout spine (♂). Antennæ scarcely so long as the body, bases of joints, from the fourth, pale testaceous; twelfth joint (♂) short and twisted. Thorax convex in the middle, without distinct tubercles, a short obtuse tubercle on each side; above dark brown, sides dingy tawny white, traversed by an indistinct dusky stripe. Elytra broad and but slightly convex; shoulders conically produced, base on each side obtusely elevated and very coarsely granulate-punctate, sides under the humeral projections also coarsely punctured, rest of the surface faintly punctured; dark brown, base dingy tawny white; each elytron beyond the middle ornamented with a distinct oblong, transverse, chalky spot. Body beneath dingy tawny; abdomen black in the middle. Legs very stout, black; tibiæ compressed; hind tibiæ (♂) dilated from the base, and obliquely truncated at the apex; claw-joint red.

Tapajos. Apparently allied to *H. fasciatus* of Thomson; but no mention is made by this author of any peculiar formation in the legs.

9. *Hypselomus simplex*, n. sp.

H. subelongatus, brunneo-fulvus, unicolor; elytris modice attenuatis, humeris conicis; antennis gracilibus, articulis basi griseis. Long. 6½–9 lin. ♂ ♀.

Rather more elongate than the allied species; but the elytra rather convex, and the third antennal joint strongly bent. Head dusky. Antennæ slender, a little longer than the body in the female, much longer in the male; basal joint strongly clavate; dark brown, bases of the joints, from the fourth, grey. Thorax bituberculate on each side the central ridge, dingy tawny brown. Elytra elongated, gradually and slightly tapering from base to apex; shoulders conical, base on each obtusely raised, finely punctured, colour uniform brownish tawny. Body beneath tawny brown; abdomen black down the middle. Legs simple, posterior tibiæ scarcely dilated in the male; black, thinly clothed with tawny pile; claw-joints black.

Ega.

10. *Hypselomus lignicolor*, n. sp.

H. subcylindricus, brunneus; thorace et pectore vittis lateralibus obliquis, elytris sutura vittisque lateralibus abbreviatis curvatis, nigris pallide marginatis; elytris compressis, sparsim punctatis, humeris paulo productis haud tuberculatis. Long. 5¼ lin. ♀.

Head tawny, spotted with dark brown. Antennæ as long as the body (♀), moderately stout, brown, unicolorous. Thorax unarmed and free from tubercles, surface smooth, brown; sides each with two oblique, blackish vittæ, the upper one margined with dull ochreous; there is also a short dusky central line near the middle of the hind margin. Scutellum blackish in the middle. Elytra nearly cylindrical, sides compressed, shoulders produced each into a slightly elevated ridge not surmounted by a tubercle; surface sparingly and finely punctured, brown, suture and several curved streaks on each side blackish, the lateral streaks margined on the upper sides with pallid brown. Body beneath brown; breast with oblique stripes, dull ochreous and blackish; basal half of abdomen dusky. Legs simple, tawny brown.

Ega. This species is much more elongate and narrow than the typical forms of the genus; it consorts, however, much better with the *Hypselomi* than with *Hesycha* or *Oncideres* (which comprehend elongated forms), having antennæ approximated on the forehead instead of widely separated at their bases. It seems to be nearly allied to *Hypselomus egens*, Erichson (Consp. Col. Peru. p. 148).

11. *Hypselomus obscurellus*, n. sp.

H. subelongatus, nigricans, griseo variegatus; antennis articulo basali apice subgloboso; thorace postice constricto; elytris elongato-trigonis, humeris conico-elevatis, obtusis. Long. 5½ lin. ♂.

Head dusky, eyes ample; forehead narrow, coarsely punctured; antenniferous tubercles unarmed. Antennæ black, base of joints grey, basal joint very abruptly clavate near the apex, subglobose, third joint very slightly curved. Thorax cylindrical, constricted behind the middle, surface very uneven, coarsely wrinkled transversely, dark brown. Elytra moderately elongated, wide at the base, and narrowed thence towards the apex; shoulders conically produced, but apex of cone obtuse and not tuberculated; surface very roughly punctured near the base, more finely so towards the middle, colour dark brown or blackish, thinly variegated with greyish pile. Body beneath tawny brown; abdomen in the middle glossy blackish, and sides spotted with black. Legs blackish, varied with tawny; hind tibiæ dilated near the apex (♂).

Obydos, Lower Amazons. Similar in size and general figure to *H. Syrinx* * (*Hesycha syrinx*, Dj. Cat. and French collections), but differing in the shape of the basal joint of antennæ and in the constricted thorax.

Genus JAMESIA, Jekel.

Jekel, Journal of Entomology, i. p. 259.

This genus is distinguished from *Hypselomus* by the basal joint of the antennæ being very gradually thickened from the base to the apex, not abruptly clavate, and by the third joint being quite straight instead of crooked. The claw-joints of the tarsi are quite as long as the three remaining joints taken together. The species have the same heavy figure and dull colours; but the elytra are much more elongated, and less trigonal. The genus is distinguished also by the large volume and subquadrate form of the eyes.

There seems to be scarcely sufficient difference to warrant the separation of *Jamesia* from *Clytemnestra* (Thoms.)†, the larger volume of the eyes being the only apparent definite character.

* *Hypselomus Syrinx.* Subelongatus, brunneus vel nigricans, elytris utrinque vittâ obscurâ obliquâ pallidiore. Caput angustum, fronte impunctatâ; tuberis antenniferis intus dente armatis. Antennæ corpore paulo longiores, articulis basi pallidioribus, articulo basali paulo incrassato. Thorax basi latus, antice angustatus, lineâ dorsali elevatâ. Elytra elongata, postice paulo attenuata, subtiliter punctata, brunnea, lineâ curvatâ medianâ obscure fulvâ; humeris prominulis, in carinam lævem curvatam desinentibus. Corpus subtus fuscum. Pedes fusci, unicolores; tibiis compressis. Long. 4½–5½ lin. ♂ ♀. *Hab.* Rio Janeiro.

† Since the early part of the genus *Hypselomus* in this memoir was in print, I have found that Perty and Serville happen to have described the types of two distinct genera under the respective names of *Hypsioma* and *Hypselomus*. The latter genus is equivalent to *Clytemnestra* of Thomson, which therefore becomes a synonym. M. Thomson, in his later work, 'Systema Cerambycidarum,' has adopted this change of nomenclature. The following rectification of synonymy is therefore necessary :—

Gen. 1. *Hypselomus*, Perty, Delect. An. Art. Bras.
    = *Clytemnestra*, Thomson, Class. des Cérambycides.
    = *Jamesia*, Jekel, Thomson, Systema Cerambycid. (section).
Gen. 2. *Hypsioma*, Serville, Ann. Soc. Ent. Fr. iv.
    = *Hypselomus*, Thoms. (Class. des Cérambl.) Bates (*ut supra*) and authors, *nec* Perty.

1. *Jamesia globifera*, Fab.

Lamia globifera, Fabricius, Syst. Eleuth. ii. 284. 15.
Hypselomus variolosus, Pascoe, Trans. Ent. Soc. n. s. v. pt. 1 (1859).

J. subelongata, sordide griseo-brunnea; thorace transverse ruguloso et acute tuberculato; elytris prope basin tuberculis globosis nigris politis et postice maculis nigris leviter impressis variegatis; capite lateribus parallelis, oculis magnis, subquadratis; antennis brunneis, maris corpore multo longioribus; pedibus simplicibus. Long. 10 lin.

Not uncommon on dead trees throughout the Amazons region; also found at Cayenne.

2. *Jamesia pupillata*, Pascoe.

Hypselomus pupillatus, Pascoe, Trans. Ent. Soc. n. s, v. pt. 1.
Jamesia bipunctata, Jekel, Journ. of Entom. i. 260.

J. subelongata, parum convexa, olivaceo-brunnea, nigro punctata; elytris medio utrinque ocellatis; maris capite infra dilatato, cornibus frontalibus magnis acutis porrectis; antennis quam corpus duplo longioribus. Long. 11 lin. ♂ ♀.

Differs from *J. globifera* chiefly by the more depressed form of the elytra, and the absence of basal elevation with globular tubercles. It may readily be recognized also by the eye-like spot on the disk of each elytron, consisting of a rounded, black, slightly impressed spot, surmounted by a white speck. The antennae are much more elongated, and the projecting angles of the antenniferous tubercles in well-developed males are very large and acute, and are directed horizontally. The base of the elytron has a few minute granulations with punctures, and the rest of the surface is sprinkled with rounded, dark-brown, slightly impressed spots, as in *J. globifera*.

Ega; not uncommon.

Genus Hesychia (Dj. Cat.), Thomson.

Thomson, Archiv. Entom. i. 187 (1857):
Fairmaire, Ann. Soc. Ent. Fr. (1859), p. 523.

This genus was first characterized, in few words, by M. Thomson in 1857; but the description subsequently published by M. Fairmaire defined more accurately its points of distinction. It agrees with *Hypselomus* in having the first joint of the antennae abruptly clavate, and the third joint curved; the curvature, however, is much less pronounced than in *Hypselomus*, and is sometimes very slight. Its other distinguishing characters are (1) the elongate, parallelogrammical, and depressed form of body, (2) the more elongated claw-joint of the tarsi, and (3) the wide separation of the antennae at their origin.

1. *Hesycha Nyphonoïdes*, Pascoe.

Hesycha Nyphonoïdes, Pascoe, Trans. Ent. Soc. n. s. v. pt. 1.

H. parallelogrammica, depressa, obscure fusca cinereo-fulva variegata; elytris medio fascia undulata, obscura, cinereo-fulva. Long. 6¼–8 lin. ♂ ♀.

Head dull brown; forehead broad, sparingly punctured; antenniferous tubercles in the male produced on their inner side into a stout pointed tooth. Antennæ in the male nearly twice the length of the body, with the apical joint greatly elongated; in the female about the length of the body, apical joint shorter than the preceding; colour dull brown or blackish. Thorax uneven above, sides with a short pointed tubercle; dull brown, speckled with black. Elytra slightly narrowed from base to apex, shoulders slightly prominent and surmounted by an obtuse shining tubercle; surface even, thickly but finely punctured, dull brown, covered with dingy tawny confluent spots, and crossed beyond the middle by a zigzag fascia of a little paler hue. Body beneath and legs blackish or dull brown.

Common on branches of dead trees at Ega. There are two closely allied species in collections from the interior of French Guiana *.

* *Hesycha jaspidea*, n. sp. *H. Nyphonoïdei* simillima, robustior, maris elytris postice magis angustatis et fronte valde cornuta. Obscura fusca; thoracis lateribus utrinque tuboris duobus obtusis armatis. Elytra humeris prominentibus, basi rugoso-punctata et inæqualia fusco-nigra, maculis sordide fulvis sparsis quarum tribus majoribus medianis in fasciam abbreviatam conjunctis. Corpus subtus fulvo tomentosum. Antennæ valde elongatæ, articulis basi griseis. Long. 8 lin. ♂. *Hab.* In Cayenna interiore (Dom. Bar).

Hesycha liturata, n. sp. Minor, brunnea, elytris litura tenui obliqua albicante. Caput fuscum, fronte punctata, tuberculis antenniferis utroque sexu intus acutis. Antennæ brunneæ, maris corpore paulo longiores. Thorax quadratus, lateribus tuberculo distincto subacuto, supra brunneus vittis tribus nigris, lateribus cinerascentibus. Elytra postice paulo angustata, apice oblique breviter truncata, humeris vix productis, obtusis; dorso punctata, brunneo et fulvo variegata, infra humeros (cum prothoracis et pectoris lateribus) nigricantia, apud medium litura tenui valde obliqua albicante. Corpus subtus et pedes brunneo tomentosa. Long. 5–6 lin. ♂ ♀. *Hab.* In Cayenna (Dom. Bar).

The following species belongs also to this genus, from its linear subdepressed form and the somewhat wide separation of the antennæ at their bases:—

Hesycha xylina, n. sp. Elongata, sordide brunnea; elytris rugoso-punctatis, fusco et griseo strigatis, humeris subuncinatis. Caput fuscum, fronte grosse sparsim punctata, tuberculis antenniferis intus dente valido curvato armatis(♂). Antennæ valde elongatæ, brunneæ, apice pallidæ, articulis (à tertio) basi testaceis, articulo 12mo acuto, curvato. Thorax supra inæqualis, inermis, brunneus. Elytra valde elongata,

2. *Hesycha maculosa*, n. sp.

H. elongata, convexiuscula, fusca, maculis numerosissimis partim
confluentibus fulvis; vertice nigro trilineato; thorace nigro maculato. Long. 8½ lin. ♂ ♀.

Head dusky, front channeled down the middle, punctured;
eyes rather elongated, margined on the inner side narrowly with
tawny; vertex tawny, marked in the middle with three parallel
black lines; antenniferous tubercles produced into a short acute
tooth on the inner side, longer in the male than in the female.
Antennæ longer by one half than the body in the male, and the
terminal joint very slender and much longer than the preceding;
in the female a little longer than the body, with the terminal
joint shorter than the preceding; basal joint abruptly clavate,
third joint scarcely perceptibly curved; colour blackish. Thorax
quadrate, surface uneven, with several impressed curved lines
and raised interspaces, sides behind the middle with an acute
tubercle; colour tawny, marked with two short black lines in
front in the middle and a spot behind them, and four spots on
each side of the disk. Scutellum black. Elytra elongate and
rather convex, slightly tapering; shoulders prominent, and surmounted by a glossy black tubercle; surface quite even and
moderately punctured, dark brown, covered uniformly with a
multitude of tawny specks, mostly confluent. Body beneath
tawny. Legs blackish.

Ega.

3. *Hesycha cretacea*, n. sp.

H. oblongo-elongata, subdepressa; elytris maculis numerosis fulvis
maculaque magna laterali cretaceo-alba. Long. 8 lin. ♀.

Head grey, margins of eyes with tawny lines, front punctured;
eyes elongated; antenniferous tubercles acute on their inner
side, vertex with three short black streaks. Antennæ a little
longer than the body, dark brown; basal joint clavate, third
joint very slightly curved. Thorax quadrate, sides each with
two large obtuse tubercles, surface with transverse furrows,
tawny mixed with grey, and spotted with black. Scutellum
black, margined with grey. Elytra oblong, a little dilated beyond the middle, slightly convex, shoulders moderately prominent; with irregular clusters of punctures arranged in lines,
black, covered with pinkish-tawny spots, partly confluent, and

---

parum convexa; humeris productis, antice curvatis, postice tuberculo
nigro armatis, quasi uncinatis; supra grosse punctata, punctis partim
confluentibus, sordide brunnea, strigis pallidis et fuscis variegata.
Corpus subtus brunneum. Pedes fusci. Long. 4½ lin. ♂. Hab. Rio
Janeiro, à D. Squires capta.

having in the middle on each side a large chalky-white spot. Body beneath dull chalky white; breasts with pinkish streaks, and abdomen spotted with black. Legs black, thinly clothed with grey pile.

Ega; rare. This handsome species, like the preceding (*H. maculosa*), approaches *Oncideres* in many of its characters, especially the elongate eyes, subconvex form of body, and scarcely curved third antennal joint; but it lacks the massive head, cylindrical form of body, and short transverse thorax of *Oncideres*, and therefore must be classed with *Hesycha*.

Genus Trachysomus, Serville.

Serville, Ann. Soc. Ent. Fr. iv. (1835).
(Char. emend.) Buquet, Ann. Soc. Ent. Fr. 1852, p. 345.

This remarkable group is distinguished from the allied genera chiefly by the elytra being disfigured by tubercular excrescences, and by the antennæ being composed of short joints reaching only three-fourths the length of the body. The head is moderately narrow, the eyes oblong (not narrow and elongated as in *Oncideres*), the basal joint of the antennæ very abruptly clavate, the third joint very slightly curved, the thorax subcylindrical, and the claw-joint of the tarsi shorter than the remaining joints taken together. The species are found closely clinging to thin woody stems of plants, and strongly resemble portions of the stems distorted by glandular prominences or galls.

Trachysomus Santarensis, n. sp.

T. Trachysomo fragifero (Kirbii) valde similis, differt colore ochraceo- vel rufo-fulvo; thorace supra ochraceo; elytris juxta scutellum utrinque spinis quatuor acutis, fasciculis singulis pilorum subapicalibus nigris linea curvata nigra communi connexis. Long. 7½ lin.

This is so closely similar in form of body and tubercular excrescences to the South-Brazilian *T. fragifer*, that it can scarcely be considered more than a local form of the same stock. It is a little broader and more robust, the thorax is less uneven on the disk, and is there of a bright yellowish-tawny colour. The two tubercles on each elytron, near the scutellum, are longer and more acute. The elytra are of a nearly uniform reddish or orange-brown hue; the subapical fascicle of hairs is a little further removed from the apex and margin of the elytra; it is connected with the corresponding fascicle posteriorly by a curved black line, and a large portion of the disk behind each basal excrescence is quite smooth.

Dry woods near Santarem.

Genus Oncideres, Serville.

Serville, Ann. Soc. Ent. Fr. (1835) iv.

The chief characters of this, the typical genus of the group, are furnished by the elongate-oblong or cylindrical form of body; the broad head and convex occiput, with consequent wide separation of the antennæ at their bases; the elongated eyes; the clavate shape of the basal antennal joint, and straight form of the third joint; the short transverse thorax; and, lastly, the great length of the claw-joint of the tarsi, which exceeds that of the three remaining joints taken together.

The species are all found on the branches of trees, which they amputate from the living tree by gnawing deeply into the bark and wood, making a ring-like incision, until the bough breaks off by its own weight. I have often seen boughs thus severed from green and living Cajú trees, and hence discovered that the best means of finding the insects was by examining the amputated portions lying on the ground in woods or the thinner parts of the forest. The object of the severance is apparently to create a supply of dead wood in which to deposit their eggs and rear the larvæ.

1. *Oncideres Callidryas*, n. sp.

O. minus convexus; thorace griseo-tomentoso; elytris basi minute granulatis, medio confertim punctatis, nigris, guttis numerosissimis carneo-griseis. Long. 10½ lin. ♂ ♀.

Head much narrower than the middle part of the thorax, clothed with pinkish-tawny pile; forehead plane, punctured; antenniferous tubercles (♂) on each side armed with longish acute teeth directed forwards; eyes oblong. Antennæ about the same length as the body in the female, twice the length in the male, black. Thorax with transverse depressions, sides each armed with a strong conical tubercle, clothed with hoary-grey pile. Scutellum and basal margin of elytra hoary grey. Elytra less cylindrical and convex than in the more typical species; shoulders prominent and surmounted by a retrocurved tubercle, base and shoulders thickly and finely granulated, middle part simply but thickly punctured, punctures becoming finer posteriorly, and disappearing before the apex; colour black, sprinkled throughout with small grey or pinkish-grey spots, some very minute, others larger; near the middle of each side the spots are whiter, and tend to aggregation. Body beneath hoary white. Legs black, thinly clothed with grey pile.

Pará, banks of the Tapajos, and Ega; one pair taken *in copulá* on a branch of a felled tree at Pará. The elytra are much more thickly spotted in the Ega examples than in those from Pará and the Lower Amazons.

2. *Oncideres Satyrus*, n. sp.

O. cylindricus, fulvo-brunneus; elytris guttis albis paucis sparsis, basi tuberculis nigris; antennis validis; thorace basi valde constricto. Long. 10–12 lin. ♂ ♀.

Head in the ♂ much narrower than the thorax, in the ♀ as wide as the widest part of the thorax, with broad plane front, colour tawny brown, a black stripe below each eye. Antennæ about the length of the body in the female, a little longer in the male, with the apical joint twice the length of the preceding; they are robust in both sexes, but the four basal joints are thicker in the ♂ than in the ♀; colour black. Thorax with transverse depressions; a conical tubercle on each side, and much constricted behind the tubercle; brownish tawny, with a fine, black, central, transverse line. Elytra cylindrical, brownish tawny, sprinkled with a small number of minute white spots; base and shoulders with a few polished rounded tubercles; rest of surface impunctate, smooth. Body beneath and legs thickly clothed with tawny pile; sides of breast chalky white.

Pará. Closely allied to *O. vomicosus*, Germar (Ins. Nov. 482), but differing greatly in the maculation of the elytra, the spots being small, few in number, and all distinct from each other.

3. *Oncideres fulvus*, n. sp.

O. oblongo-subcylindricus; thorace postice haud constricto, guttis nigris quinque discoidalibus in linea transversa dispositis, tuberculo parvo laterali; elytris modice elongatis, valde convexis, lævibus, guttis parvis albis sparsis, prope basin tuberculis utrinque circa duodecim nigris. Long. 11 lin. ♀.

Closely resembles *O. Satyrus*; but the body is proportionately shorter and broader in the female than in the corresponding sex of that species; the thorax is shorter, and shows no constriction near the base; the elytra are uniformly convex and impunctate, and there are very few tubercles near the base, only two conspicuous ones on each side of the scutellum, and a small number under each shoulder. The colour is entirely ochreous tawny, with the exception of five small spots placed in a transverse row across the thorax, the black elytral tubercles and a small number of widely separated, but tolerably uniformly distributed, white specks over the elytra. The antennæ are somewhat darker, and and there is a very distinct oblong chalky spot on each side of the breast.

Tapajos.

4. *Oncideres Diana*, Olivier.

Lamia Diana, Oliv. Ent. 67. p. 107. f. 168.

O. subcylindricus, griseus; elytris quarta parte basali dense ac minute tuberculata, parte apicali lineis tenuissimis furcatis nigris, medio guttis sparsis nigris; thorace linea transversa nigra; fœminae capite luto fulvescente; maris capite angusto, fusco, inermi. Long. 8–11 lin. ♂ ♀.

This species is distinguished by the basal portion of the elytra being thickly covered with small glossy-black tubercles, of which one at the hinder part of the humeral prominence is much larger than the rest. The tuberculated area ceases abruptly behind, and the disk of the elytra has only a very few scattered and slightly elevated black specks, which towards the apex subside into simple spots, not raised at all from the smooth surface. The general colour is pale ashy grey (white beneath); the apical part of the elytra has a few fine black lines in the form of a double or treble fork joined at the base. The male differs greatly in width of head from the female, but the antennæ scarcely differ in proportionate length or stoutness; they are, however, more nearly approximated at their bases by one-half in the male than in the female, which gives to a male insect an appearance quite foreign to the genus. The male specimen before me has a finely reticulated black patch across each elytron at the tips of the forked lines, of which there is only a trace in one of the female examples.

Pará, and at Santarem on the Tapajos.

5. *Oncideres crassicornis*, n. sp.

O. subcylindricus, postice utroque sexu attenuatus, fulvo-brunneus; elytris basi tuberculis diversis sparsis instructis, postice punctis impressis rufescenti-brunneis in lineis furcatis ordinatis; maris antennis basi valde incrassatis, capite bicornuto. Long. 9–10 lin. ♂ ♀.

Head not much wider in the female than in the male, brownish tawny, with the usual black stripe below each eye; antenniferous tubercles in the male dentiform on each side. Antennæ dark brown, simple in the female, one-half longer than the body in the male, with the basal and third joints much thickened, especially the latter. Thorax impressed transversely, and furnished on each side with a tubercle; colour brownish tawny. Elytra narrowed to the tip in both sexes, tawny brown, inclining towards ashy near the middle; the basal part raised in the middle, and studded with a moderate number of scattered tubercles, differing greatly in size, and all glossy black; from the middle to the apex there is a number of shallow punctures covered each with a reddish-brown spot and arranged in forked lines. Body beneath and legs clothed with tawny-brown tomentum.

Ega, and banks of the Tapajos.

6. *Oncideres dignus*, n. sp.

O. cylindricus, fuscus; thorace tuberculis quinque in linea transversa ordinatis; elytris prope basin tuberculis magnis globosis
utrinque sex nigris, postice guttis numerosis albis. Long.10 lin. ♂.

Head (♂) moderately narrow; forehead very narrow, being
encroached upon by the voluminous eyes, which are oblong and
reach very nearly to the extremity of the muzzle; antenniferous
tubercles unarmed; colour dark brown. Antennæ nearly twice
the length of the body, black, basal joint gradually thickened
from base to apex, rest of the antennæ tapering to the tip.
Thorax longer and narrower than in the typical species of *Oncideres*; lateral tubercles small, obtuse, and black; in a line with
them is a row of five similar glossy-black tubercles lying across
the middle of the thorax; colour dark brown. Elytra cylindrical, clear dark brown, impunctate; middle of base with six
very prominent glossy-black tubercles, arranged in two rows;
besides these, there are ten or twelve smaller tubercles on each
side, three of which are on the shoulder: the rest of the elytra
smooth, and ornamented with a number of small clear white
spots, distributed regularly and widely apart over the surface.
Body beneath and legs dark brown.

This handsome species was very rare, at Ega, Upper Amazons.

7. *Oncideres pulchellus*, n. sp.

O. minor, cylindricus, griseo-brunneus; elytris cinereo maculatis,
dimidio basali tuberculis rotundatis, dimidio apicali maculis impressis, nigro-nitidis. Long. 6½ lin. ♀.

Head and thorax of same breadth; head ashy brown, with a
streak down each side of the front tawny; buccal organs and
circuit of the mouth red. Antennæ a little longer than the
body, dark brown. Thorax ashy brown, with three shining-
black tubercles in a triangle on the disk, and two smaller ones
on each side, the outermost of which is in the position of the
ordinary lateral tubercle. Elytra cylindrical, obtuse behind,
ashy brown, varied with a small number of equal-sized and equidistant pale ashy spots, and with a number of scattered shining
round spots, those over the basal half covering large rounded
tubercles of small elevation, and those towards the apex shallow
impressions; the tubercles are not crowded near the base or
shoulders, but are widely dispersed. Body beneath and legs
light brown; sides of breast with an ashy patch.

Ega; rare.

8. *Oncideres Cephalotes*, n. sp.

O. magnus, robustus, convexus, postice attenuatus, cinereo-brunneus;
elytris prope basin dense, pone basin sparsim tuberculatis, tuber-

culis ovatis, obliquis et postice elevatis; thoracis tuberculis lateralibus elongatis, fronte magna, latissima, nuda, punctulata. Long.
15 lin., lat. capitis 4¼ lin. ♀.

Head brown; front naked, coriaceous, punctured, black; eyes
moderate, reaching little more than halfway down the forehead;
vertex very convex. Antennæ rather shorter than the body (♀),
tapering to the apex, basal joint curved; colour brown. Thorax
twice as broad as long, a little narrowed behind the lateral tubercles, which are long and spiniform; surface dull ashy brown,
with a central transverse black line. Elytra massive, narrowed
to the apex, convex, especially in the middle of the basal part on
each side; shoulders prominent and oblique, with a conspicuous
tubercle at their hinder angles; colour ashy brown, paler near
the middle, and covered with small, oblong, raised, scale-like
tubercles, which are very crowded and strongly elevated at their
posterior ends near the base, much scattered and very slightly
elevated near the middle, and arranged in rows, simply as spots,
near the apex. Body beneath ashy white; legs ashy brown.
Ega*.

Genus EUDESMUS, Serville.

Serville, Ann. Soc. Ent. Fr. iv. (1835) p. 82.

This well-marked genus resembles *Oncideres* in its cylindrical
form of body. Its distinguishing character is derived from the
bulbous ovate shape of the third antennal joint in the male.

---

* The following new species have lately been received from entomological
travellers in South America:—

Oncideres limpidus. Cylindricus, fusco-nitidus; elytris fulvo-ochraceo
irroratis. Caput (♂) modice angustatum, fronte punctata, ochracea,
vitta infraoculari nigra; tuberculis antenniferis intus prominulis,
acutis. Antennæ corpore longiores, nigro nitidæ; articulo basali
distincte clavato, articulis tertio et quarto infra dense ciliatis. Thorax
postice angustatus, tuberculis lateralibus modice productis, nigris;
supra ochraceo-brunneus, linea nigra transversa, ante medium fascia
rufo-fulva. Elytra cylindrica, fusco-nitida maculis numerosissimis
discretis tomentosis ochraceo-fulvis; juxta basin tuberculis globosis
paucis; deinde leviter granulata, humeris confertim tuberculatis.
Corpus subtus fulvo-tomentosum. Pedes nigricantes, femoribus
fulvo tomentosis. Long. 10 lin. ♂. Hab. in Bahia Brasiliæ, a Dom.
Reed lecto.

Oncideres Bouchardii. Cylindricus, cano-griseus; elytris nigro punctatis et maculis majoribus rotundatis fulvis sparsis. Caput latum,
griseum, maris paulo angustius, tuberculis antenniferis intus vix prominulis. Antennæ griseæ; articulo basali gradatim incrassato, nigro,
maris valde rugoso. Thorax griseus, linea transversa nigra. Elytra
convexa, vage punctata, cano-grisea, maculis rotundatis carneo-fulvis
conspersa, punctis nigris; prope basin tuberculis numerosis globosis.
Corpus subtus pedesque cano tomentosa. Long. 10–11 lin. ♂ ♀.
Hab. in Sta. Martha Novæ Granatæ, a Dom. Bouchard copiose missus.

The females of some of the species resemble *Oncideres* very
closely; and almost the only feature by which their generic
position may be recognized is the peculiar dark patch, streaked
with paler colours, which exists on the apical part of the elytra
of all the species. The head is broad, very little broader in the
females than in the males; but the forehead is not so plane or
so much elongated as in *Oncideres*. The basal joint of the antennæ forms a smooth ovate club; the thorax is relatively a little
longer than in *Oncideres*; the elytra are free from ridges and
tubercles, and are obtusely rounded at the apex; the claw-joint
of the tarsi is moderately elongated, and is about equal in length
to the remaining three.

1. *Eudesmus rubefactus*, n. sp.

E. cylindricus, convexus, rufescens; thorace nigro-lineato; elytris
dimidio basali griscscente, apice utrinque macula magna ovata
saturatiore strigis nigris et griseis ornata. Long. 7½–9 lin. ♂ ♀.

Head reddish tawny, vertex streaked with black; front plane,
coarsely punctured, dingy grey; eyes oblong, one-half the length
of the front; antenniferous tubercles in the male acute on their
inner sides. Antennæ about the length of the body, reddish
tawny; apices of joints, from the fourth, blackish. Thorax
cylindrical, of same width as the head, very uneven, especially
on the sides, where the inequalities rise to broad, obtuse tubercles; colour pinkish red, centre with two black lines continuous
with those on the vertex, sides each with two or three much-
broken and oblique lines. Scutellum and basal margin of elytra reddish, spotted with black. Elytra cylindrical, convex,
abruptly declivous near the apex; surface uneven, with faintly
raised lines, thickly punctured, especially towards the base,
basal half occupied by a large, triangular, common, dingy-grey
patch; on this follows a belt of pale greyish red, which broadens
greatly on the lateral margins; the apical portion of each elytron
is occupied by a dark, neatly limited, oval patch, streaked longitudinally with black, tawny red, and grey. Body beneath and
legs reddish brown; breast ashy in the middle.

Ega, clinging to dead boughs of trees; rare.

2. *Eudesmus caudalis*, n. sp.

E. cylindricus, depressiusculus, cinereo-brunneus; thorace postice
fusco notato; elytris dimidio basali grisco-fusco, apice utrinque
macula magna ovata nigricante fulvo strigata, medio cinereo fasciata. Long. 5½–6 lin. ♂ ♀.

Very closely allied to *E. rubefactus*, and scarcely differing in
the disposition of the colours and markings of the elytra. The
latter, however, are much more depressed; and the insect is of a

dull ashy-brown hue, and much narrower and smaller. The forehead is uneven, punctured, and of a dull slaty hue; the third antennal joint in the male is much less swollen than in *E. rubefactus*, and therefore more elongate, and fusiform rather than ovate in shape. The thorax is uneven and obtusely tuberculated on the sides, but is destitute of longitudinal lines, except two very short ones near the base. The elytra are of the same grey leaden hue over their basal halves, and have a pale belt beyond the middle; but the latter does not expand on the margin. The dark apical streaked spot has an ashy transverse streak across the middle.
Also found at Ega.

8. *Eudesmus posticalis*, Guérin.

Eudesmus posticalis, Guérin-Méneville, Icon. Règne Animal, p. 248.

E. cylindricus, subdepressus, brunneus; thorace dorso valde inaequali immaculato, tuberculis lateralibus parvis; elytris medio fascia obliqua grisea, deinde brunneis griseo et griseo-brunneo strigatis, ante apicem signatura nigra griseo marginata; antennis brunneis, articulis (duobus basalibus exceptis) basi testaceis; maris articulo tertio valde inflato, ovato. Long. 6½ lin. ♂.

"D'un gris-brunâtre couvert d'un duvet très-court et très-fin d'une couleur cendrée, surtout en dessous, sur les côtés du corselet et au milieu des élytres, où ce cendré blanchâtre forme une bande crochue en arrière, terminée en pointe près de la suture et précédant une tache arrondie d'un brun plus foncé, en arrière de laquelle on voit une petite tache allongée blanche et deux ou trois petites lignes noirâtres. Antennes d'un gris brun, avec la base du troisième article et des suivants d'un jaune roussâtre pâle, une petite pointe avancée à la saillie du front sur laquelle s'insèrent les antennes. Pattes courtes et fortes, d'un gris brun dessus, cendrées en dessous. Long. 14, lat. 5 mill.—Brésil intérieur." (Guérin-Méneville, l. c.)
My example was found at Ega.

4. *Eudesmus servilatus*, n. sp.

E. elongatus, depressus, fulvo-brunneus; thorace supra vittis sex nigris; elytris ultra medium dilatatis, plaga laterali infra humeros, linea basali strigisque ante apicem fuscis, vitta curvata laterali cinerea; fronte abbreviata, oculis magnis subconvexis. Long. 6¼ lin. ♀.

Head slightly convex on the forehead, with short muzzle; eyes very large, broad, and somewhat convex, reaching very nearly to the edge of the epistome; vertex bright tawny, and marked with a semicircular figure of a blackish-brown hue. Antennæ rather longer than the body (♀) and stout, ochreous brown, base of joints (from the fourth) pallid. Thorax convex, but depressed near the hind margin; lateral tubercle small, conical; colour above bright tawny, with six blackish-brown vittæ; sides ashy, with a broader and paler dusky stripe. Scutellum pale tawny ochreous. Elytra dilated a little behind the middle, depressed, and thickly punctured (except towards the apex), rusty tawny, with a few short ashy streaks and a number of dark-brown strigæ a little behind the middle, the innermost of which runs near the suture to the apex: the basal half of the suture is broadly margined with dusky, and there is a short blackish stripe on each side near the scutellum, and a broad patch of similar hue beneath each shoulder, on the upper edge of which is an ashy streak, which continues in a curved line to the lateral margin, and then to the apex. Body beneath ashy; sides of breast and abdomen dark brown. Legs reddish; femora and tibiæ each with a blackish ring round the middle.

I met with the female only of this remarkable species, which differs so much from the other *Eudesmi* in the shortness of the muzzle. If the male, when discovered, should be found not to possess the swollen third antennal joint, the species will have to be removed from this genus. It was found at Ega.

Genus XYLOMIMUS, nov. gen.

Body cylindrical, narrow. Head vertical, or slightly inclined backwards; muzzle moderately elongated; sides rounded; forehead very slightly convex; eyes small, lower lobe nearly circular. Antennæ moderately distant at their bases, with inner side of antenniferous tubercles prominent and angular; basal joint dilated almost from the base, and forming a thick, oblong club, with the lower edge slightly waved; third joint one-third longer than the first, and also thickened nearly from the base, continuing of the same thickness to the apex, furnished on the underside with a fringe of long bristles; fourth joint slender, slightly thickened in the middle, and about one-half the length of the third; fifth to seventh joints each about one-half the length of the fourth, slender (rest wanting). Thorax cylindrical, longer than broad, and deeply wrinkled transversely; lateral tubercles inconspicuous. Elytra linear, obtusely rounded at the apex, surface free from excrescences; pro- and meso-sterna plane. Legs very short, thighs clavate, tibiæ broad; claw-joint of the tarsi as long as the remaining joints taken together.

The species on which this genus is founded presents, from its shape and style of coloration, a striking resemblance to a fragment of a slender decayed branch.

Xylonimus baculus, n. sp.

X. angustatus, cylindricus, thorace transversim crebre ruguloso; elytris stria impressa suturali, apice singulatim obtuse rotundatis; corpore supra brunneo, lateribus obscure ochraceo; elytris pone medium fascia lata flexuosa brunnea ochraceo lineata; antennis brunneis, articulo quarto flavo. Long. 5¼ lin. ♂!

Head dingy ochraceous, front uneven, punctured; vertex and occiput ample, brown, streaked with rusty ochreous; antenniferous tubercles slightly prominent on their inner sides, and leaving a small semicircular notch between them. Antennæ with the first and third joints dark brown varied with ochreous, bristly, fringe of the third also dark brown, fourth joint yellow, fifth, sixth, and seventh rusty brown. Thorax cylindrical, surface covered with numerous, irregular, transverse wrinkles; lateral tubercles small, conical, dark brown in the middle, with three indistinct rusty-brown vittæ; sides each with an ochreous vitta, below which is a broader brown vitta. Elytra linear, shoulders not prominent, apex of each obtusely rounded; surface slightly uneven, plane towards the base and more convex beyond the middle, punctured (except near the apex) and marked with an impressed stria near the suture; colour rusty ochreous, with a broad common brown vitta over the suture from the base to beyond the middle, and a broad irregular brown fascia (lineated with rusty brown) at the termination of the vitta, the space near the apex having an irregular ochreous spot followed by a similarly shaped brown spot. Body beneath light brown; sides of prothorax and breast with an ochreous-white vitta; abdomen streaked with ochreous white. Legs clothed with pale tawny-brown pile.

Found on a slender dead branch of a tree in the forests of the Tapajos.

Genus Ecthœa, Pascoe.

Pascoe, Trans. Ent. Soc. n. s. iv. p. 244 (1858).

Syn. *Talasius,* Buquet, Thoms. Arcana Naturæ, p. 99 (1859).

This remarkable genus is distinguished from the allied groups by many well-marked features, which have been well described by the authors above quoted. I myself met with female examples only, and have not been able to examine the opposite sex, which bears one of the chief marks of the genus—namely, four horn-like projections from the forehead. The body is large and cylindrical; the head very broad, and remarkable (besides the horned forehead of the male) for the great convexity of the crown, which rises very much higher than the base of the antennæ, and descends perpendicularly from its front edge towards the tubercles which support those organs. The elytra are broad and square at the apex, and each one is deeply sinuated in the middle, so as to form two projections or lobes. The antennæ are rather slender, in the female as long as the body, with the basal joint tumid on one side at the apex, and the third joint slightly curved.

My specimens differ in colour from the one figured by M. Buquet; but I believe them to be referable to the *E. quadricornis* of Olivier. The *Trachysomus faunus* of Erichson (Consp. Peru. p. 148) seems to be quite a distinct species of this genus.

Ecthœa quadricornis, Olivier.

Cerambyx quadricornis, Oliv. Ent. iv. p. 97, pl. 20. f. 158.
Talanxus quadricornis, Buquet, Thoms. Arc. Nat. p. 100, pl. 5. f. 6.

The female example now in my collection, and which I found at Ega, is 9¼ lines in length, the head being 2¼ lines in width. The upper part of the forehead is yellow, brown near the crown, where it is marked with three black spots; the lower part is of a blackish olive-colour, the line of demarcation between the two colours being a transverse carina, from which in the male rise the two lower frontal horns. The thorax is very uneven on each side, one of the elevations near the anterior part of the disk on each side forming an acute tubercle; the colour above is rusty ochreous, the hind part having two blackish lines, which are severally continuous with the rounded velvety black spots on the elytra, on each side of the scutellum. The elytra are of a light green hue, except on the apical fourth, where there is a large ashy-ochreous spot, streaked with dark brown, very similar to the streaked apical spots in the genus *Eudesmus*. The underside of the prothorax and breast is greenish ashy. The legs are green, varied with greenish ashy. These green and rusty-ochreous hues, combined with the rugged surface of the insect, give it very much the appearance of a mossy fragment of wood, when it is seen clinging close to a dead bough, as is the habit of the creature.

Genus TÆNIOTIA, Buquet.

Buquet, in Thomson's 'Arcana Naturæ,' p. 45.

Like many other generic groups of Longicorns, the present one is recognizable rather by a similar general form and coloration than by definite structural characters. The species are cylindrical or linear and depressed in shape, and exhibit a dark-brown or black curved mark towards the apex of the elytra, preceded by a pale-ashy or greenish patch, and succeeded by fulvous strigæ nearer the apex. The possession of this characteristic mark points to a near relationship with *Eudesmus* and *Ecthœa*; but some species answer very well to the definition of

the genus *Hesycha*, as far as structure is concerned. All the
species, however, are more linear in form than the *Hesychæ*, and
the antennæ in nearly all are more nearly approximated at their
bases. The head is variable in width, and the forehead is somewhat convex in the middle; the latter is in most species clothed
with pale-coloured tomentum. The antenniferous tubercles, in
the broader-headed species, have prominent and sometimes
cornuted inner angles. The antennæ themselves are slender
and setaceous, in the males often twice the length of the body;
their basal joint is clavate, and the third joint, with few excep-
tions, a little curved. The thorax is cylindrical and uneven,
never short and broad. The elytra are linear, obtusely rounded
at the apex, free from centro-basal elevations and tubercles; the
shoulders are prominent and acute, and curved anteriorly. The
legs are moderately short, the thighs clavate, the claw-joint robust, as is universal in the Oncideritæ, and equal in length to
the three remaining taken together.

The *Trestoniæ*, like the other genera of the present group,
are found on branches of trees, clinging closely and gnawing the
bark and surface-wood.

1. *Trestonia Chevrolatii*.

Trestonia Chevrolatii, Buquet, Thoms. Arc. Nat. p. 46.

T. elongata, subdepressa; capite lato, tomento flavescente dense vestito, maculis duabus verticis alterisque frontalibus nigris, genis et
gula nigricantibus; antennis basi distantibus, brunneis, tuberculis
antenniferis intus modice productis acutis (♀); thorace obscure
fusco-grisescente, supra transverse ruguloso sulcis duobus transversis juxta marginem posticum distinctioribus; elytris postice
paulo attenuatis, dorso depressis, humeris subconicis, granulatis,
disco bicostato lineaque elevata suturali, punctatis, griseis, ante
apicem utrinque plaga curvata nigra (antice albo marginata), dein
fulvo-brunneis macula subapicali pallida; corpore subtus pedibusque viridi-griseis, abdomine ferrugineo-brunneo, segmentis tribus
posterioribus lateribus ochraceis. Long. 10 lin. ♀ .

One example, taken at Ega, and named as above, from the
typical specimen formerly belonging to M. Chevrolat. It would
be impossible to determine the species from the meagre description of M. Buquet.

2. *Trestonia ramuli*, n. sp.

T. elongato-oblonga, postice (♂) angustata, subdepressa, fusca, fulvo
variegata; elytris medio macula magna laterali viridi-cinerea
postice dentata et fusco marginata, intra apicem macula distinc-
tiore fulva; tuberculis antenniferis distantibus, intus utroque sexo
prominulis acutis; antennis corpore paulo longioribus, articulo
tertio curvato. Long. 6–6½ lin. ♂ ♀ .

Head moderately broad, forehead punctured, dingy brown

varied with tawny; antenniferous tubercles with their inner angles in both sexes prominent, acute, conical, and distant from each other somewhat widely. Antennæ very little longer than the body in either sex, dark brown; joints paler at the base; third joint rather strongly bent in the middle. Thorax sub-cylindrical, widest in the middle, convex, transversely depressed near the hind margin, very uneven above, and obtusely tuber-culose on the sides; dark brown, varied with rusty tawny. Ely-tra with prominent conical shoulders, and gradually narrowed towards the apex (much less so in the female than in the male), surface scarcely convex, simply punctured (except near the apex), dark brown, minutely varied with rusty tawny, and having on each side in the middle a large, oblique, greenish-ashy spot, widest on the margin : this spot is bordered posteriorly by a broadish, flexuous, blackish streak ; and close to the apex there is a tawny spot, larger and clearer in colour than the other tawny marks. Body beneath and legs clothed with olivaceous-ashy tomentum.

On dead branches, Ega.

8. *Trestonia albilatera*, Pascoe.

Hesycha albilatera, Pascoe, Trans. Ent. Soc. n. s. v. pt. 1. 25.

T. elongato-oblonga, apicem versus paulo attenuata, subdepressa, fusca, fulvo minute varia ; capite latiusculo, fronte ochracea, tuber-culis antenniferis intus in lobulos erectos oblongos productis (♂) ; elytris utrinque plaga maxima laterali (fere ad basin extensa) cana, postice nigro marginata. Long. 6½ lin. ♂ .

Similar, in its elongate-oblong subdepressed form of body and general colour, to *T. ramuli*, but differs in the elytra being much less prominent at the shoulders, and not attenuated, except from very near the apex; the pale lateral spot, too, is much larger and whiter, extending from behind the middle to the shoulders. The thorax is cylindrical and very uneven on its surface, as in *T. ramuli*; but it has two transverse impressed lines near the hind margin, and a distinct conical lateral tubercle, much behind the middle. The forehead is clothed with dense tomentum of a pale ochreous hue. The underside of the body is ashy, with a broad rusty-tawny stripe down the middle of the abdomen. The antennæ are very slender and twice the length of the body in the male; the terminal joints are greatly elongated, and the third with a scarcely perceptible bend.

Ega, on branches of trees.

4. *Trestonia coarctata*, n. sp.

Trestonia terminata, Buquet, Thoms. Arc. Nat. p. 47, pl. 5. f. 3?

T. cylindrica, cinereo-fusca, fulvo varia, vertice coarctato; antennis basi valde approximatis, articulo basali elongato, apice abrupte clavato; elytris crebre punctatis, apice nigris, fulvo lituratis. Long. 4½–6 lin. ♂ ♀.

The form of body and situation of the dark apical spot (close to the apex of the elytra) in this species so closely resemble the same features in the figure above quoted of *T. terminata*, that it is not unlikely the specimens here treated of belong to that species. I cannot, however, reconcile the description of the colours given by M. Buquet with my insects; and the figure is as uncertain in this respect as the description. His words are, "Couleur générale d'un gris-verdâtre mélangé de blanc et parfois de jaunâtre sur le devant de la tête, sur les bords latéraux du prothorax et sur la partie inférieure des élytres." The head in all my specimens is of a pale ashy hue, with a dark-brown spot on the upper part of the forehead between the eyes. The elytra as well as the thorax are dark brown, clothed with thinnish ashy pile, and sometimes varied with tawny, and becoming of a paler ashy hue near the dark apical spots. The thorax has a number of large scattered punctures, and the elytra are thickly punctured, except at the extreme apex. The antennæ are closely approximated at the base, the bases of the tubercles being separated only by the impressed line on the vertex; the angles of the tubercles are not produced. The antennæ are more than twice the length of the body in the male, the apical joint being twice the length of the preceding, and of great tenuity; in the female they are but little longer than the body, but the apical joints are very slender and more elongated than is usual in the female sex of Longicorn insects; the basal joint is as long as the third, and clavate at the apex.

Found, rather commonly, on slender branches on the banks of the Tapajos, and also at Ega.

Genus PEDITROX, nov. gen.

Body subcylindrical. Head moderately narrow; face plane, inclined obliquely backwards; eyes ample, convex; antenniferous tubercles with their inner angles produced. Antennæ elongated, simple; basal joint gradually thickened from the base; third joint straight, one-fourth longer than the first, fringed beneath with fine hairs. Thorax subcylindrical, uneven, sides armed with prominent, acute lateral tubercles. Elytra cylindrical, free from ridges and tubercles; apex rounded. Legs moderate; thighs clavate; claw-joint of tarsi greatly elongated, longer than the three remaining joints taken together.

This new genus, founded on one species only, is very closely allied to *Trestonia*, differing, in structural characters, chiefly in the gradually thickened basal joint of the antennæ. The characteristic feature in the coloration of the elytra of *Trestonia* is entirely absent, the colours being dull and uniform. In form of body and head, the species described below resembles much *Trestonia terminata* and *T. coarctata*.

Peritrox denticollis, n. sp.

P. subcylindrica, paulo convexa, fuliginosa; elytris maculis tomentosis fulvo-brunneis adspersis; capite inter antennas profunde impresso; thorace transverse ruguloso, lateribus acute tuberculatis. Long. 5 lin. ♂.

Head sooty black, coarsely punctured on the forehead and crown, and deeply grooved between the antenniferous tubercles, which are closely approximated at their bases, and have their inner edges produced into short ear-like lobes. Antennæ blackish, shining. Thorax subcylindrical, surface uneven, and marked with a few sharp transverse wrinkles, besides two impressed lines parallel to the hind margin; lateral tubercles conical, acute; colour sooty brown. Elytra cylindrical, narrowed only very near the apex, the latter rounded; surface thickly punctured, except near the apex, sooty brown, sprinkled with spots formed of dingy-tawny tomentum. Body beneath and legs pitchy, thinly clothed with ashy pile.

Santarem, on a dead branch: one example.

Genus PACHYPEZA, Serville.

Serville, Ann. Soc. Ent. Fr. (1835) iv.

The forehead, muzzle, and eyes in this genus resemble much the same features in *Oncideres*; but the crown is narrower and more depressed between the antenniferous tubercles. The body is elongate, but narrower than in any species of *Oncideres*. The antennæ have their joints beneath (including the basal joint) clothed more or less densely with longish hairs. The thorax is cylindrical, about as long as broad, and covered above with transverse wrinkles. The pro- and meso-sterna are extremely narrow. The legs are short and stout, the femora clavate, the tibiæ very short and compressed, and the tarsi have the clawjoint, although elongated, much less robust and shorter than in *Oncideres*.

Pachypeza lanuginosa, n. sp.

P. cylindrica, robusta, fusco-cinerea; capite latiore; antennis distantibus, articulis sex basalibus infra pilis tenuibus dense vestitis; elytris prope basin confertim et subtiliter granulato-punctatis. Long. 9½–10 lin. ♂ ♀.

Head rather broad, forehead between the antenniferous tubercles depressed; eyes large, oblong, ashy tawny. Antennæ a little longer than the body in the male, about the same length in the female; terminal joints shorter than the median ones, last joint short and pointed; basal and five succeeding joints densely clothed beneath with very fine hairs; colour ashy brown. Thorax scarcely so long as broad, surface closely wrinkled, many of the wrinkles not continuous; colour ashy brown. Elytra cylindrical, convex; shoulders somewhat prominent; basal fourth of the surface studded with small, regular granulations, accompanied by punctures; finely punctured in the rest of their surface; colour ashy brown, deflexed sides paler. Body beneath and legs tawny brown; base of abdomen on each side, and hind legs, sooty brown.

Ega and S. Paulo, Upper Amazons, on slender woody stems.

Genus CACOSTOLA (Dej. Cat.), Fairmaire.

Fairm. Ann. Soc. Ent. Fr. (1859), p. 532.

This genus, imperfectly characterized by M. Fairmaire, comprises a number of small-sized linear insects, closely allied to *Hesycha* and *Trestonia*, but distinguished by their narrow forms, obscure coloration, and especially by their much shorter heads, the muzzle being very little prolonged beyond the lower margin of the eyes. The antennæ are moderately distant at their bases, their supporting tubercles having a conical projection on their inner sides; they are slender, filiform, naked, and very little longer than the body; their first joint forms a smooth club, their third joint is in some species curved, and their terminal joint is at least as long as the preceding. The thorax is short and cylindrical, with a scarcely perceptible prominence in the middle of each side, and the surface punctured, not wrinkled transversely. The elytra are linear, obtusely rounded at their apices, and their surface is free from ridges and tubercles. The legs are short, the thighs clavate, and the claw-joint of the tarsi longer than the remainder taken together. The sterna are narrow, the pro- and mesosterna of equal width, and simple. The species are found, like the *Trestonia*, clinging to slender decaying branches of trees.

1. *Cacostola simplex*, Pascoe.

Pachypeza simplex, Pascoe, Trans. Ent. Soc. n. s. v. pt. 1. p. 44.

C. linearis, griseo-fusca; thorace elytrorumque lateribus griseo lineatis; capite latiusculo; antennis articulo tertio subrecto. Long. 4½–5 lin. ♂ ♀.

Head moderately broad; forehead uneven, and, with the

u

vertex, punctured, tawny-grey. Antennæ distant at the base,
supporting tubercles with their inner edges prominent; filiform,
but somewhat tapering to the extremity, dark brown, bases of
joints grey; third joint scarce perceptibly curved. Thorax of
the same width as the head, cylindrical, scarcely longer than
broad; lateral tubercle inconspicuous; surface coarsely but
sparingly punctured, greyish brown, dorsal line and two obscure
lateral streaks grey. Elytra linear, coarsely punctured (more
thickly so towards the base), and with faint longitudinal elevations on the disk, brown, sides in some examples paler; disk
with one or more oblique grey vittæ. Body beneath and legs
greyish brown; abdomen variegated with brown and grey.

Tapajos and Upper Amazons, also Cayenne. Examples from
Cayenne and the Tapajos are much darker than those from the
Upper Amazons.

2. *Cacostola flexicornis*, n. sp.

C. lineariis, castaneo-fusca, obscura; capite angustiore; thorace brevi;
elytris creberrime punctatis; antennis tenuibus, articulo tertio
valde curvato. Long. 3½ lin. ♂ ♀.

Head small; forehead with a deeply impressed longitudinal
line, punctured, coarsely pubescent; vertex coarsely punctured;
antenniferous tubercles with a small conical projection on their
inner sides. Antennæ rather slender, dark brown, with the
bases of the joints pale testaceous; third joint strongly bent;
terminal joint in the male half as long again as the preceding.
Thorax short, lateral prominences conspicuous, surface closely
punctured, dark rusty brown. Elytra linear, very closely and
equally punctured from base to apex, dark rusty brown. Body
beneath and legs dingy ashy; abdomen variegated.

Slender dead twigs, Santarem.

Genus AMPHICNÆIA, nov. gen.

Body small, linear. Head very short, vertically; upper portion of the eyes encircling the base of the antennæ; but the
reniform lobe of considerable width, and not attenuated as
in the eyes of the genus *Dorcasta**; lower lobe convex,
prominent; forehead convex. Antenniferous tubercles very
short, oblique, and unarmed: antennæ filiform, stout, clothed
with short hairs, the joints beneath fringed with long and
straight hairs; first joint moderately short, thickened nearly

* The upper, reniform portion of the eyes in *Dorcasta* is very narrow,
and, in the middle, attenuated. This is a step towards the total disappear-
ance of the upper lobe, which is a distinguishing feature of *Spalacopsis*,
Newm., a genus closely allied to *Dorcasta*.

from the base; third joint straight. Thorax cylindrical, sides without tubercles, surface punctured. Elytra linear, apex rounded, surface punctured throughout. Legs moderately elongated; thighs clavate; claw-joint of tarsi about as long as the three remaining joints taken together. Sterna narrow, simple.

This genus forms a portion of a small group—including *Dorcasta*, *Aprosopus*, and *Spalacopsis* (= *Eutheia*, Guér.)—which differs from all the foregoing in the form of the head and in the shortness of the antenniferous tubercles.

1. *Amphicnæia lineata*, n. sp.

A. brevis, sublinearis, fusco-nigra, thoracis vittis tribus, scutello et elytrorum vittis duabus lateralibus griseis; elytris longe setosis, crebre punctatis, apice subobtuse rotundatis. Long. 2¼ lin.

Head very short in front; forehead thickly punctured throughout. Antennæ filiform, rather thick, black. Thorax very thickly punctured, convex; dorsal line and a lateral vitta on each side greyish. Scutellum grey. Elytra sublinear, moderately narrowed towards the apex, and rounded at the tips; surface thickly punctured throughout, and clothed with longish stiff hairs; blackish brown, with two tawny-ashy vittæ on each side approximating towards the base. Body beneath and legs rusty, shining, thinly clothed with greyish pile.

Ega; common on dead twigs.

2. *Amphicnæia pusilla*, n. sp.

A. minuta, testaceo-fusca; thorace punctato, griseo trivittato; elytris setosis, punctatis, testaceo-fuscis, sutura lateribusque obscurioribus; antennis pedibusque ferrugineis. Long. 1½ lin.

Head rusty brown, forehead punctured, vertex and occiput thickly punctured. Antennæ rusty red, sparingly setose, basal joint rather thick, forming an ovate club. Thorax evenly punctured throughout, rusty brown, the dorsal line and a broadish vitta on each side grey. Scutellum grey. Elytra linear, punctured throughout, testaceous, suture and sides rusty brown. Body beneath and legs pale ferruginous.

Santarem.

Closely allied to *A. lineata*, but distinguished by its smaller size and different coloration*.

* A third species occurs at Rio Janeiro, in South Brazil:—

A. lyctoides. Linearis, fusco-ferruginea; corpore supra crebre passim punctato. Antennae infra sparsim hirsutæ. Elytra linearia, glabra, punctis sublineatim ordinatis. Corpus subtus et pedes fusco-ferruginea, glabra; episternis, pectore segmentisque abdominalibus medio grosse punctatis. Long. 1½ lin. *Hab.* in Rio Janeiro.

Genus ALETRETIA, nov. gen.

Body elongate-elliptical. Head short, vertically; forehead convex; eyes not prominent; upper or reniform lobe moderately broad and reaching the centre of the crown, so that the eyes above are separated only by the longitudinal line of the vertex. Antenniferous tubercles short, unarmed: antennæ stout, a little longer than the body, and tapering towards the apex, fringed beneath with long and fine hairs; basal joint moderately short and thickened almost from the base. Thorax cylindrical, lateral tubercles very small. Elytra narrowed towards the apex, the tips obliquely and briefly truncated. Legs moderately elongated, tarsi narrow, claw-joint stout and as long as the three remaining joints taken together.

The form and clothing of the antennæ, shape of claw-joint, and general habit show this genus to be closely allied to the preceding, notwithstanding the numerous points of difference.

Aletretia inscripta, n. sp.

A. elongato-elliptica, nigra; thorace vittis quinque, elytris utrinque vittis quatuor (juxta basin et pone medium interruptis) fulvo-griseis, spatio nigro mediano elytrorum lineola transversa fulvo-grisea. Long. 3¼–4½ lin. ♂ ♀. ·

Head clothed with greyish or tawny pile, not visibly punctured; central line deeply impressed; eyes nearly touching on the vertex. Antennæ one-third longer than the body, dark brown, pubescent, fringed with long fine hairs beneath. Thorax cylindrical, rather broader in the middle, and having on each side a minute tubercle; surface punctured throughout, black, clothed with fine grey pile, and marked with five greyish-tawny vittæ. Scutellum tawny grey. Elytra narrowed towards the apex, the tip briefly and squarely truncate, with the outer angle prominent; surface deeply but sparsely punctured towards the base, faintly so and glossy towards the apex, black, the basal half with four light-brown vittæ (the second one from the suture alone reaching the base), and the apical part with a number of short streaks of the same colour, the intermediate black space having on each elytron a transverse wedge-shaped line. Body beneath and legs clothed with light-brown pile.

Upper and Lower Amazons; on dead twigs.

Genus DORCASTA, Pascoe.

Pascoe, Trans. Ent. Soc. n. s. iv. p. 264.

In this genus the body is much more elongated than in any of the preceding, being narrow and linear, but tapering towards the apex of the elytra. The head has an elongated crown, or, in other words, is prolonged horizontally; and the forehead in the typical species is directed obliquely towards the edge of the prosternum. The upper reniform lobe of the eyes is very narrow. The antennæ are not longer than the body, and are closely approximated at their bases; but the antenniferous tubercles are not elevated or armed; the antennal joints are short, thick, and setose, the bristles on the under surface being longest; the basal joint is thickened from the base, and of equal breadth thence to the apex. The legs are short and stout, and the claw-joint of the tarsi is about equal in length to the remaining joints taken together. The elytra are briefly sinuate-truncate, and dentate at the apex.

1. *Dorcasta oryx*, Pascoe.

Dorcasta oryx, Pasc. Trans. Ent. Soc. n. s. iv. p. 264.

D. sublinearis vel attenuato-elliptica, fusca, griseo tomentosa; capite thoraceque vitta laterali lineaque dorsali fulvis; elytris utrinque fulvo trilineatis, apice oblique sinuato-truncatis, angulis acutis; corpore toto setoso; capite elongato, infra valde retracto. Long. 3¼ lin.

Abundant on dry twigs in hedges, Santarem. The *Hypopsis dasycera* of Erichson (Schomburgk's Reise in Brit. Guiana, vol. iii.) is evidently a *Dorcasta* closely allied to *D. oryx*, if not the same species.

2. *Dorcasta ignea*, n. sp.

D. linearis, grisea, capite thoraceque lineis duabus dorsalibus, regione scutellari et elytrorum vitta lata curvata fusco-nigris; capite elongato, infra retracto; elytris striato-punctatis, subcostatis, apice oblique sinuato-truncatis, angulis externis valde productis crassis obtusis: corpore haud setoso. Long. 4 lin.

Head prolonged above and retracted beneath, as in *D. oryx*; forehead clothed with tawny-grey pile; vertex and occiput dingy tawny, lineated with black; upper reniform lobe of the eyes extremely attenuated in the middle; vertex punctured. Antennæ about as long as the body, clothed with short setæ; basal joint oblong, angular; colour blackish, bases of joints greyish. Thorax convex in front; surface punctured, dingy tawny; sides each with a light-grey line; centre with two flexuous blackish lines extending to the head and meeting on the crown. Elytra free from setæ, slightly tapering from base to apex, the latter obliquely sinuate-truncate, with the outer angles produced into vertically thickened lobes; surface with coarse punctures arranged in lines, some of the interstices subcostate; colour dingy grey; the scutellar area and a broad streak, curving from each shoulder to the suture and subapical margin, dark brown. Body beneath and legs dingy brown.

Dry twigs, Santarem.

3. *Dorcasta occulta*, n. sp.

D. cylindrica, postice subobtusa, griseoceus, brunneo variegata, regione scutellari fusco-nigra; elytris juxta apices abrupte declivibus, apice breviter suboblique sinuato-truncatis, angulis acutis; capite infra minus retracto. Long. 2¾ lin.

Head less elongated and less retracted beneath than in the typical species, clothed with dingy greyish tomentum; central line deeply impressed; upper lobe of eyes attenuated. Antennæ thick, filiform, sparsely clothed with short bristles, longer underneath; basal joint oblong, angular. Thorax convex, sparingly punctured, tawny grey, with whitish streaks on the sides. Elytra cylindrical, subobtuse and abruptly declivous near the apex, the latter briefly sinuate-truncate, both angles slightly produced and acute; surface free from bristles, coarsely punctured, partly in lines, dingy grey-tawny, with brownish spots, a large patch over the scutellar area, and sometimes a curved spot on each side, in the middle, dark brown. Body beneath and legs tawny ashy.

Santarem, on dry twigs.

4. *Dorcasta cænosa*, n. sp.

D. cylindrica, postice subobtusa, griseo-fusca; thoracis lateribus et elytrorum maculis cinereis; elytris apice oblique sinuato-truncatis, angulis prominulis acutis; capite infra minus retracto. Long. 1¾ lin.

Head less elongated and retracted beneath than in the typical species, rusty-brown, clothed with dingy-grey tomentum; central line deeply impressed; upper lobe of eyes attenuated in the middle, the extremity, on the crown, raised. Antennæ filiform, clothed throughout with short setæ; basal joint thickened abruptly from the base, oblong: colour dingy brown. Thorax subcylindrical, slightly tumid in the middle; surface punctured, rusty grey-brown, sides pale ashy. Elytra linear, narrowed a little before the apex, the latter obliquely sinuate-truncate, with both angles acute; surface coarsely punctured, greyish rusty brown, with an ashy streak near each shoulder, and a discoidal ashy line divided into spots by brown specks. Body beneath and legs rusty brown.

Santarem, on dried twigs.

Group *Hippopsinæ*.

Genus MEGACERA, Serville.

Serv. Ann. Soc. Ent. Fr. iv. p. 43.

Megacera agrees with *Hippopsis* in the greatly elongated form of body, and in the long setiform antennæ, more than twice the length of the body, and fringed with fine bristles or hairs beneath. It differs, according to Serville, in the vertical instead of retracted inclination of the face, and in the elytra being squarely instead of obliquely truncated or pointed at the apex. I find, on the examination of a series of species, that these two characters do not go together, some species having the head of a *Megacera* with the elytra of an *Hippopsis*. One of the following species described under *Megacera* (*M. prælata*) has, however, a facies quite distinct from *Hippopsis*, owing to the greatly swollen posterior orbits of the eyes and absence of lineation in the colours of the thorax and elytra. In general form it much resembles *M. vittata* of Serville, the type of the genus.

1. *Megacera prælata*, n. sp.

M. linearis, parallelogrammica, olivaceo-cinerea; capite, thorace, elytrorum basi et antennis obscurioribus; antennis longissimis; capite verticali, orbita oculorum incrassato; thorace transversim valde rugoso; elytris sinuato-truncatis, angulis prominulis acutis. Long. corp. 9 lin., antenn. 28 lin.

Head with vertex moderately elongated and subconvex, punctured; face short, nearly vertical, clothed with dark olive-ashy tomentum; posterior orbit of eyes thickened and prominent. Antennæ more than three times the length of the body, blackish, scantily clothed with olivaceous tomentum. Thorax cylindrical, anterior and posterior transverse sulcus well marked, the intermediate part of the dorsal surface covered with coarse transverse rugæ; dark olivaceous. Elytra linear, very slightly narrowed close to the apex, the latter transversely sinuate-truncate, both angles faintly prominent; surface finely punctured towards the base, light olivaceous ashy, smooth, base a little darker. Body beneath and legs clothed with smooth olivaceous-ashy tomentum.

One example on a slender branch in the forest, Ega.

2. *Megacera apicalis*, n. sp.

M. linearis, postice perparum angustata, griseo-nigra; capite pone oculos tumidulo, lateribus lineisque duabus verticis antice convergentibus fulvis; thorace et elytris utrinque fulvo trivittatis, vitta interiore elytrorum juxta basin attenuata, vittis omnibus ante apicem in fasciam griseo-fulvam terminatis, ipso apice nigro, sinuato-truncato, angulis acutis. Long. 5½–7 lin.

Head with vertex moderately prolonged; face short, slightly retracted; black, clothed with thin grey pile, sides and two coronal vittæ converging in front tawny; vertex coarsely but sparingly punctured; sides somewhat tumid behind the eyes. Antennæ nearly three times the length of the body, basal joints densely fringed beneath; colour blackish, thinly clothed with grey pile. Thorax cylindrical, a little narrowed in front, surface coarsely punctured, the punctures here and there running into rugæ; greyish black, with six tawny vittæ. Elytra linear, very slightly narrowed from base to apex, the latter transversely sinuate-truncate, both angles acute; surface thickly punctured, except near the apex, greyish black; each elytron marked with three tawny vittæ, the innermost one of which is very narrow near the base, and all terminate in a broad, subapical, tawny-ashy belt, which is succeeded by a black belt occupying the apex. Body beneath and legs grey; sides of breast with two tawny streaks.

Ega, on slender branches.

8. *Megacera rigidula*, n. sp.

M. linearis tenuis, postice sensim attenuata, griseo-nigra; capite lateribus vittisque duabus verticis cinereo-fulvis; thorace grosse sparsim punctato, vittis sex, et elytris utrinque vittis tribus cinereo-fulvis, vittis duabus lateralibus elytrorum ante apicem terminatis. Long. 4¼ lin.

Head with vertex moderately prolonged, face short, slightly retracted; black, clothed with grey pile, covered with large punctures; sides and two convergent vittæ on the vertex ashy tawny. Antennæ rust-coloured. Thorax cylindrical, covered with large scattered punctures, some of which are confluent, and marked with six tawny-ashy vittæ. Elytra slender, gradually narrowed from base to apex, the latter sinuate-truncate, with both angles produced and acute, the external one most so; surface coarsely punctate-striate to the apex, greyish rusty black, each elytron with three ashy-tawny vittæ, all thickest towards the base (the lateral one furcate), and the two lateral ones terminating before the apex in an ashy spot. Body beneath and legs grey, the tomentum more dense on the sides of the body.

Santarem.

Genus HIPPOPSIS, Serville.

Serville, Encyel. Méthod. x. p. 336.

As already observed in the remarks under the head of *Megacera*, this genus is remarkable for the very elongated narrow form of body, and equally elongated hair-like antennæ, which are fringed with fine hairs beneath, at least the basal joints. The body is not linear, as in *Megacera*, but is gradually attenuated posteriorly, the elytra having their apices prolonged into a point. The degree to which this prolongation of the elytral tips is carried varies in the different species, and offers a good mark for distinguishing some of them. In some, namely those which approach *Megacera*, the elytra are simply very obliquely sinuate-truncate at the apex, both angles of the truncature being acute, but the external one greatly prolonged. In others the external angle is still further prolonged, and the sutural one only just perceptible. This feature is carried out to greater lengths in other species, in which the truncature is so extremely oblique as to be imperceptible, the elytra then appearing to be terminated each in a long, fine point.

The species of *Hippopsis*, like all other Oncideritæ, are parasitic on the slender branches of trees. They choose, however, the most slender twigs, and cling to them so closely by their short stout legs and elongated claws as to be difficult of detection. All that I have seen possess the same style of coloration —a ground-colour black or brown, clothed with extremely fine grey pile, and marked with tawny or dingy grey stripes extending over head, thorax, and elytra, the diversities of which sometimes form good specific characters.

1. *Hippopsis truncatella*, n. sp.

H. linearis, fusca, capite, thorace et elytris utrinque vittis tribus testaceo-griseis; capite thorace latiore, pone oculos sensim angustato; elytris paulo ante apices attenuatis, apice utrinque oblique sinuato-truncatis, angulo interiore prominulo acuto, exteriore late producto, vittis griseis duabus interioribus ante apicem conjunctis. Long. 4½ lin.

Head broader than the thorax, curvilinearly narrowed behind the eyes; face strongly retracted; eyes prominent; brown, face clothed with thick greyish pile; vertex coarsely punctured, and, with the sides, marked with six greyish vittæ, the two central ones of which gradually converge on the crown, and the four others traverse the deflexed sides of the neck and cheeks. Antennæ slender, basal joint gradually thickened from base to apex; colour rusty brown. Thorax narrower than the head or elytra, cylindrical, coarsely punctured, brown, marked on each side with three greyish vittæ, the lowermost of which is continuous along the sides of the breast. Elytra scarcely perceptibly narrowed from the shoulders to near the apex, thence rapidly narrowed; the apex truncated a little obliquely, the truncature incurved near the sutural angle, which is produced and acute, the outer angle being broad and also acute, but moderately produced; surface thickly punctured, partly in lines, brown, and marked on each elytron with three broad, greyish vittæ, the two inner ones of which unite before the apex, and the lateral one interrupted at the shoulder, under which is a small grey streak. Body beneath and legs clothed with fine greyish tomentum.

Pará and Lower Amazons.

x

2. *Hippopsis griseola*, n. sp.

H. linearis, fusca griseo-suffusa; thorace elytrisque utrinque vittis tribus, collo vitta lata, vertice lineis duabus parallelis testaceo-cinereis; capite pone oculos tumidulo, deinde angustato, vittis elytrorum omnibus ante apicem commixtis; elytris apice acuminatis, divaricatis. Long. 4¾ lin.

Head a little broader than the thorax, tumid behind the eyes, then rather abruptly narrowed; face strongly retracted; brown, rather thickly clothed with grey pile, side of the neck with a broad ashy vitta, vertex with two narrower vittæ parallel up to the eyes. Antennæ rusty brown, basal joint gradually thickened from base to apex. Thorax cylindrical, surface having very large confluent punctures, brown, clothed with fine grey pile, and marked with six testaceous-ashy vittæ. Elytra linear to near the apex, thence gradually narrowed, each elytron ending in a point, the sutural side of which is nearly straight, the outer side a little incurved, hence giving an outward turn to the pointed apices; surface punctured, partly in lines, punctures fainter near the apex, brown, clothed with grey tomentum, and marked on each elytron with three testaceous-ashy vittæ, all of which coalesce at a distance from the apex. Body beneath and legs thinly clothed with greyish pile, sides of breast and abdomen streaked with denser tomentum.

Santarem.

3. *Hippopsis clavigera*, n. sp.

H. linearis, tenuis, fusca, vertice vittis quatuor geminatis, thorace et elytris utrinque vittis tribus griseis; corpore toto grosse punctato; antennis articulo basali apice clavato. Long. 2¾ lin.

Head broader than the thorax, gradually narrowed behind the eyes, beneath strongly retracted; forehead elevated at the summit a little above the level of the crown; antenniferous tubercles suborbicular and prominent; eyes lateral, nearly round, slightly emarginated near the base of the antennæ, but not extending in a reniform lobe upon the vertex; the latter closely punctured, marked with four greyish stripes united in pairs posteriorly; face clothed with greyish hairs. Antennæ very slender, capilliform, scantily fringed with long hairs; basal joint slender, somewhat abruptly clavate towards the apex. Thorax cylindrical, evenly and thickly punctured; brown, marked with six greyish vittæ. Elytra linear, gradually tapering, more quickly so nearer the apex, which is moderately prolonged and pointed, without truncature; surface closely punctured from base to apex, brown, marked with three broad greyish stripes. Body beneath coarsely but evenly punctured throughout, and, with the legs, thinly clothed with greyish pile.

This singular little species occurred only at Santarem, on the Lower Amazons.

4. *Hippopsis prona*, n. sp.

H. linearis, elongata, fusca, nitida, collo vitta lata laterali, vertice lineis duabus, thorace et elytris utrinque vittis tribus testaceo-griseis; capite infra valde retracto, supra quadrato; elytris leviter oblique truncatis, acutissimis. Long. 5 lin.

Head above quadrate, the lateral outline behind the eyes being nearly straight; face elongated and very strongly retracted, tending towards the horizontal position, clothed with greyish hairs, and deeply impressed on the summit between the antennæ; vertex coarsely punctured, having a shining, raised dorsal line, brown; sides each with a broad vitta, and vertex with two stripes, greyish. Antennæ piceous, finely and densely fringed, basal joint gradually thickened from base to apex. Thorax cylindrical, covered with large even punctures; rusty brown, marked with six tawny-grey stripes. Elytra much elongated, four and a half times the length of the thorax, linear, gradually narrowed, and near the apex more quickly narrowed; the latter prolonged into an acute point; the inner side of the prolongation formed by an oblique truncature, the sutural angle of which is distinct; surface punctured in distinct rows, punctures indistinct towards the apex, brown, shining, marked on each elytron with three testaceous-grey stripes, the inner two of which unite at the apex; the middle stripe is fainter and greyer than the other two, and is interrupted towards the base. Body beneath faintly punctured, piceous, and, with the legs, clothed with thin, grey pile.

S. Paulo, Upper Amazons.

5. *Hippopsis fractilinea*, n. sp.

H. elongato-fusiformis, fusco-nigra, collo vitta laterali, vertice lineis duabus, thorace et elytris utrinque vittis duabus fulvis, vitta interiore elytrorum mox pone medium fracta; thorace supra transverse ruguloso; elytris valde acuminatis. Long. 5–10 lin.

Head narrower than the middle part of the thorax, and constricted midway between the eyes and the hind margin; face very short, moderately retracted, clothed with fulvous pile, central line deeply impressed; antenniferous tubercles with their inner margin dentate; vertex having a few large punctures in the middle, and a shining central line impressed posteriorly; dark brown, sides each with a stripe, vertex with two narrow converging lines fulvous. Antennæ greatly elongated, black. Thorax narrowed in front, and constricted near its hind margin, surface transversely punctate-rugose; brownish black, shining, surface with two tawny lines, sides each with one similar line continuous with a streak on the side of the breast. Elytra tapering from base to apex, each elytron ending in a straight point, the sutural edge being also nearly straight; surface shining brown-black, punctured (except towards the apex), and marked on each with two tawny vittæ, the inner one of which is severed after the middle, the severed ends oblique and running parallel for a short distance; suture towards the base and disk marked with faint silky grey lines. Body beneath shining black, clothed with fine silky greyish pile; abdomen with three tawny stripes. Legs black, clothed with silky tawny pile.

Common on dead branches of trees at Ega.

Subtribe DÉSMIPHORITÆ.

Group *Exocentrinæ*.

Genus EXOCENTRUS, Mulsant.

Mulsant, Coléopt. de France, Longicornes, p. 162.

Exocentrus is a well-known genus of wide distribution, and comprising a number of small Lamiaires, of ovate or oblong form of body with thorax armed on each side with a distinct acute spine. The antennæ are not much longer than the body in the most slender species, and are generally setose; the basal joint is of moderate length, forming an elongate club thickened almost from the base. The claw-joints of the tarsi are elongated but slender, and the claws are widely divergent. The genus may be known from all the genera of Acanthocinitæ by the sockets of the anterior thighs being open or angulated on their outer edges.

1. *Exocentrus striatus*, n. sp.

E. oblongus, convexus, fusco-ferrugineus, griseo sparsim pubescens; antennis corpore paulo longioribus, pubescentibus; oculis magnis, supra fere contiguis; thorace pone medium spina valida longa armato; elytris striato-punctatis; pedibus testaceo-ferrugineis, femorum clavis fuscis. Long. 3¼ lin.

Head rather narrow; sides occupied by the voluminous eyes, which also almost meet on the vertex; muzzle below the eyes short but rectangular; rusty brown, clothed with hoary pile. Antennæ filiform, a little longer than the body, clothed with laid pubescence, rusty brown, bases of joints reddish; basal joint of nearly equal thickness throughout, gradually narrowed near the base. Thorax subquadrate, very little narrowed behind, each side, behind the middle, armed with a long, stout, slightly curved spine; surface thickly punctured and sparsely clothed with recumbent shining hoary pile. Elytra oblong, convex, a little narrowed towards the apex, the latter rounded; surface punctured in rows, except about the suture near the base, where they are very closely punctured ; the scant hoary pile lies in lines along the interstices ; colour rusty brown. Body beneath rusty brown, thinly clothed with shining hoary pile. Legs moderately elongate, pale reddish ; thighs strongly clavate, clubbed part blackish. Santarem, on slender dry twigs.

2. *Exocentrus nitidulus*, n. sp.

E. oblongus, convexus, fusco-ferrugineus, nitidulus pube sparsa brevissima cinerea vestitus ; antennis corpore dimidio longioribus, thorace utrinque spina recta armato, supra postice linea transversa impresso ; elytris punctatis, punctis apud discum sublineatim ordinatis. Long. 2–2¾ lin.

Head convex in front ; central line deeply impressed ; muzzle narrowed below the eyes ; the latter moderately large, distant on the vertex ; rusty brown, clothed with ashy pubescence. Antennæ half as long again as the body, nearly naked, ferruginous. Thorax subquadrate, constricted behind the spines, the latter stout, very acute, and straight ; surface closely punctured, dark rusty, scantily clothed with ashy pubescence. Elytra oblong-ovate, convex, very thinly clothed with short, shining, cinereous hairs, thickly punctured, the punctures on the disk partly arranged in rows ; colour rusty, in some examples with a brassy tinge. Body beneath dark rusty, scantily clothed with ashy hairs. Legs dark rusty, thighs abruptly clavate.
Santarem, on slender dry twigs.

Genus BLABICENTRUS, nov. gen.

Body oblong-ovate, convex, clothed with longish stiff hairs. Head small ; muzzle narrowed below the eyes ; the latter large and nearly approximating on the vertex. Antennæ filiform or setaceous, a little longer than the body, clothed with stiff hairs ; basal joint narrowed towards the base. Thorax tumid on each side in the middle, but quite destitute of spine. Elytra oblong-ovate, convex, rounded or briefly and obliquely truncated at the apex. Legs moderately elongated ; thighs abruptly clavate ; tarsi rather narrow and shorter than the tibiæ even in the hind legs ; claw-joint elongated, claws divergent.

1. *Blabicentrus hirsutulus*, n. sp.

B. oblongo-ovatus, convexus, undique setosus, brunneus, nitidulus ; elytris maculis elongatis griseis lineatim ordinatis, apice rotundatis. Long. 3 lin.

Head dingy brown, clothed with coarse light-brown pubescence and with longish stiff hairs ; central line faintly impressed ; eyes

simple. Antennæ very little longer than the body (? ♀), rusty
red, scantily clothed with longish stiff hairs. Thorax equal in
width to the head, much narrower than the elytra, convex above,
very slightly tumid on the side in the place of the missing lateral
spine, faintly constricted posteriorly ; rusty brown, shining,
sparsely pubescent, and clothed with a few longish stiff hairs. Elytra elongate-ovate, rounded at the tip ; surface punctured in rows
and bristly with dark-coloured hairs, brown, pubescence greyish
except near the base, and forming several rows of short linear spots
separated by dark-brown specks. Body beneath and legs dark
brown, the latter clothed with long, stiff hairs.

Banks of the Tapajos, on dead twigs.

2. *Blabicentrus angustatus*, n. sp.

B. angustatus, ellipticus, minus convexus, fusco-ferrugineus ; thorace
medio utrinque distincte tumido, deinde angustato ; antennis elytrisque setosis, his apice oblique breviter subobtuse truncatis.
Long. 2¾ lin.

Head rusty brown, impunctate, scantily clothed with greyish
pubescence ; eyes moderate, distant on the vertex. Antennæ setaceous, half as long again as the body (? ♂), scantily clothed
with fine bristles, rusty brown. Thorax broader than the head
in the middle, thence sinuate-angustate to the base ; surface very
slightly convex, smooth, rusty brown, shining, scantily clothed
with very fine pubescence. Elytra scarcely broader than the
middle part of the thorax, narrowed towards the apex, which is
briefly and obliquely truncated ; surface very slightly convex,
marked with a few scattered punctures, and clothed throughout with
longish and rather fine erect hairs, rusty brown, with fine greyish
pubescence arranged in lines. Body beneath and legs rusty
brown, the latter partially clothed with fine hairs.

Santarem, on dead twigs.

Genus EAIOPSILUS, nov. gen.

Body elongate-oblong or sublinear, clothed throughout with
long woolly hairs. Face short and rather broad ; muzzle a little
dilated below the eyes ; eyes small, widely distant on the vertex ;
crown broad and not depressed between the antenniferous tubercles, the latter scarcely prominent. Antennæ scarcely so long
as the body, filiform ; basal joint short and thick, attenuated at
the base ; third and fourth joints together as long as all the succeeding joints, which are each very short. Thorax subquadrate,
each side armed in the middle with a short conical tubercle.
Elytra elongate-oblong, rounded at the apex. Legs short ; thighs

clavate; tarsi short and broad, basal joint triangular; claw-joint elongated, slender, claws widely divergent and simple.

Eriopsilus nigrinus, n. sp.

E. elongato-oblongus, fuliginosus, nitidus, capillis longis ubique vestitus, supra grosse punctatus. Long. 3 lin.

Head broad, forehead closely and finely punctured and with an impressed central line, vertex coarsely punctured, black shining. Antennæ a little shorter than the body, thickly clothed throughout with long and fine woolly hairs of a blackish colour; second and third joints elongated and equalling in length the succeeding joints taken together. Thorax sooty black, shining, coarsely punctured, and clothed with long blackish hairs. Elytra elongate-oblong, rounded at the tip, coarsely punctured, the punctures becoming shallower towards the apex, sooty black, shining, clothed with long blackish hairs. Body beneath punctured, black, clothed with dark-greyish hairs. Legs black, thickly clothed with dark hairs.

S. Paulo, Upper Amazons.

Genus Omosarotes, Pascoe.

Pascoe, Journal of Entomology, vol. i. p. 131.

The remarkable insect which constitutes this genus is much more elongated in form even than the preceding (*Eriopsilus*); yet its essential characters show that its true place is amongst the series of genera composing the Exocentrine group—a position already accorded to it by Mr. Pascoe (Trans. Ent. Soc. 3rd ser. vol. iii. p. 55). In the form of the head it does not differ much from *Eriopsilus* or even *Exocentrus*, the face being moderately broad and the muzzle slightly dilated and quadrate below the eyes; but the antenniferous tubercles are more conspicuously developed and the vertex depressed between them. The antennæ are nearly as long as the body; the basal joint forms a smooth, elongato-pyriform club, the third and fourth joints are much elongated, and the succeeding joints abbreviated, the fifth being only half the length of the fourth; but what is remarkable in them is their clothing, the long fine hairs which exist scantily on the joints being changed into very long and rather stiff bristles at the apices of the joints; the third joint is thickened towards the apex, beneath. The thorax is oblong, very convex and almost gibbous in the middle, and constricted before and behind; in the middle of each side is a very distinct and sharp tubercle. The elytra are scarcely longer than the head and thorax taken together, and are remarkable for a very long pencil of hairs surmounting the prominent centro-basal ridges, besides an acute carina extending from the prominent shoulders half-way down the sides of each elytron. The legs are rather elongated, the thighs clavate, and the tarsi very short. The insect in general form and colour resembles certain species of *Mallocera* or *Ibidion* in the Cerambycide section of Longicornes.

Omosarotes singularis, Pascoe.

Omosarotes singularis, Pasc. Journal of Entomol. vol. i. p. 131, pl. 8. f. 5.

O. elongatus, niger, antennis podibusque nigro hirsutis ; capite et thorace subtiliter strigosis, hoc antice griseo-sericeo ; elytris pube tenuissima griseo-sericea vestitis, lateribus fasciaque pone medium nigerrimis, pedibus piceo-rufis. Long. 4½ lin.

I met with two examples of this insect, namely on a slender branch of a tree in the forest at S. Paulo, Upper Amazons.

Genus SCOPADUS, Pascoe.

Pascoe, Trans. Ent. Soc. n. s. vol. iv. p. 100.

This genus resembles *Omosarotes* in its elongate shape and Cerambyceideous aspect; but its antennæ are much elongated, filiform to their apex, and nearly naked. The groove of the anterior tibiæ, which is the invariable character of the Lamiaires, is scarcely perceptible, so that, were it not for the vertical face, square muzzle, and pointed palpi, it might be doubted whether the genus would not have its true place amongst the Cerambycidæ; the groove, however, on careful examination, is seen to be present. The legs are elongated ; the thighs very abruptly clubbed, the tibiæ slender and linear, and the tarsi short, with the basal joint triangular. The anterior and middle coxæ are globular, the sterna very narrow, and the anterior sockets angulated on their outer side. As in *Omosarotes*, the elytra have raised centro-basal ridges surmounted by a pencil of hairs; on the outer side of each ridge lies an oblique linear depression, extending from the inner side of the prominent shoulder to the middle of the suture.

Scopadus ciliatus, Pascoe.

Scopadus ciliatus, Pasc. Trans. Ent. Soc. n. s. iv. p. 100, pl. 22. f. 5.

Sc. elongato-oblongus, rufescens, capite et pronoto nigris opacis, elytris dimidio apicali purpureo-nigro velutino ; thorace supra convexo tuberoso lateribus utrinque tuberculo acuto armatis. Long. 5 lin.

On stem of dead tree, Ega ; three examples.

Genus ESMIA, Pascoe.

Pascoe, Trans. Ent. Soc. n. s. vol. i. p. 44.

Like the three preceding genera, the present has an elongate form of body. The antennæ are a little longer than the body, and have the basal joint and the third and basal half of the fourth joints thickened and densely clothed with hairs; the fifth joint has also a dense patch of hairs on its upper surface; the third and fourth joints are greatly elongated; the rest of the antennæ, body, and legs are clothed less densely with shorter hairs. The front of the head is vertical, and the muzzle quadrate. The thorax is short, subquadrate, and armed on each side with a tubercle. The legs are moderately short, the tarsi short and rather broad, the claws divergent.

Esmia turbata, Pascoe.

Esmia turbata, Pascoe, Trans. Ent. Soc. n. s. vol. i. p. 44.

E. sublinearis, saturate castanea, subnitida, breviter hirsuta, punctata, linea laterali tetina corporis, altera per thoracem et suturam elytrorum extensa lineolisque discoidalibus elytrorum flavis. Long. $3\frac{1}{4}$ lin.

Ega, on slender branches; rare.

Group Tapeininæ *.

Genus TAPEINA, Serville.

Serville, Encycl. Méthod. x. p. 545.

Body oblong, extremely depressed, clothed with erect hairs. Head broad and short, the lower part not being prolonged below the eyes, and the front edge of the crown in the female either forming a transverse ridge a little above the labrum or sloping to the epistome, and in the male elongated laterally into projections of various forms according to the species. Antennæ longer than the body, stout, setaceous. Thorax transverse oval. Elytra rounded at the tip. Legs moderately short; thighs clavate; tarsi short and broad; claws divergent.

The species forming this curious genus are found underneath close-fitting bark of trees, after they have been felled or uprooted in the forest. They share this peculiar habitat with the flattened Cucujidæ, Nitidulidæ, Histeridæ, and others, all of which form together a somewhat extensive insect-fauna suited to these confined habitations.

* This group was placed provisionally under the Superditæ, in the synopsis previously given of the Lamiaires. A more accurate examination has convinced me that it has closer affinities with the members of the Desmiphoritæ. The Tapeinæ, in fact, appear to be abnormally flattened forms of Exocentrinæ.

Y

1. *Tapeina dispar*, Serville.

Tapeina dispar, Serv. Encycl. Méthod. x. p. 546.
—— bicolor, id. (♀).
—— dispar, Thomson, Archives Entomolog. i. p. 42, pl. 7. fig. 4 a, b.

T. castaneo-rufa, capite thoraceque supra nigris nitidissimis, antennis nigris; armatura frontali maris elongata transversa, plana, apice utrinque obtuso truncato, margine superiore medio dentato. Long. 3½-4 lin. ♂ ♀.

Generally distributed in the forests of the Amazons.

2. *Tapeina eroctifrons*, Thomson.

Tapeina erectifrons, Thoms. Archives Entomol. i. p. 43, pl. 7. f. 2 a.

T. nigra, nitida; armatura frontali maris elongata transversa, angustata, concava, apice utrinque rotundato, margine superiore subrecto, margine inferiore utrinque angulato-dilatato. Long. 4-4½ lin. ♂ ♀.

Generally distributed throughout the forests of the Amazons.

Group *Compsosomina*.

Genus COMPSOSOMA, Serville.

Serville, Ann. Soc. Ent. 1835, p. 55.

This well-known and handsome genus of Lamiaires, by its compact, thick, oval forms, reminds one of the Anisocerinæ and Hypsiomæ. The group has been placed in the neighbourhood of the Hypselominæ by Mr. Pascoe, and M. Thomson sees a resemblance between the genus *Ærenea* (belonging to the Compsosominæ) and *Gymnocerus*. *Compsosoma* and its associated genera, however, differ from the Anisocerinæ by the tarsal claws, which are scarcely divergent, and from the Hypselominæ by the shortness of the claw-joint. The hairy clothing of body and antennæ, and the form of the head, gradually rounded off or sloping from the occiput to the epistome, are also characters which distinguish the Compsosominæ from the Anisocerinæ and the Onciderites, to which *Hypsioma* and *Hypselomus* belong. Although the lower part of the head, or muzzle, of some species resembles, in its square form, that of the Anisocerine group, yet this is evidently an inconstant character in the Compsosominæ; for other species (e. g. *Compsosoma Muiszechii*) have almost precisely the same form of muzzle as the Desmiphoritæ, to which group I consider, notwithstanding the difference in the general form of the body, the Compsosominæ belong. This form of head is utterly foreign to the Anisocerinæ and the Onciderites.

1. *Compsosoma Mniszechii*, Thomson.

Compsosoma Mniszechii, Thoms. Archiv. Entom. i. p. 74, pl. 9. f. 4.

C. oblongo-ovatum, crassum, convexum, hirsutum, grosse punctatum, elytris nigro-tuberculatis; thorace fuliginosa, vitta lata cinereo-fulva; elytris humeris rotundatis, plaga humerali fuliginosa (fulvo tincta), deinde utrinque vitta lata obliqua cinereo-fulva, parte postica fuliginosa, medio fulvo-sericea, suturaque cinerea; pectore utrinque plaga cretacea; antennis filiformibus, hirsutis. Long. 7 lin.

I found a few examples only of this fine species, on the slender stem of a young tree, in the forest at Ega, Upper Amazoné. The lower part of the face is extremely short, scarcely extending below the eyes; the latter are large and convex.

2. *Compsosoma terrenum*, Pascoe.

Ærenea terrena, Pascoe, Trans. Ent. Soc. n. s. vol. i. p. 25.

C. parvum, ovatum, obscure fulvum, undique breviter setosum; capite parvo, infra oculos brevissimo, contracto; antennis grossis, filiformibus, corpore paulo brevioribus, fuscis; elytris humeris subfalcatis, maculis duabus nigris utrinque basalibus; abdomine plagis duabus basalibus nigris. Long. 3¼ lin.

S. Paulo, Upper Amazons.

Genus TESSARECPHORA, Thomson.

Thoms. Archiv. Entom. i. p. 77.

The chief differences which M. Thomson assigns as distinguishing this genus from *Compsosoma* are the swollen and densely hirsute third and fourth joints of the antennæ, and the elevated shoulders and centro-basal ridges of the elytra. To them may be added the convexity of the front part of the head, and the extension of the lower part considerably below the narrow, oblong and scarcely convex eyes. The *Compsosomæ*, so far as at present observed, are found in their perfect state only on woody stems or trunks of trees; *Tessarecphora arachnoïdes* I found only on the foliage of Mimosa trees.

Tessarecphora arachnoïdes, Thomson.

Tessarecphora arachnoïdes, Thoms. Arch. Entom. i. p. 77, pl. 9. f. 10 *a, b*.

T. ovata, nigra, nitida; capite coriaceo, opaco; antennis articulis 6°–7° albis et albo hirsutis, 8°–11° fere nudis; thorace et elytris lineolis reticulatis cinereis, his carinis centrobasalibus conico-elevatis et longe penicillatis, humeris falcatis et valde oblique elevatis. Long. 4 lin.

I found this exquisite little insect only in the forest of Obydos, in the month of March, on the foliage of Mimosa trees.

Genus ÆRENEA, Thomson.

Thomson, Archives Entom. i. p. 298.

This genus is closely allied to *Compsosoma*, but differs in several points, admitting of clear definition. The antennæ are destitute of the dense fringe which exists in *Compsosoma*, and are furnished with scattered hairs. The face is broad and plane, and the muzzle quadrate and prolonged below the eyes. The mesosternum has a conical horizontal projection in front; and the prosternum is longitudinally convex or keeled, and sometimes vertical on the posterior face. The general form of body and the structure of the legs and tarsi are very similar to the same features in *Compsosoma*.

1. *Ærenea albilarvata*, n. sp.

Æ. breviter ovata, fulvo-brunnea, fronte fascia lata cinerea albo marginata, thorace lateribus castaneo-fuscis, elytris prope apicem fascia lata curvata grisea; antennæ parce breviter setosæ articulo basali clavato, articulis tertio et quarto longitudine æqualibus; elytris pedibusque breviter setosis. Long. 4½ lin.

Head broad, upper part of the forehead with a curved impressed line on each side besides the central longitudinal line, face plane and broad; colour tawny; the face crossed by a broad belt of milky-white tomentum, margined with lines of denser white, and extending up the face of each antenniferous tubercle. Antennæ a little longer than the body, sparingly clothed with short bristles, tawny brown; basal joint forming an oblong pyriform club; third and fourth joints about equal in length. Thorax quadrate, convex and tubercular above, and marked with a few punctures, tawny; sides dark chestnut-brown. Elytra short, ovate, shoulders obtusely rounded but slightly falcate; surface punctured and beset with short bristles; tawny, basal margin edged with dark brown; a broad curved grey fascia on each at a short distance from the apex. Body beneath and legs reddish, clothed with greyish-tawny pile.

Forests of the Tapajos.

2. *Ærenea cognata*, Pascoe.

Ærenea cognata, Pascoe, Trans. Ent. Soc. n. s. vol. i. p. 25.

Æ. ovata, breviter griseo setosa, purpureo-brunnea, fronte plana, griseo tomentosa; antennis rufescentibus, fere nudis; thorace supra tuberoso et cum occipite fulvo; scutello fulvo; elytris brevibus, convexis, punctatis, humeris falcatis, purpureo-brunneis, marginibus lateralibus fulvis, fasciaque obliqua grisea; pedibus testaceo-rufis, cinereo variegatis; abdomine piceo-nigro, nitido. Long. 6 lin.

Ega; Upper Amazons.

Group *Desmiphorinæ*.

Genus DESMIPHORA, Serville.

Serville, Ann. Soc. Ent. Fr. iv. 62.

Desmiphora is distinguished from the neighbouring genera by
the numerous tufts of hair arising from the thorax and elytra,
and the long hairy clothing of its body and limbs. The body is
elongate-oblong or linear, with the apex of the elytra obtusely
rounded. The head is small and retracted, with sloping crown,
very short face and muzzle, and large eyes. The antennæ
are stout, about as long as the body, tapering to a point,
with short thick basal joint narrowed at the base, elongated
second and third joints, and progressively abbreviated remaining
joints. The thorax has an acute prominent tubercle on each
side in the middle. The legs are stout, thighs not clavate; tarsi
with short triangular joints, and fine divergent claws.

The *Desmiphora* are found clinging to slender decaying branches
of trees, and are numerous in species in Tropical America.
Some of them resemble, in their colours and tufted forms, decayed
fragments of wood covered with minute cryptogamic plants.

1. *Desmiphora fasciculata*, Oliv.

Lamia fasciculata, Olivier, Ins. p.67, t. 17, f. 131; Fab. Ent. Syst. i. 2. 281.
268; Syst. El. ii. 299.

D. oblongo-elongata, fusco-nigra, capite, thorace articulisque basalibus
antennarum fulvo hirsutis et penicillatis, articulo tertio apice infra
dilatato; elytris utrinque pone medium fulvo plagiatis, basin et
apicem versus nigro penicillatis, undique breviter setosis, et griseo
hirsutis; pedibus nigris, fulvo variegatis; tibiis extus dense setosis; corpore subtus nigro nitido; abdomine utrinque fulvo
plagiato. Long. 8–9 lin. ♂ ♀.

Ega; Upper Amazons.

2. *Desmiphora cirrosa*, Erichs.

Desmiphora cirrosa, Erichs. Consp. Ins. Col. Peruan. p. 147.

D. oblongo-elongata, brunnea, capite fusco, vertice fusco bipenicillato; antennis fulvo-brunneis, hirsutis; thorace supra plaga magna,
postica brunnea, parte antica et lateribus sordide albis albo penicillatis; elytris utrinque prope basin fusco penicillatis, postice et
abdomine albo strigatis et penicillatis. Long. 6 lin. ♂ ♀.

Generally distributed throughout the forests of the Amazons;
also found in South Brazil near Rio Janeiro.

8. *Desmiphora senicula*, n. sp.

D. cylindrica, brunnea, griseo hirsuta, vertice bipenicillato; antennis
obscuris; thorace disco cristis duabus elongatis parallelis fulvo-

brunneis; elytris antice simplicibus, postice sordide albo strigatis
et fasciculatis; abdomine cinereo-fulvo lanuginoso; pedibus rufescentibus, cinereo dense hirsatis. Long. 4 lin.

Head dark-brown, coarsely pubescent, vertex with two short
erect pencils of dark-brown hair. Antennæ blackish brown,
densely pubescent, and clothed besides with long, coarse, brown
hairs. Thorax brown; disk with two parallel lines of tawny-
brown hairs. Elytra moderately punctured, dingy brown, pubescent, and clothed with long hairs; base with one or two short tubercles on each side, but without tufts of long hairs; apical part
ashy and marked with whitish streaks and tufts of whitish hairs.
Body beneath and especially the abdomen densely clothed with
woolly tawny pile; legs reddish and clothed with woolly pubescence.

Forests of the Tapajos.

4. *Desmiphora elegantula*, White.

Desmiphora elegantula, White, Cat. Longic. Brit. Mus. ii. p. 401.

D. cylindrica, ferrugineo-castanea, longe hirsuta, nitida; thorace et
elytris grossissime punctatis, illo disco cristis duabus parallelis et
lateribus fulvis, his utrinque prope basin unipenicillata, apice albo
strigatis et penicillatis; corpore subtus tenuiter cinereo pubescente. Long. 2½–3 lin.

Forests of the Tapajos.

5. *Desmiphora multicristata*, n. sp.

D. elongato-oblonga, fulvo-testacea, undique longe hirsuta; antennis
gracilibus; thorace convexo, crebre punctato, tripenicillato; elytris grosse punctatis, utrinque prope basin cristis tribus densis
elongatis parallelis, prope apicem penicillis tribus, testaceo-fulvis;
corpore subtus subnudo; pectore abdomineque lateribus nigricantibus. Long. 4½ lin.

Head coarsely punctured, brown, clothed with pale-tawny pubescence, the forehead and vertex having numerous long and
erect pale hairs. Antennæ rather longer than the body, slender,
the joints being much longer and thinner than in the other
species; third joint rather strongly curved; fourth less curved; all
the joints pale testaceous tawny, shining, and clothed throughout with long pale hairs. Thorax convex, surface even, coarsely
punctured, tawny-pubescent, and clothed with erect hairs; disk
on each side and front margin each with a thin pencil of hairs.
Elytra oblong, coarsely punctured, especially towards the base,
tawny testaceous, shining; each elytron towards the base with
three rather long parallel crests of dense hairs all of equal height,
and towards the apex with three thin pencils of similar hairs arranged in a row across the elytron. Body beneath and legs pale,

tawny testaceous, shining, clothed with long pale hairs; sides of breast and basal segments of abdomen black*.

Forests of Obydos, Lower Amazons.

Group *Pogonocherinæ.*

Genus Prymnosis, nov. gen.

Body elongate, plane above, and clothed with short, fine, erect hairs. Head small, depressed on the crown between the antenniferous tubercles, prolonged some distance below the eyes, and contracted at the occiput behind the eyes. Antennæ filiform, nearly twice the length of the body, and clothed throughout with fine, stiff hairs, longest on the underside of the joints; the basal joint elongate, nearly as long as the third, the third a very little longer than the fourth, and the rest very slightly diminishing in length. Thorax oblong, and armed on each side with a stout, porrect and acute spine. Elytra plane above, shoulders armed with a short spine, tapering thence to the apex, which is truncated, with the external angles prolonged each into a spine. Legs moderately elongated, thighs slightly clavate, tarsal joints triangular, claws divergent. Mesosternum narrowed and elevated behind; sockets of anterior coxæ widely angular externally.

Prymnosis bicuspis, n. sp.

P. elongata, postice attenuata, supra plana, punctata, fusco-castanea, vertice, thorace et scutello linea dorsali flava; elytris apice trun-

---

* The following new species, sent from South Brazil by Mr. Squires, belong also to this genus:—

Desmiphora ornata. Elongato-oblonga, fulvo-ochracea, nigro lineata et variegata. Caput grosse punctatum, fulvo hirsutum, vertice nigro bipenicillato. Antennæ corpore breviores, robustæ, hirsutæ, dimidio basali rufo, apicali nigricante. Thorax fulvo-ochraceus, lateribus utrinque nigro trilineatis, disco postice brunneo; juxta marginem anticum penicillis tribus porrectis quarum una antica fusca, alteræ duæ posticæ fulvo-ochraceæ. Elytra grosse punctata, fulvo-ochracea, pone medium annulo communi nigro, fasciaque subapicali alba; singulis cristis setosis parum elevatis, una prope basin, altera longe ante apicem. Corpus subtus fulvo villosum, pectore nigro. Pedes fulvo-testacei. Long. 4½ in. *Hab.* in Rio Janeiro, a Dom. Squires lecta.

Desmiphora venosa. Elongata. Caput fuscum, fulvo hirtum, punctatum. Antennæ robustæ, pilosæ, fulvo-brunneæ, articulis supra nigris. Thorax niger, nitidus, crebre foveolatus, lateribus fulvo plagiatis, dorso fusco bipenicillato. Elytris juxta basin et latera crebre foveolata, fusca, medio plana vix punctata, cinereo-brunnea, apice fusco maculata; singulis penicillis grossis fuscis, decumbentibus, una prope basin, altera apicem versus, lateribus et parte postica lineis flexuosis elevatis cinereo tomentosis. Corpus subtus et pedes testacea, cinereo villosa. Long. 3½ lin. *Hab.* in Rio Janeiro, a Dom. Squires lecta.

catis, angulis externis spinosis divaricatis, supra sublineatim punctatis, cinereo confluenter maculatis; pedibus rufo-testaceis. Long. 4 lin.

Head very coarsely punctured, black, depressed between the antenniferous tubercles; occiput constricted and marked with a yellow central vitta, which is continuous over the thorax to the scutellum. Antennæ nearly twice the length of the body, clothed sparingly throughout with fine, stiff hairs; reddish. Thorax oblong, armed on each side with a stout spine, surface very coarsely punctured, dark castaneous, with a yellow central line. Scutellum yellow. Elytra narrowed in a straight line from base to apex; shoulders armed with a small spine; apex truncated, external angles produced each into a long slightly diverging spine; surface punctured partly in lines, dark castaneous, sprinkled with grey confluent spots, which leave an oblique belt about the middle spotless. Body beneath shining, thinly pubescent; thoracic segments coarsely punctured, black; abdomen reddish, faintly punctured. Legs testaceous red, clothed with fine hairs.

Santarem and Ega.

Genus ESTHLOGENA, Thomson.

Thomson, Systema Cerambyc. p. 107.

In this genus the body is elongated, subdepressed, and parallelogrammical or slightly narrowed behind, with the apex of the elytra more or less truncated, and sometimes dentate. The hairy clothing usual in this group is, in some of the species of *Esthlogena*, short and bristly. The head is small, with no depression between the antenniferous tubercles; the face is short and convex, and very slightly prolonged below the eyes. The thorax is armed on each side with a conical tubercle. The legs are moderately long and stout, with subclavate femora and short, triangular tarsal joints. The claws are only semidivergent—a character which, together with the more elongated body and linear elytra, distinguishes this genus from *Estola*, to which it is very closely allied. The antennæ, as in *Estola*, are scarcely longer than the body, hairy, with the fourth joint a little longer than the third, and the remaining joints becoming gradually and slightly shorter. The basal joint is short and thick, and narrowed at the base.

1. *Esthlogena pulverea*, n. sp.

E. elongata, angustata, postice paulo attenuata, breviter setosa, cinereo-ochracea; antennis corpore haud longioribus, articulis apice fuscis; thorace supra sparsim punctato, vittis sex obscuris brunneis, lateribus utrinque tuberculo lato apice spinoso; elytris sparsim punctatis, apice breviter sinuato-truncatis, angulis externis productis, cinereo-fuscis, sutura maculaque apicali cinereo-ochraceis. Long. 4½ lin.

Head small, clothed with laid ashy-ochreous tomentum. Antennæ about as long as the body, clothed sparingly with stiff hairs, ashy ochreous, tips of the joints, from the fourth, blackish. Thorax as wide as head and elytra, slightly uneven on the surface; sides each with a broad dentiform prominence in the middle; disk marked with a few scattered punctures; ashy ochreous, disk with two, and sides each with two, obscure brownish vittæ. Elytra elongate, narrowed before the apex, which is briefly sinuate-truncate, with the outer angles dentiform; surface clothed with fine bristles, marked with a few widely scattered punctures, ashy brown, with the suture and a spot near the apex ashy ochreous. Body beneath and legs ashy ochreous.

Santarem.

2. *Esthlogena mucronata*, n. sp.

E. elongata, postice paulo attenuata, dense breviter hirsuta, castaneofusca; pedibus piceo-rufis; elytris confertim cinereo confluenter maculatis, punctato-striatis, apice sinuato-truncatis, angulis externis spinosis; tibiis dilatato compressis. Long. 5–6 lin.

Head dark pitchy, thinly pubescent, and marked with large scattered punctures. Antennæ about as long as the body, setose, pitchy red, becoming darker towards the apex, with the joints pale ashy. Thorax closely covered with large deep punctures, leaving a smooth longitudinal dorsal space scored by an impressed line; lateral prominence small, dentiform; surface thinly clothed with ashy pubescence, forming faint lines. Elytra elongate, subdepressed, tapering behind; apex sinuate-truncate, with the outer angles spiniform; surface very thickly clothed with erect hairs springing from punctures arranged in lines; the colour is dark blackish chestnut, shining and varied throughout with cinereous confluent specks. Body beneath castaneous, thinly clothed with ashy pile. Legs reddish; intermediate and posterior tibiæ broad and compressed from base to apex.

Ega, on dead branches.

8. *Esthlogena sulcata*, n. sp.

E. elongata, subdepressa, postice paulo attenuata, undique breviter setosa, nigro-castanea; capite, thorace et scutello cinereo-fulva vittatis; elytris cinereo confluenter maculatis, punctato-striatis, striis postice fortiter impressis, apice truncatis, angulis externis spinosis; pedibus rufo-castaneis. Long. 7 lin.

Head very coarsely punctured; forehead strongly convex; vertex with two ashy-tawny stripes. Antennæ about as long as the body, castaneous, clothed with grey pubescence; apices of the joints, from the fourth, black. Thorax broadened in the middle, and having on each side a distinct acute tubercle; surface covered with scattered punctures, leaving a smooth space along the middle; clothed with ashy pubescence arranged in vittæ, the central vitta (continuous to the scutellum) tawny. Elytra elongated, closely covered with short bristles (like the rest of the body); apex squarely truncated, with the external angles produced into an acute tooth; surface punctate-striate, the striæ more deeply impressed posteriorly and the interstices costate, dark blackish castaneous, covered with small confluent spots of grey tomentum. Body beneath blackish, thinly clothed with grey pile. Legs reddish; tibiæ simple.

Santarem, dead branches of trees.

4. *Esthlogena linearis*, n. sp.

E. linearis, dense longe hirsuta, fusco-castanea; thorace foveolato, linea dorsali lævi, lateribus breviter spinosis; elytris lineatim punctatis, cinereo irroratis, apice sinuato-truncatis, angulis vix productis; antennis pedibusque testaceo-rufis, illis articulis apice obscurioribus. Long. 3½ lin.

Head coarsely punctured, clothed with tawny-brown pubescence. Antennæ as long as the body, reddish testaceous; apices of the joints, from the fourth, darker. Thorax sparsely covered with large and deep punctures, leaving a smooth dorsal line; sides each with a small acute spine; colour blackish chestnut. Elytra linear, narrowed close to the apex, the latter sinuate-truncate, with the angles acute, but not distinctly produced; surface clothed with long, stiff hairs very dense towards the apex, punctured in rows, dull castaneous, sprinkled with greyish confluent spots. Body beneath dull reddish brown; legs testaceous red; tibiæ simple.

Santarem. There are two undescribed species, closely allied to this, found in the province of Rio Janeiro*.

* *Esthlogena obtusa.* Elongata, parallelogrammica, setosa, nigro-castanea, griseo irrorata. Caput grosse punctatum. Antennæ dense setosæ, rufo-testaceæ, articulis a tertio basi pallidioribus. Thorax grosse punctatus, niger, linea dorsali lævi et interstitiis griseo pubescentibus; tuberculo laterali apice unguiculato. Elytra linearia, apice obtuse truncata, lineatim punctata, interstitia lævia, nigro-castanea, griseo confluenter maculata. Corpus subtus nigrum, nitidum, sparse tomentosum. Pedes rufi. Long. 5 lin. *Hab.* in Rio Janeiro (D. Squires).

Esthlogena prolixa. Elongatissima, linearis, sparsim setosa, nigra, fusco-griseo tomentosa. Caput sparsim punctatum. Antennæ nigræ. Thorax supra subplanus, punctis magnis paucis notatum, tuberculis lateralibus brevibus, latis, obtusis. Elytra apice recte truncata, angulis externis spinosis; supra sparsim punctata, postice costata. Corpus subtus et pedes nigra. Long. 6 lin. *Hab.* in Rio Janeiro.

Genus ESTOLA, Fairmaire.

Fairmaire, Ann. Soc. Ent. Fr. 1859, p. 524.

This genus is very closely allied to *Esthlogena*, the shape of the head, form and proportion of antennal joints, clothing of body, and general appearance offering no points of difference worthy of mention. The body, however, is less elongated, the elytra being shorter and subtrigonal. The tarsal claws in all the species that I have examined are fully divergent—a character which will at once distinguish the present genus from the preceding.

1. *Estola basivittata*, n. sp.

E. elongato-oblonga, postice attenuata, setosa, brunneo-fulva; thorace basi utrinque maculis duabus, elytris singulis basi macula rotundata, nigro-velutinis; pedibus rufescentibus. Long. 3–4½ lin.

Head thickly punctured and clothed with tawny-brown pubescence. Antennae as long as the body, fringed beneath with stiff hairs; reddish, joints from the third tipped with dusky, eighth joint white, tipped with dusky. Thorax slightly narrowed at the base; sides each with a conical, acute tubercle; surface punctured, setose, tawny brown, base on each side with two velvety blackish spots margined with ashy. Elytra narrowed from base to apex, the latter rounded; surface setose, punctate-striate, punctures elongated; uniform tawny brown, base of each with a rounded, velvety, purplish-black spot. Body beneath blackish, clothed with fine grey pile, and setose; legs reddish, setose.

Forests of the Tapajos.

2. *Estola variegata*, n. sp.

E. elongato-oblonga, postice attenuata, setosa, nigro griseo et fulvo laete variegata. Long. 4 lin.

Head coarsely punctured, setose, black, varied with fulvous spots and spotted with grey behind the eyes. Antennae as long as the body, sparingly setose, dark reddish, bases of the joints testaceous; eighth joint whitish, tipped with brown. Thorax slightly narrowed behind; lateral tubercles large, with apex acute and slightly recurved; surface setose, coarsely punctured, but leaving small smooth interspaces, black, varied with clear, large, fulvous spots. Elytra tapering from base to apex, the latter rounded; surface setose, punctured in lines; third interstice costate behind, minutely varied with black, clear fulvous, and grey, the last colour prevailing along the suture, and a light fulvo-testaceous spot lying across the suture towards the apex. Body beneath black, thinly clothed with grey pile; legs reddish, varied with greyish and fulvous.

Ega.

3. *Estola lineolata*, n. sp.

E. elongato-oblonga, postice attenuata, setosa, fusca, griseo-fulvo variegata; antennis pallide annulatis; thorace basi utrinque lineola obliqua griseo-fulva; elytris punctato-striatis, apice angustatis, obtusis. Long. 3–1½ lin.

Head coarsely and irregularly punctured, blackish, thinly clothed with coarse tawny-brown pubescence; in brightly coloured individuals obscurely variegated. Antennæ fringed beneath with stiff hairs, dull reddish or testaceous; apices of all the joints dusky, sometimes variegated with grey; eighth joint greyish testaceous, tipped with dusky. Thorax very coarsely punctured; lateral tubercles acute; surface setose, dingy brown, sometimes varied with dull reddish, clothed with scanty tawny-brown pubescence, the base at each side having a short, thin, pale line running obliquely towards the disk, and in fresh examples surrounded by blackish. Elytra tapering to the apex, which latter is narrow and obtuse, almost truncated; surface setose, coarsely punctate-striate, with the third interstice costate before the apex, dingy brown or blackish or partially dull reddish, more or less varied with tawny spots, in fine examples minutely varied with blackish and tawny. Body beneath dingy black; legs reddish, thighs and tibiæ varied with black.

Banks of the Tapajos, common. Also found at Cayenne, and existing in some French collections under the names of *Hebestola annulicornis* and *Lepricurii*. I have a specimen also which was taken by Mr. Squires at Rio Janeiro, where several other species are found allied to this, three of which have truncated elytra *.

* *Estola truncatella.* Elongato-oblonga, parce setosa, nigro-fusca, griseo obscuro tomentosa. Caput angustum, punctatum. Antennæ ciliatæ, fuscæ, articulo octavo albo annulato. Thorax parvus, spina laterali acuta; supra crebre punctatus, griseo-fuscus, unicolor. Elytra elongato-oblonga, prope apicem angustata, apice oblique truncata, angulis haud productis; supra tenuiter setosa, griseo-fusca, grosse punctato-striata, interstitio tertio postice costato. Pedes nigri, griseo pilosi. Long. 4½ lin. *Hab.* in Rio Janeiro.

Estola acricula. Elongata, postice attenuata, parce setosa, cinereo-fulva, nigro punctata. Caput punctatum. Antennæ corpore breviores, testaceæ, articulis apice brunneo variegatis, articulo octavo testaceo. Thorax grosse punctatus, spinis lateralibus longiusculis acutis. Elytra punctato-striata, cinereo-fulva, nigro punctata, apice oblique truncata, angulis externis breviter spinosis. Long. 4½ lin. *Hab.* in Rio Janeiro.

Estola varicornis (Dj. Cat.). Elongato-oblonga, postice vix attenuata, setosa, nigrina, griseo obscuro variegata. Caput punctatum, inter antennas valide concavum. Antennæ ciliatæ, nigræ, articulis basi pallido testaceis; articulo octavo testaceo, apice nigro. Thorax crebre grosse punctatus, tuberculis lateralibus brevibus acutis. Elytra punctato-striata (interstitio tertio postice acute costato), nigrina, griseo obscure variegata, apice oblique truncata, angulis internis rotundatis, externis distinctis. Corpus subtus nigrum. Pedes nigri, griseo variegati, tarsis rufescentibus. Long. 3–4 lin. *Hab.* in Rio Janeiro.

4. *Estola porcula*, n. sp.

E. oblongo-ovata, hispida, obscure brunnea, griseo confluenter maculata; antennis testaceo annulatis; elytris antice confuse, postice
sublineatim punctatis, apice obtusis. Long. 2–3½ lin.

Head thickly punctured, blackish, clothed with coarse greyish
pubescence and rigid hairs. Antennæ dusky; fourth, sixth, eighth,
and tenth joints ringed with pale testaceous. Thorax convex,
thickly punctured, setose, and clothed with dull-greyish tomentum;
lateral tubercles small, acute. Elytra oblong, scarcely narrowed
behind, apex obtusely rounded; surface closely setose, minutely
varied with dingy grey and dusky brown, punctured, the punctures confused except towards the apex, where they are partly
arranged in rows. Body beneath and legs dusky, clothed with
coarse greyish pile.

Lower Amazons, at Santarem and Villa Nova, on dead twigs.

Genus EPECTASIS, nov. gen.

Body greatly elongated, narrow, cylindrical, clothed throughout with erect, fine hairs. Head small, face convex, vertex depressed between the bases of the antennæ; eyes reniform, rather
distant on the crown. Antennæ as long as the body, filiform,
clothed both above and beneath with long and fine hairs; basal
joint short and thick, but narrowed at the base; third joint considerably shorter than the fourth, the following joints gradually
and successively shorter. Thorax elongate, cylindrical; lateral
tubercles nearly obsolete. Elytra elongated, cylindrical, apex
obliquely truncated. Legs short; thighs scarcely clavate, basal
joint of the posterior tarsi cylindrical, as long as the second and
third taken together; claws semidivergent. Sterna narrow,
plane.

The chief points of distinction between this genus and the two
preceding are the elongated cylindrical form of body, the hairy
antennæ, both above and beneath, and the unarmed thorax. The
insect known in collections under the MS. name of *Euteles lurida*,
might be included in it, as it offers most of the characters, with
the exception of the fourth antennal joint not exceeding in length
the third.

Epectasis attenuata, n. sp.

E. elongata, cylindrica, hirsuta, obscure castaneo-fusca; antennis
piceo-rufis, articulo terminali dimidioque penultimi pallide testaceis; thorace crebre punctato, medio late cinereo-fusco vittato;
elytris grosse confuse punctatis, prope apicem cinereo plagiatis,
apice oblique valde truncatis. Long. 4 lin.

Head small; face convex, hairy, and clothed with dingy-greyish pubescence, punctured. Antennæ dull pitchy red, basal half of tenth joint and the whole of the eleventh greyish testaceous, bases of several preceding joints also greyish. Thorax cylindrical, elongate, sides slightly conical in the middle; surface closely punctured, dull blackish castaneous, middle with an obscure dull-ashy vitta. Elytra elongate, cylindrical, hirsute, covered with large punctures, dull chestnut-brown; apex with a greyish patch and obliquely truncated. Body beneath and femora blackish, tibiæ and tarsi reddish, hirsute, and clothed with dingy-ashy pubescence.

Ega, on a dead twig.

Group *Apomecynina.*

Genus AGENNOPSIS, Thomson.

Thomson, Archives Entom. i. p. 302.

This genus is tolerably well known to students of the Longicornes under the name of *Talæpora* of Dejean's catalogue. The body is of an elongate-elliptical shape with obtusely rounded elytra, the apex of which is adorned in most of the species by a black spot, margined anteriorly with pale ashy, the pale streak existing in those species which are destitute of the black spot. The antennæ, as is usual in the Apomecyninæ, are much shorter than the body, and filiform, with the terminal joints much abbreviated, and the third of great relative length. The thorax is unarmed, the head small, with rounded vertex and forehead and retracted face. The claws of the tarsi are short and scarcely divergent.

1. *Agennopsis pygæa*, n. sp.

A. elongato-elliptica, brunnea; thorace grosse vage punctato, lateribus chnereo-brunneis; elytris vage punctatis, nigro cinereoque obscure irroratis, apice macula rotundata communi nigro-velutina antice cano marginata. Long. 3½-5½ lin. ♂ ♀.

Head retracted beneath, sprinkled with large punctures, and clothed with tawny-brown pubescence. Antennæ about half the length of the body in the female, two-thirds the length in the male, filiform; third joint as long as the three following taken together, dark brown. Thorax narrowed anteriorly and rounded on the sides, marked with large evenly distributed punctures, which leave a narrow impunctate dorsal space; colour brown, sides each with a broad ashy-brown vitta. Elytra considerably broader than the thorax at the base, scarcely widened beyond the middle, then narrowed to the apex; surface smooth and marked with scattered punctures not arranged in lines; colour light brown, obscurely speckled with dusky and pale ashy, apex ornamented with a rounded velvety black spot, narrowly margined anteriorly with ashy white. Body beneath and legs dingy brown; abdomen with a black spot on each side of the second to the fourth segments.

Santarem, Lower Amazons. Also found at Rio Janeiro.

2. *Agennopsis cordida*, n. sp.

A. elongato-elliptica, brunnea; thorace grosse vage punctato, lateribus cinereo-brunneis; elytris lineatim punctatis, interstitiis subcostatis, ante apicem utrinque lineola transversa cinerea. Long. 4 lin. ♀.

Head marked throughout with very large punctures. Antennæ about half the length of the body, dingy brown. Thorax slightly narrowed anteriorly and scarcely rounded in the middle; surface thickly marked with large punctures, leaving no smooth dorsal line; brown, sides each with a broad ashy-brown vitta. Elytra considerably broader than the thorax, scarcely widened beyond the middle, then narrowed to the apex; surface punctured in rows from base to apex, with some of the interstices elevated; colour brown, obscurely spotted with black and ashy; apex concolorous, and near the apex on each elytron a short oblique ashy line. Body beneath and legs ashy brown; abdomen with a black spot on each side of the second to the fourth segments.

Santarem.

3. *Agennopsis cylindrica*, n. sp.

A. elongata, cylindrica, obscure fusca; capite thoraceque lateribus fulvis; elytris lineatim punctatis, prope apicem linea transversa flavescente. Long. 4 lin.

Head irregularly punctured, clothed with yellowish-tawny pubescence. Antennæ black, three basal joints (except the apex of the third) tawny. Thorax cylindrical, covered with coarse, large punctures; dark brown, sides tawny. Elytra linear, singly rounded at the apex; surface punctured in rows, with a mixture of large punctures; dull brown, with a straight transverse yellowish line near the apex, the space between the line and the apex studded with large black punctures. Body beneath coarsely punctured, dark grey; legs blackish.

Santarem.

Subtribe SAPERDITÆ.

Group *Callianæ*.

Genus EUMATHES, Pascoe.

Pascoe, Trans. Ent. Soc. n. s. iv. p. 251; Journal of Entom. i. p. 354.

The characters of this genus are well defined by Mr. Pascoe, in the Journal of Entomology as above referred to. Its position is not so well ascertained. The form of the tarsal claws (widely divergent, with a broad, acute tooth at the base) points to an affinity with the Callianæ; and as I think this feature outweighs in importance the dissimilarity of general form and facies, I have placed the genus in the Callianæ group, rather than amongst the Pogonocherinæ, with which it agrees in some points. The body is elongate-oblong, narrowed behind, depressed above, and beset with short bristles. The head is short, the crown, in profile, not forming an angle with the forehead, and the face very little prolonged and narrowed below the eyes, which latter are large and convex. The thorax has a distinct acute tubercle on each side in the middle. The elytra are singly rounded at the apex. The antennæ are half as long again as the body, filiform, and setose, the basal joint short and forming an oblong club, the third joint a little longer than the fourth, and the rest very gradually decreasing in length. The legs are moderately long, the thighs slightly clavate; the tarsi moderately short, with the claws, as before mentioned, armed each at the base with a large, acute tooth.

Eumathes Amazonicus, n. sp.

Eu. elongato-oblongus, supra planus, setosus, viridi-cinereus, obscure fusco maculatus; elytris dense et confuse punctatis; maris tarsorum posticorum articulo primo valde elongato. Long. 5—5¼ lin.

Head coarsely punctured, clothed with grey pubescence. Antennæ dingy grey. Thorax irregularly punctured on its surface, light-greenish ashy, obscurely varied with dusky; lateral tubercles small, acute. Elytra slightly narrowed behind, plane above and free from costæ, rather thickly but irregularly covered with small punctures, especially on the basal half, and clothed with short bristles; pale-greenish ashy, obscurely varied with dusky spots of various sizes. Body beneath and legs clothed with ashy pubescence. First joint of the hind tarsi in the male as long as the remaining joints taken together.

Ega. I am indebted to Mr. Alexander Fry for pointing out the differences between this species and its near relative *Eumathes undatus* (Pascoe) of Southern Brazil. The great length of the basal joint of the posterior tarsi in the male, and the closer punctation of the elytra, are the chief distinguishing characters.

Genus CHALCOLYNE, nov. gen.

Closely allied to *Gryllica*, Thoms. (Classif. des Cérambyc. p. 120), but differs in the thorax being armed on each side with an acute spiniform tubercle. Body oblong, clothed with short, stiff hairs; elytra subtrigonal, rounded at the tip. Head with long, slightly retracted face; mouth projecting; palpi elongate, pointed; eyes ample both above and beneath, and nearly approximating on the crown ; antenniferous tubercles distinct, divergent. Antennæ scarcely so long as the body, stout, the joints simple and gradually tapering to the apex, basal joint thickened gradually from base to apex. Thorax subcylindrical, finely wrinkled transversely, sides each armed with an acute spiniform tubercle. Legs moderately elongated, thighs clavate, middle tibiæ simple on their outer edge ; tarsi about half the length of the tibiæ, broad, not compressed ; basal joint in all the feet short, triangular ; claw-joint slender, projecting beyond the third joint to an extent equal to the length of the third joint ; claws widely divergent and strongly curved, furnished at the base on the inner side with a broad square enlargement. Prosternum narrow, simple ; mesosternum rather broad, bituberculated, and vertically inclined anteriorly.

Chalcolyne metallica, Pascoe.

Onocephala(?) *metallica*, Pascoe, Trans. Ent. Soc. n. s. iv. (1858).

C. oblonga, nitens, nigro-ænea, breviter fusco setosa ; elytris viridi-scencia, striato-punctatis ; thorace subcylindrico, elytris multo angustiore, antice leviter angustato, supra transverse rugoso, lateribus utrinque tuberculo acuto armatis. Long. 5 lin. ♂ ?

Found only at Ega, Upper Amazons, on the stem of a slender tree in the forest. The insect is very similar in form to *Gryllica flavo-pustulata*, Thoms., but differs not only in the spinose thorax and metallic colours, but in the basal joints of the antennæ not being compressed.

Genus EUMIMESIS, nov. gen.

Body oblong, above plane, clothed with short, stiff hairs. Elytra oblong, broadly rounded at the tip. Head with long, slightly retracted face ; mouth somewhat projecting ; palpi elongate, pointed ; eyes ample, but distant on the vertex ; antenniferous tubercles distinct, divergent. Antennæ short ; basal joint oblong-quadrate, compressed ; second joint rather abruptly dilated from the middle ; third joint curved and dilated at the apex ; fourth with the upper edge enlarged into a short foliaceous expansion ; remaining joints very short, simple. Thorax subcylindrical, thickly punctured, sides each armed with an acute spiniform tubercle. Legs moderately elongated, thighs clavate, middle tibiæ simple on their outer edge, tarsi short and uncompressed, claw-joint slender and short ; claws divergent and strongly curved, furnished at the base on their inner side with a broad tooth. Prosternum narrow, simple ; mesosternum much broader, bituberculate, steeply inclined anteriorly.

This genus, as will be seen by the above description, harmo-

z

nizes with *Chalcolyne* in the majority of its characters. Mr.
Alexander Fry, who has paid especial attention to the Saperditæ and their allies, having examined my specimens, is inclined
to think that the insect on which I have founded the genus
Chalcolyne is a male individual of a species of *Eumimesis*. The
great difference in the antennæ, in the absence of positive evidence of identity, forbids, however, the fusion of the two forms
into one genus.

Eumimesis heilipoides, n. sp.

E. speciebus *Heilipi* generis Curculionidarum simillima, oblonga,
   fusco-ferruginea, dense breviter setosa; thorace utrinque vitta lata,
   elytris vitta lata basali et macula magna subapicali sordide albis.
   Long. 6 lin. ♀?

Head dark red, hispid and thinly clothed with whitish recumbent pile. Antennæ dark red, fifth joint and apices of third to
eleventh joints black, bases grey. Thorax subcylindrical, a little
narrowed in front; sides each armed with a small acute tubercle,
thickly punctured, rusty brown, each side marked with a broad
tawny-white vitta. Elytra oblong, broadly rounded at the apex,
surface in the middle depressed and very closely punctured, the
sides over the tomentose whitish parts sparsely punctured, over
the naked parts closely so; from the base to beyond the middle of
each runs a tawny-white stripe, thickest in the middle, and
within the apex is a similarly coloured rounded spot composed
of dense tomentum, the edges of the elytra and a large triangular spot between the vitta and the apical patch being dark
and shining. Body beneath and legs rusty red, sprinkled with
grey tomentum. The whole body clothed with short erect hairs.
St. Paulo, Upper Amazons.

This insect, from its colour and form, bears a most deceptive
resemblance to many species of *Heilipus*, a genus of Curculionidæ.

Genus Hastapis, Buquet.

Buquet, in Thoms. Archives Entom. i. p. 338.

In this genus the body is oblong, slightly convex, and beset
with short bristles. The head is moderately short, depressed
between the antenniferous tubercles; the eyes are rather small.
The antennæ are about the length of the body, and clothed
above and beneath with short, stiff hairs. The lateral tubercles
of the thorax are acute and spiniform. The elytra are rounded
at the apex, and depressed in the middle. The mesosternum
is prominent in front. The thighs are clavate, the tarsi short
and broad, with a broad tooth at the base of each claw.

Hastatis galerucoides, n. sp.

H. oblonga, breviter setosa, fulvo-brunnea, vertice thoracisque lateribus cinereis; elytris marginibus lateralibus lineaque longitudinali discoidali pallide testaceis; antennis nigris, articulis 3°–6ª apice dilatatis, angulis productis. Long. 5 lin. ♀.

Head brown, partly clothed with yellowish-ashy pubescence, which forms two divergent stripes on the vertex. Antennæ a little shorter than the body (♀), black, clothed with short bristles; third to sixth joints gradually dilated at the apex, with the apical angles produced. Thorax clothed with dense tawny-brown pubescence, sides each with a broad ashy vitta, lateral tubercles large and acute. Elytra oblong, obtuse at the apex; surface clothed with short bristles, finely punctate-striate, depressed along the suture, tawny brown, with the lateral and apical margins and a line from base to apex terminating at the sutural angle pale testaceous. Body beneath dusky castaneous; mesosternum with two tubercles in front. Legs pale-reddish testaceous, with a large black spot on the outer side of the middle and posterior femora.

Santarem.

Genus CALLIA, Serville.

Serville, Ann. Soc. Ent. Fr. iv. (1835) p. 60.

The species composing this well-known genus are all of small size and of the most diversified colours—some being metallic, and others resembling species of various other families of Coleoptera. The antennæ are filiform, with the joints from the third (inclusive) gradually and proportionally decreasing in length. The tarsal claws have a broad and acute tooth at their base.

1. *Callia fulvocincta*, n. sp.

C. oblonga, setosa, chalybea, nitida; elytris violaceis, cano tomentosis, basi fascia lata fulvo-aurantiaca. Long. 3 lin. ♀.

Head glossy steel-blue, thinly clothed with hoary tomentum; front with a deeply impressed longitudinal line. Antennæ dark metallic blue, setose. Thorax short (much shorter than in the allied *C. axillaris*), glossy steel-blue, smooth, convex. Scutellum steel-blue. Elytra oblong, setose, punctured, violaceous, obscured with fine hoary tomentum; base with a broad tawny-orange fascia, broadest a little before the lateral margin. Body beneath and legs steel-blue.

Santarem, flying over masses of dried branches.

2. *Callia chrysomelina*, Pascoe.

Callia chrysomelina, Pascoe, Trans. Ent. Soc. n. s. v. p. 34.

C. oblonga, postice paulo dilatata, setosa, nigra; capite, thorace, an-

tennarum articulo basali (apice excepto) femoribusque anticis et intermediis (geniculis exceptis) laete ferrugineis; elytris crebre punctatis azureis; corpore subtus chalybeo. Long. 3½ lin. ♀.
Ega, dry twigs.

3. *Callia criocerina*, n. sp.

C. oblongo-elongata, setosa, nigra nitida; capite, thorace, antennarum articulo basali (apice excepto) femoribusque anticis et intermediis (geniculis exceptis) flavis; elytris elongatis, crebre punctatis, violaceis. Long. 3 lin. ♂.

Head and mouth, except the tips of the palpi, yellow. Antennæ a little longer than the body, bluish black; basal joint of the antennæ, except the extreme base and the apex, yellow. Thorax glossy yellow; lateral tubercles large, obtuse at their apex. Scutellum yellow. Elytra elongate-oblong, parallel-sided, setose, thickly punctured, violet. Body beneath and legs black; anterior and middle femora, except their apices, yellow.

S. Paulo, Upper Amazons.

4. *Callia halticoïdes*, n. sp.

C. elongata, setosa, nigra; thorace (margine postico excepto) ferrugineo; antennis articulis tribus terminalibus albo-testaceis; femoribus anticis (geniculis exceptis) abdominisque lateribus flavotestaceis. Long. 2¼ lin. ♂.

Head small, deeply impressed down the middle, shining black, except the margin of the epistome, which is pale testaceous. Antennæ scarcely longer than the body, black, extreme bases of the joints and the whole of the three terminal joints whitish testaceous. Thorax very short, transverse; lateral tubercles very acute, red, hind margin black. Scutellum black. Elytra elongate, linear, setose, thickly punctured, partly in rows, black. Body beneath and legs black; anterior femora in the middle, and intermediate femora on one side, pale testaceous; sides of abdomen testaceous.

Ega, Upper Amazons.

5. *Callia lycoïdes*, n. sp.

C. elongata, setosa; capite thoraceque flavis, lateribus nigro vittatis; elytris fulvo-flavis, plaga quadrata communi basali lineola prope basin marginali et plaga magna apicali nigris; antennis nigris, articulis tribus terminalibus flavis; femoribus (apice exceptis) et tibiis basi pallide testaceis. Long. 2¼ lin. ♂.

Head small, tawny yellow, with fine golden pubescence, sides behind the eyes each with a dusky stripe. Antennæ not longer than the body, black; three terminal joints pale yellow. Thorax somewhat elongated, rusty yellow, shining, and clothed with fine golden pubescence; lateral tubercles broad, but acute;

disk obtusely tubercular, with a dusky stripe on each side. Scutellum tawny yellow. Elytra elongate, regularly punctate-striate (punctures large), tawny yellow; a quadrate patch over the scutellar region, a basal marginal streak, and a broad fascia at the apex black. Body beneath black; legs black; thighs, except their apices and the bases of the tibiæ, yellow testaceous. S. Paulo, Upper Amazons.

6. *Callia cleroïdes*, n. sp.

C. sublinearis, postice paulo ampliata, setosa, nigra; capitis lineolis, thoracis vitta laterali antennarumque annulo magno mediano fulvo-flavis. Long. 3½ lin. ♂.

Head small, deeply impressed in the middle, black, a line down the centre of the crown and one on each side, and the lower part of the face, tawny. Antennæ as long as the body, black, apex obscurely rufescent; apical half of the fourth and nearly the whole of the fifth joint clear tawny yellow. Thorax elongated, lateral tubercle small, conical; surface coarsely punctured and tubercular, black; sides each with a broad golden-fulvous vitta. Scutellum black. Elytra elongated, a little dilated at the apex; surface setoso, closely punctured, partly in rows, black; lateral edges near the base obscurely rufescent, and an indistinct streak from the shoulder down each side dull tawny. Body beneath clothed with silvery-grey tomentum. Legs black; femora at the base rufescent.
Ega.

The preceding series of species, mimicking respectively various types of Coleoptera, do not exhaust the variety of dress which the *Calliæ* put on. I have a small species in my collection, from Rio Janeiro, which presents the style of coloration of certain species of Lampyridæ*.

Genus PRETILIA, nov. gen.

Closely allied to *Callia*, but differs in the thorax being unarmed on the sides. This part of the body is short, convex, and rounded, the sides being tumid instead of having the distinct conical tubercle. The eyes are short and convex, their reniform

---

* *Callia lampyroïdes*. Elongato-oblonga, depressa, setosa, fusco-nigra, testaceo marginata. Caput breve, nigrum, læve, ore testaceo marginato. Antennæ corpore multo breviores, parce ciliatæ, nigræ, articulis basi pallide testaceis. Thorax subquadratus, supra tuberosus interstitiis grossissime punctatis; niger, lateribus litura rufo-testacea, breviter tuberculatis. Elytra elongato-oblonga, apice rotundata, setosa, supra punctato, fusco-nigra, lateribus late testaceo marginatis. Corpus subtus nigrum. Pedes nigri; coxæ et femora pallide-testacea, his nigro maculatis. Long. 3¼ lin. ♀. Hab. in Rio Janeiro (Squires).

undivided shape distinguishing the species from the Tetraopinæ, to which they are allied by the form of the thorax. The body is linear and setose. The antennæ are filiform and longer than the body in both sexes; the third joint is much elongated, and half as long again as the fourth, the remainder being filiform and slender to the apex. The pro- and meso-sterna are both very narrow. The legs are moderately elongated, and the tarsal claws have a large tooth at the base.

I am indebted to Mr. Alexander Fry for pointing out the chief distinguishing characters of this genus.

Pretilia telephoroïdes, n. sp.

P. linearis, setosa; capite flavo-ferrugineo, occipite nigro nitido; thorace rufo, pube aurea tecto; elytris nigris vel fulvo-brunneis, apice nigris, pube fulvescente vestitis, punctato-striatis, apice obtusis; pedibus testaceis, tarsis fuscis; pectore abdomineque nigris, griseo tomentosis, hoc lateribus fulvo-testaceis; antennis nigris, basi ferrugineis, articulis 5° et 6° flavis. Long. 3–4½ lin. ♂ ♀.

Head depressed between the antenniferous tubercles; face, cheeks, and palpi reddish yellow; crown and occiput shining black. Antennæ black; basal joint, except the apex, reddish yellow; fifth and sixth joints (sometimes also the apex of the fourth) pale yellow. Thorax short, rounded, convex; sides tumid, reddish yellow, clothed with golden pubescence. Scutellum black. Elytra linear, obtuse at the apex, punctate-striate, setose, purplish black or tawny brown, gradually becoming black towards the apex, clothed with a changing tawny pubescence. Breast and abdomen dusky, clothed with griseous pile; abdomen brownish testaceous on the sides. Legs reddish yellow; tarsi dusky.

Pará and Lower Amazons.

Group Astatheinæ.

Genus PHÆA, Newman.

Newman, Entomologist, p. 13.

Syn. Lamprocleptes, Thomson, Arch. Entom. i. 377.

The chief character which distinguishes this genus from *Tetraopes* (the chief American representative of the group Astatheinæ) is the form of the tooth of the claws. The tooth in *Tetraopes* is long and acute, running parallel to the claw itself, but much shorter; in *Phæa* it is very broad and short, adhering only to the base of the claw, as in the Callianæ. The eyes, as in the rest of the Astatheinæ, are completely divided. The body is more or less elongate and linear.

Phæa coccinea, n. sp.

P. linearia, brevis, coccinea, pube pallida sericea vestita ; femoribus apice, tibiis, tarsis et antennis (basi exceptis) nigris. Long. 3½ lin.

Head as broad as the middle part of the thorax, bright red ; eyes moderately prominent, black. Antennæ about as long as the body, filiform, hirsute, black, basal half of the first joint red. Thorax constricted near the front and hind margins, surface strongly elevated and smooth in the middle, clothed with long erect hairs ; bright red. Elytra linear, bright red, clothed with fine pale silky pubescence (visible only in certain lights), and with erect hairs, strongly punctate-striate, the punctures fainter and more confused towards the apex. Body beneath and thighs yellowish red ; apex of thighs, tibiæ, and tarsi black.

Santarem.

Group *Amphionychinæ.*

[The Amphionychinæ are distinguished from the Phytœciinæ (both having bifid claws) by the sides of the elytra having a longitudinal carina extending from the shoulders.]

Genus LYCIDOLA, Thomson.

Thomson, Systema Cerambyc. p. 125.

The proposer of this genus has omitted to state the essential characters which distinguish it from *Spathoptera* and *Hemilophus.* These are furnished by the peculiar width of the sterna, especially of the prosternum, which is as broad as, or a little broader than, the mesosternum. The prosternum in *Spathoptera* is much narrower than the metasternum, and in *Hemilophus* it is reduced to a mere thread, almost concealed by the large coxæ. *Lycidola* is moreover distinguished from *Spathoptera* by the dilatation of the elytra commencing almost from the shoulders, by the breadth and shortness of the head, and the transverse thorax. The genus is founded on *Saperda palliata,* Klug (Entom. Bras. Specimen alterum, pl. 42. f. 11).

Lycidola simulatrix, n. sp.

L. nigra, breviter setosa ; capite et thorace vitta laterali communi fulva ; elytris apud medium fascia alba diaphana, apice singulatim rotundatis ; femoribus basi flavo-testaceis. Long. 5–6 lin.

Head short and broad, the face extending a short distance below the eyes, and not dilated ; black, face reddish ; occiput on each side with an oblique fulvous stripe. Antennæ black ; third joint one-fourth longer than the fourth, cylindrical ; the fourth a little dilated ; both densely hairy ; the remaining joints shorter than the third and fourth taken together, and sparingly setose.

Thorax considerably broader than long, coarsely punctured, except on the disk, which is smooth, deep black; sides each with a fulvous stripe. Scutellum black. Elytra dilated almost from the shoulders, and quite abruptly, at the apex singly rounded; disk punctured, and having on each three longitudinal carinæ, the two outer of which are united before the apex, and the inner one abbreviated; expanded sides shagreened and traversed by a flexuous carina; colour wholly deep black with a violet tinge, except a white diaphanous belt across the middle, interrupted at the suture. Body beneath and legs black; basal part of thighs testaceous yellow.

Var. Base of each elytron with a small fulvous spot in continuation of the thoracic stripe; lateral edge of the elytron also fulvous near the base (approaching *L. palliata*, Klug). Tapajos.

The typical form not uncommon at Ega, on leaves. The var. found only on the banks of the Tapajos.

Genus SPATHOPTERA, Serville.

Serville, Ann. Soc. Ent. Fr. 1835, p. 50.

Body elongated, dilated behind; facies of the genus *Lycus.* Head somewhat prolonged on the vertex; face elongated and dilated below the eyes. Thorax short, a little narrower than the head. Elytra dilated from beyond the middle, apex briefly emarginated. Legs short; claws bifid. Prosternum narrower than the mesosternum. Antennæ about the length of the body, or a little shorter; basal joint greatly elongated, gradually and slightly thickened from base to apex, ciliated; third and fourth joints greatly elongated, hairy and ciliated beneath, sometimes very thickly ciliated; following joints short and sparingly setose. The lateral carina of the elytra is thick and prominent, and extends from the shoulder to the apex.

1. *Spathoptera capillacea*, n. sp.

S. elongata, postice dilatata, nigra; capite thoraceque vitta laterali fulva, fronte rufescente; elytris macula angulari humerali fasciaque lata pone medium fulvis; antennis articulis tertio et quarto haud dilatatis, infra pilis longis densissimis nigris vestitis. Long. 6 lin.

Head coarsely punctured; vertex elongated, shining black, with a fulvous vitta on each side behind the eye; face dull reddish, clothed with scant tawny pile. Antennæ a little shorter than the body, black; fifth and sixth joints reddish; third and fourth joints greatly elongated, neither of them thickened, but furnished on their under surface with a dense fringe of long thick hairs. Thorax coarsely punctured, shining black, with a fulvous stripe on each side. Elytra with their dilatation commencing a little before the middle, at first very gradual, at about two-thirds their length abruptly dilated; apex of each rounded, and offering a small triangular emargination; surface finely setose; disk closely punctured, and with two very fine raised lines, united before the apex (where alone they are distinct); dilated margins (outside the strong lateral carina) shagreened and traversed, to the apex, by a nearly straight carina; colour black, with a basal spot on each shoulder bent towards the suture, and a broad fascia beyond the middle, fulvous; the edges both of the humeral mark and the fascia irregular. Body beneath and legs black; coxæ and thighs beneath pale testaceous. Ega.

2. *Spathoptera mimica*, n. sp.

S. elongata, postice dilatata, fulva, capite vitta laterali nigra, thorace lateribus maculaque triangulari dorsali nigris; elytris nigris, macula humerali angulata fasciaque lata pone medium fulvis; antennis breviter hirsutis, nigris, articulis quinto et sexto testaceis, quarto incrassato. Long. 6–7½ lin.

Head fulvo-testaceous, punctured; sides behind the eyes with a black stripe; vertex elongated, convex. Antennæ wholly clothed with shortish hairs, black; fifth and sixth joints pale testaceous; fourth joint dilated. Thorax with a few large punctures and an elevated dorsal line fulvous; deflexed sides, and a triangular dorsal spot with the apex scarcely reaching the anterior margin, black. Elytra elongated, the dilatation commencing very gradually before the middle, and at two-thirds the length more abrupt; at the apex singly rounded and faintly emarginated; surface finely setose, closely punctured, and with two indistinct raised lines united before the apex; dilated margins (outside the lateral carina) shagreened, and traversed by a raised line from base to apex; colour black, with a basal spot on each shoulder (bending towards the suture), and a broad fascia beyond the middle, fulvous; the edges both of the humeral spot and the fascia jagged. Body beneath fulvo-testaceous; sides of breast black, and abdomen with two rows of brown spots. Legs black; coxæ and inside of femora testaceous.

Ega; found only on leaves of trees in the deep forest.

Genus HEMILOPHUS, Serville.

Serville, Ann. Soc. Ent. Fr. 1835, p. 50.

The chief difference existing between this genus and *Amphionycha* resides in the antennæ, which in *Hemilophus* are formed almost the same as in *Spathoptera* and *Lycidola*: that is to say, the third and fourth joints are disproportionately elongated, occupying together, in some species, nearly one-half the total

2 A

length of these organs; they are, besides, thickened and densely hirsute, sometimes ciliated. Both genera have a strongly elevated lateral carina, bifid claws, and very narrow prosternum. The elytra in *Hemilophus* are sometimes a little dilated before the apex, but in *Amphionycha* never show any trace of dilatation.

Hemilophus fasciatus, n. sp.

H. elongatus, sublinearis, ante apicem paulo ampliatus; capite fulvo, vitta laterali maculaque triangulari occipitali nigris; thorace nigro, vitta utrinque laterali fulva; elytris nigris, macula cuneiformi humerali fasciaque recta mediana fulvis; antennis nigris, articulo quinto basi rufo, articulis quarto et quinto paulo incrassatis, dense breviter setosa. Long. 5 lin. ♂.

Head tawny yellow, with a triangular spot on the occiput and a stripe behind each eye black; forehead convex and marked with a deeply impressed line. Antennæ a little longer than the body (♂); black, with the base of the fifth joint reddish; basal joint clothed with longish hairs; third and fourth joints together longer than the whole of the following joints, thickened, linear, densely clothed with short hairs; remaining joints clothed sparingly with very short hairs. Thorax coarsely punctured, leaving smooth spaces on the disk, and having a deep transverse impression behind; black, with a fulvous vitta on each side of the upper surface. Elytra nearly linear, being very slightly dilated a little before the apex, the latter, on each elytron, presenting a very shallow emargination with a short spine at its outer side; surface densely punctured, partly in lines, and with several interstices slightly raised, black; a straight humeral spot, pointed behind, the basal part of the lateral edges, and a straight fascia about the middle fulvous. Body beneath tawny yellow; sides of thorax and breast and middle of the abdominal segments black. Legs black, base of thighs yellow.
Ega.

Genus Tyrinthia, nov. gen.

This genus includes a number of species which agree with *Hemilophus* in the great length and dense clothing of the third and fourth (or, at least, the third) antennal joints, but differ in the absence of a distinct continuous lateral carina from the elytra. The vertically deflexed sides of the elytra form with the disk, in section, a distinct angle; but the carina is not apparent, except for a short distance from the shoulders.

I have adopted the name that the group bears in the rich collection of Mr. Alexander Fry.

1. Tyrinthia capillata, n. sp.

T. elongata, setosa, nigra; capite fulvo-flavo, supra nigro, vitta laterali fulvo-flava, inter antennas profunde indentato ; thorace utrinque vitta laterali fulva ; elytris elongatis, juxta apicem angustatis, apice singulatim rotundatis et brevissime emarginatis, supra punctato-striatis, macula humerali cuneiformi vittaque lata mediana fulvis ; antennis nigris, ultra medium annulo lato flavo, articulo tertio longissimo, ciliato. Long. 5 lin. ♂.

Head coarsely punctured, forehead convex, mouth projecting, vertex deeply depressed between the bases of the antennæ, tawny yellow, the crown and occiput and a stripe behind each eye black. Antennæ as long as the body, black, with the apical half of the fourth, the whole of the fifth, and the base of the sixth joints yellow; basal joint elongate, gradually thickened and fringed with very long, fine hairs ; third joint nearly as long as the whole of the succeeding joints taken together, not thickened, but furnished beneath with a continuous fringe of very long hairs ; fourth joint not much longer than the fifth, and destitute of fringe. Thorax coarsely punctured, leaving smooth spaces on the disk, behind deeply impressed ; fulvous, with a broad central and lateral vittæ black. Elytra linear, except very near the apex, where they are narrowed, the apex itself being narrow and apparently entire, but showing, on close examination, a very shallow emargination and minute tooth ; disk regularly and rather deeply punctate-striate, black, a wedge-shaped basal spot and a broad median vitta fulvous. Body beneath black ; sterna and centre of the breast bright testaceous yellow. Legs black, base of thighs testaceous yellow.
S. Paulo, Upper Amazons.

2. *Tyrinthia scissifrons*, n. sp.

T. elongata, linearis, setosa, fuliginoso-nigra ; fronte, vitta laterali thoracis lineolaque laterali elytrorum fulvo-testaceis ; femoribus basi articuloque quinto antennarum rufo-testaceis ; antennis articulis tertio et quarto biciliatis ; fronte (maris) tumida, conica, apice fissa. Long. 4 lin. ♂.

Head testaceous yellow, vertex and occiput black ; upper part of the forehead (♂) conically produced and cleft at the apex, and antenniferous tubercles armed on the inner side with a conical prominence. Antennæ as long as the body, black, with the fifth joint reddish ; basal joint on its upper side abruptly thickened, hairy ; third and fourth joints together longer than the whole of the remaining joints, slightly thickened and furnished beneath with two fringes of long and fine hairs. Thorax coarsely punctured, and with a smooth dorsal line, black, a narrow stripe on each side pale testaceous. Elytra linear, singly rounded and entire at the apex ; surface very closely punctured and furnished with three obtuse costæ, dull black ; lateral edge

2 A 2

and carina near the base dull testaceous. Body beneath dull black; base of thighs reddish testaceous.

Banks of the Tapajos and Ega, Upper Amazons. Mr. Fry informs me that the peculiar bilobed prominence of the head is found in the males of some Rio Janeiro species. *Hemilophus frontalis* of Guérin-Méneville (Ins. rec. par Osculati, n. 265) belongs to this genus.

Genus Isomerida, nov. gen.

This new genus is distinguished from *Hemilophus* by the antennal joints decreasing in length in regular proportion from the third joint to the apex, and by the fringe of hairs on their under surface existing in uniform density on all the joints. The only difference between *Isomerida* and *Amphionycha* lies in the shortness of the antennæ, which are not longer than the body, even in the males, and decrease greatly in thickness from the third joint to the apex.

I have adopted the name under which the genus stands in the collection of Mr. Alexander Fry.

1. *Isomerida albicollis*, Castelnau.

Hemilophus albicollis, Laporte de Castelnau, Animaux articulés, ii. p. 488.

I. elongata, linearis, postice paulo angustata, tenuiter setosa; capite thoraceque rufo-testaceis, cano interdum dense tomentosis; elytris punctatis, interstitiis duobus elevatis, apice truncatis, rufo-testaceis plus minusve fuliginosis, vel totis nigris; abdomine nigro, segmentis tertio et quarto dense cano tomentosis; antennis nigris, articulis basi testaceis. Long. 4½–5½ lin. ♂ ♀.

This common species is very variable in its coloration, and there is only a small proportion of examples which exhibit the white hue of the thorax, and these only in the dried state; in life, the thorax is always red. The truncature of the elytra is straight and offers a short tooth at the exterior angles.

It is found on the leaves of trees, and is a common and generally distributed insect throughout the Amazonian forests.

2. *Isomerida ruficornis*, n. sp.

I. robustior, elongata, linearis, postice haud angustata, tenuiter setosa, nigra; capite, thorace, antennis (apice exceptis) et pedibus (femoribus supra exceptis) rufis; elytris apice truncatis, angulis externis dentatis; abdomine segmentis tertio et quarto dense cano tomentosis. Long. 6 lin. ♂.

Head entirely red, depressed between the eyes. Antennæ stout, as long as the body, finely fringed beneath; third joint one-third longer than the fourth, the following becoming very gradually shorter; red, with the three apical joints tinged in the middle with dusky. Thorax thinly clothed with pale silky tomentum, visible only in certain lights; red, prosternum and circuit of the acetabula blackish. Scutellum black. Elytra slightly dilated a little before the apex, the latter straightly truncated, with the outer angles slightly produced; surface punctured and marked with one faintly raised line besides the lateral carina; deep black, shining. Breast and abdomen black; third and fourth ventral segments densely clothed with pale silky tomentum. Legs red, upper side of femora black.

Fonte Boa, Upper Amazons.

Genus Amphionycha (Dej. Cat.), Thomson.

Thomson, Archiv. Entom. i. p. 311.

The numerous species which compose this genus agree in the possession of long filiform antennæ, with the joints more or less densely fringed with fine hairs, but never partially thickened, clothed, or tufted; the third joint is more or less disproportionately elongated. The body is variable in shape, but is generally elongated and linear, in some species greatly elongated, in others much shorter and oblong. All have well-developed lateral carinæ on the elytra; the apices of the latter are variable, being in some species broadly truncated and toothed, in others briefly truncated, and in some species rounded and entire.

1. *Amphionycha Diana*, Thomson.

Amphionycha Diana, Thoms. Classif. des Cérambyc. p. 65.

A. elongata, postice paulo attenuata, castaneo-rufa, occipite fascia brevi et macula laterali, thorace vitta laterali et macula postica, elytris fascia communi subbasali maculisque utrinque tribus posterioribus cretaceo-albis; prothorace pectoreque lateribus cretaceo plagiatis; pedibus fulvo-testaceis; antennis ciliatis, articulo tertio modice elongato; elytris breviter truncatis. Long. 6½ lin. ♀.

This very handsome species occurred only in the forests of the Tapajos. It is found also in the interior of French Guyana.

2. *Amphionycha seminigra*, n. sp.

A. elongata, parallelogrammica, ferrugineo-testacea; antennis, elytrorum dimidio postico, pedibus posticis, tarsis omnibus et abdomine nigris; thorace tuberoso; elytris late truncatis, angulis productis. Long. 5 lin.

Head coarsely punctured, testaceo-ferruginous. Antennæ longer than the body (♂?), finely fringed to the apex; joints all slender, third double the length of the fourth; black. Thorax with three large, smooth tubercles on the disk, and one on each side, red; margins marked with a few very large punctures. Elytra parallelogrammical, broadly truncated, with both angles of the truncature produced into sharp teeth; surface closely punctured, the punctures and also the lateral carina ceasing abruptly at three-fourths the length of the elytron; black, basal third rusty testaceous. Body beneath reddish testaceous; hind part of the breast dusky; abdomen black. Legs reddish testaceous; tarsi and the hind legs black.

S. Paulo, Upper Amazons.

3. *Amphionycha nigripennis*, n. sp.

A. elongata, parallelogrammica, ferruginea; elytris, tarsis apicibusque tibiarum nigris; thorace tuberoso; elytris late truncatis, angulis productis; antennis parce setosis. Long. 6½ lin. ♀.

Head broad, muzzle dilated and having prominent angles, testaceous red, shining, and marked with a few shallow punctures. Antennæ shorter than the body, slender and tapering to the extremity, very sparingly setose; third joint nearly twice the length of the fourth; testaceous red. Thorax with a large elevated rounded tubercle on the disk, and a large obtuse one on each side, constricted near the anterior and posterior margins; bright testaceous red, marked with a very few shallow punctures. Scutellum bright testaceous red. Elytra parallelogrammical, broadly truncated at the apex, with both angles of the truncature produced and acute; surface closely punctured, the punctures as well as the lateral carinæ ceasing abruptly before the apex; deep black, suture near the scutellum red. Body beneath and legs testaceous red; apical part of the abdomen, tarsi, and apices of the tibiæ black.

Ega.

4. *Amphionycha miniacea*, n. sp.

A. elongata, parallelogrammica, glabra, rufa; elytris nigris, medio castaneo-rufis, utrinque maculis quatuor suturæque rufis; antennis nigris, articulis tertio quartoque rufis; thorace postice paulo dilatato; elytris truncatis, angulis externis valde productis, internis dentatis. Long. 4½–5½ lin. ♂♀.

Head bright red, marked with large, distinct, scattered punctures. Antennæ a little longer than the body in the ♂, shorter in the ♀, sparingly setose, black; third, fourth, and sometimes also the fifth, joints reddish testaceous; third joint one-fourth longer than the fourth. Thorax marked with very large scattered punctures, red; sides behind the middle dilated. Elytra parallelogrammical, depressed above; apex broadly truncated, with the external angle of the truncature much elongated, and the sutural angle produced into a point; surface closely punctured, dark red on the disk, shining black on the sides, glabrous, each elytron with four elongate patches, and a streak down the middle part of the suture, of dense bright-red tomentum (pallid in dried examples); one spot is near the scutellum, another underneath the shoulder, a third a little before, and a fourth a little after the middle. Body beneath and legs red.

I took numerous specimens of this beautiful species on the leaves of a tree in the forest at Obydos, Lower Amazons. In life the red colour is of a clear vermilion hue.

5. *Amphionycha megalopoides*, n. sp.

A. brevis, oblonga, flavo-testacea; capite lato, fronte nigra, bipenicillata, occipite nigro, bifasciato; thorace postice transverse sulcato; elytris singulis maculis duabus nigris; antennis rufo-testaceis, articulo quarto flavo, articulis 5ᵃ-11ᵃ fuscis. Long. 4½ lin. ♂.

Head broad, pale testaceous, clothed with fine pubescence and long pale hairs; face much narrowed below the eyes; forehead, near each eye, furnished with a cluster of long, black hairs; occiput with a black vitta behind each eye. Antennæ a little longer than the body, fringed with long scant hairs; third joint nearly twice the length of the fourth; basal joints reddish testaceous; fourth joint yellow, the rest dark brown. Thorax widened behind, and marked with a transverse sulcus near the hind margin; pale testaceous, opake. Scutellum dusky. Elytra short and broad, oblong, slightly narrowed behind, apex rounded; lateral carina thick and flexuous; surface punctured towards the base; disk with two slightly raised lines, pale yellowish testaceous, clothed with fine silky tomentum; a triangular spot over the shoulder and a round one near the suture, towards the apex, black. Body beneath and legs testaceous; breast with a black belt.

Santarem. Resembles in form and colouring certain species of *Megalopus* (family Phytophaga).

6. *Amphionycha Sapphira*, n. sp.

A. elongata, angustata, postice sensim attenuata; nigra, fronte, vitta coronali, vittis lateralibus thoracis lineisque quatuor elytrorum cæruleis; his disco bicostatis, apice sinuato-truncatis, basi macula magna aurantiaca; antennis corpore longioribus, robustis, filiformibus, nigris, dense ciliatis. Long. 5½ lin. ♂.

Head a little broader than the thorax, deeply impressed on the crown, clothed with pale-blue tomentum; occiput coarsely punctured, black, naked except on the pale-blue tomentose vittæ. Antennæ one-fourth longer than the body, stout, filiform, black, densely fringed to the apex; third joint elongated. Thorax elongated, cylindrical, uneven, broadest in the middle, black, coarsely punctured; sides each with a broad vitta of clear light blue, the black parts naked. Scutellum black. Elytra narrow, elongated, tapering from base to apex, the latter briefly sinuate-truncate, with both angles produced and acute; disk coarsely punctured, except near the apex and along the two slightly raised lines; lateral carina straight; colour blue black, shining, with the suture, a line along the disk, and lateral margins pale blue; a rounded orange-coloured spot at the base of each elytron. Body beneath and legs clothed with fine blue-grey pubescence.

I met with one example only of this remarkable species, at Ega, on the Upper Amazons, on a leaf.

7. *Amphionycha cephalotes*, Pascoe.

Amphionycha cephalotes, Pascoe, Trans. Ent. Soc. n. s. vol. iv. p. 250.

A. modice elongata, linearis, rufescens; elytris lateribus fuscis, apice suturaque antice cinereo sericeis; capite lato, convexo; thorace postice strangulato; elytris lineatibus, supra planis, punctato-striatis, apice rotundatis; antennis corpore paulo longioribus, longe ciliatis, nigris, articulis tribus vel quatuor terminalibus flavis; tibiis extus fuscis. Long. 4 lin.

Found at Ega, Upper Amazons, and on the banks of the Tapajos, on foliage.

8. *Amphionycha megacephala*, n. sp.

A. linearis; capite valde convexo, nigro, polito; antennis nigris; thorace nigro, lateribus vitta castanea testaceo plagiata, marginis postici lineola et scutello albis; elytris supra planis, crebre punctatis (apice excepto), basi fulvo-brunnois, medio nigris, apice cinereo-sericeis. Long. 4½ lin.

Head large and convex both above and in front; mandibles large, strongly curved; glossy black, lower part of the face greyish tomentose; cheeks with a small white spot under each eye. Antennæ a little longer than the body, filiform, finely fringed, black; third joint about twice the length of the fourth. Thorax cylindrical, uneven, marked above with a few large punctures, black, sides each with a broad tawny-chestnut stripe, in which is a paler spot; anterior margin with two small spots; hind margin in the middle with a short white line. Scutellum white. Elytra linear, apex rounded; surface plane, closely punctured (except near the apex); colour tawny brown near the base, black across the middle, ashy tomentose towards the apex, the colours not sharply defined. Body beneath black; breast and base of abdomen glossy tawny red; sides of the mesosternum with a white spot. Legs black.

Ega. There is another species of large-headed *Amphionycha*

found on the Isthmus of Panamá, in which this part assumes still larger proportions*.

9. *Amphionycha concinna*, White.

Phœbe concinna, White, Proc. Zool. Soc. 1856, p. 408.

A. linearis, capite lato, albo, fronte bicorni; thorace postice angustato, convexo, albo, supra plaga magna postica colore lavandulæ, disco maculis tribus lævibus nigris; elytris linearibus, apice truncatis (angulis externis productis acutis), colore lavandulæ, apice fuscia lata cretaceo-alba fusco bimaculata; corpore subtus cretaceo-albo, sternis fuscis; abdomine, pedibus et antennis rufo-testaceis, his longe ciliatis, corpore duplo longioribus. Long. 5½–6 lin. ♂.

Ega, Upper Amazons.

10. *Amphionycha bicornis*, Oliv.

Saperda bicornis, Olivier, Entom. t. iv. 68, 27, pl. 4, f. 46.

A. linearis, cretaceo-alba; thorace maculis octo, elytris singulis apice maculis tribus, griseis; abdomine, antennis pedibusque rufo-testaceis; capite lato, fronte bicorni; antennis corpore duplo longioribus, longe ciliatis. Long. 5 lin. ♂.

Forests of the Tapajos.

11. *Amphionycha testacea*, n. sp.

A. cylindrica, setosa, testacea, pube fulvescente sericea induta, thoracis marginibus pallidioribus; elytris disco abdomineque basi fuscescentibus; antennis tenuiter longe ciliatis, nigris, articulo basali (apice excepto) rufo, articulis quarto et quinto (apicibus exceptis) flavis; thorace antice angustato. Long. 3½ lin.

Head small, pale testaceous, crown darker; face convex, prominent; upper and lower lobes of the eyes connected by a very slender thread. Antennæ a little longer than the body, furnished with a scanty fringe of long straight hairs; basal joint red, except at the apex, which, together with the second and third joints, is deep black; third joint about one-third longer than the fourth, the latter (except the apex) and the basal half of the fifth pale yellow, the rest black. Thorax narrowed in front, and broadest in the middle; surface (except the disk) marked with large punctures, reddish testaceous, anterior and lateral borders

---

* *Amphionycha capito.* Robusta, linearis, nigra, nitida; thorace flavo, macula discoidali nigra. Caput magnum, convexum, corpore latius, nigrum, grosse punctatum. Antennæ corpore longiores, nigræ, ciliatæ, articulis sex terminalibus flavis. Thorax capite angustior, postice paulo constrictus, tomento flavo dense vestitus, macula quadrata discoidali nigra. Elytra brevia, linearia, supra plana, punctata, apices versus lævia, nigra, nitida, apice macula cinerea tomentosa. Corpus subtus et pedes nigra, femoribus anticis et intermediis flavis. Long. 4¼ lin. ♂. Hab. in Panamá.

paler. Elytra very briefly truncated at the apex; lateral carinæ vanishing considerably before the apex; surface punctured in lines, clothed with pale silky pubescence, brown testaceous, paler anteriorly. Body beneath and legs testaceous yellow, basal three-fourths of the abdomen blackish brown.

Ega.

12. *Amphionycha roseicollis*, n. sp.

A. brevior, linearis, nigra, subsericea; fronte, antennis (apice exceptis), corpore subtus, et pedibus flavo-testaceis; abdomine apice nigro; thorace (basi excepta) læte roseo, elytrorum lateribus et apicibus rufo-testaceis; unguiculis simplicibus. Long. 3½ lin. ♂.

Head as broad as the elytra; face yellow and densely pubescent; vertex black, naked, coarsely punctured. Antennæ one-third longer than the body, furnished with a scanty fringe of straight hairs, yellowish testaceous, sixth to eleventh joints dusky; basal joint subclavate; third joint about one-fourth longer than the fourth. Thorax with an obtuse prominence in the middle on each side, and narrowed behind; surface pale, and clothed with silky pink pubescence, hind border black, coarsely punctured, lateral prominences pale. Elytra linear, sinuate-truncate at the apex, with both angles prominent; lateral carina obsolete before reaching the apex, and accompanied in that part by a lower carina, parallel to it but not reaching the middle of the elytra; surface punctured, black, with changeable greyish pubescence; lateral margins reddish; apex testaceous. Body beneath and legs yellowish testaceous; apex of the abdomen blackish.

Ega; one example. The claws in this species are simple: it ought therefore to rank amongst the group Saperdinæ, if the evidence were complete that this is not a sexual character in this instance. As only one example exists of the species, its true position cannot at present be decided.

Group *Phytœciinæ*.

Genus ERANA, nov. gen.

Body cylindrical. Head rounded, scarcely depressed between the eyes, the latter with the upper and lower lobes connected. Antennæ moderately elongated, filiform, setose, and beneath ciliated; third joint much longer than the fourth, the remaining joints gradually diminishing in length. Thorax short, cylindrical. Elytra cylindrical, obtuse at the apex, and rounded at the sides, the discal portion not being separated from the lateral by an elevated line. Legs somewhat short, tarsal claws bifid.

I have adopted this genus from the collection of Mr. Alexander Fry, to whom is due the credit of having first detected its distinctness from *Amphionycha* and *Hemilophus*. It embraces numerous tropical American species, including *Saperda triangularis* (Germar), *S. læta* (Newman), and others.

Erana cincticornis, n. sp.

E. cylindrica, nigra, pilosa, fronte et vitta laterali thoracis albo sericeis; antennis nigris, articulis tertio et quarto basi dense setosis, quinto et quarto apice albis; elytris apice conjunctim rotundatis, angulis suturalibus spinosis. Long. 3½–4 lin.

Head convex above; front and cheeks clothed with silky whitish pubescence; vertex naked, black, coarsely punctured. Antennæ a little longer than the body, ciliated (except near the apex), black, the fifth and apical half of the fourth joints white; the third and basal half of the fourth joints appear to be thicker than the rest of the antennæ, owing to their dense clothing of short hairs. Thorax transversely depressed near the apex; surface clothed with very long and fine but erect hairs, centre part black; sides each with a pale vitta, emitting a short branch in the middle. Elytra cylindrical, apex rounded, with the sutural angles each armed with a short spine; surface clothed with erect hairs which are longest near the base, thickly punctured, dull black. Body beneath black, thinly clothed with grey pile; sides of breast and abdomen pale. Legs yellowish; tarsi and apices of tibiæ black.

Ega and S. Paulo, Upper Amazons.

Group *Saperdinæ*.

Genus AMILLARUS, Thomson.

Thomson, Archives Entom. i. p. 312.

In this very distinct genus of Saperdinæ the body is elongate linear, and, in the males, narrowed behind. The eyes are hemispheric, with a narrow angular emargination for the reception of the antennæ, the latter being greatly elongated (twice the length of the body), with very long and gradually thickened basal joint. The legs are moderately elongated, together with the tarsi. But the most characteristic peculiarity of structure is the form of the claw-joint of the tarsi and of the claws. In both sexes the claw-joint is longer than the second and third joints taken together; but in the males it is also rather abruptly dilated and thickened beneath from a short distance beyond the base. The claws are nearly straight, compressed, and scarcely divergent.

Amillarus mutabilis, n. sp.

A. elongatus, linearis, breviter parce setosus, fulvo-rufus, pectore medio et abdomine plumbeo-nigris, antennarum articulo basali nigro; tarsis tibiisque posticis fuscis; maris elytrorum parte postica, fœminæ elytris totis plumbeo-nigris. Long. 4–6 lin. ♂ ♀.

Head tawny red, vertex marked with a few shallow punctures and a smooth central line. Antennæ with a scanty fringe of short stiff hairs, reddish; basal joint and tips of other joints black. Thorax narrower than the head, broadest in the middle, constricted behind, marked with a few shallow punctures, tawny red. Scutellum reddish. Elytra tapering in the male, nearly linear in the female; apex obliquely truncated, with the outer angles dentate; surface smoothly punctured, partly in lines; colour in the males tawny red, with the posterior part more or less black, with pale silky pile; in the females wholly black, with pale silky pile. Body beneath tawny red; centre of breast and abdomen almost entirely black, with silvery silky pile. Legs tawny red, tarsi and posterior tibiæ dusky.

Abundant at Santarem on the leaves of shrubs, borders of woods. The species seems to differ from the New Granada form which has been described by M. Thomson under the name of *A. apicalis*.

ADDENDA.

The following species were accidentally omitted in treating of the genera to which they belong:—

Subtribe ACANTHOCINITÆ.

Group *Acanthoderinæ*.

Genus OREODERA.

18. *Oreodera (Anoreina) biannulata*, n. sp.

O. oblongo-ovata, convexa, fulvo-brunnea; thorace lateribus tumidis obtusis; elytris apice singulatim rotundatis, supra tenuiter punctatis fulvo-brunneo et fuliginoso variegatis, lateribus apud medium macula fulvo-brunnea annulo cinereo-albo circumcincta; antennis setosis, fuscis, articulis basi testaceis. Long. 2¾ lin.

Head clothed with tawny-brown tomentum, impressed between the antennæ; eyes distant on the vertex. Antennæ longer than the body, clothed beneath with numerous stiff hairs; basal joint reddish, the rest dark brown, with bases of joints pale testaceous. Thorax short, transverse, nearly as broad in the middle as the base of the elytra; sides tumid, obtuse; surface clothed with a mixture of tawny-brown and dark-brown tomentum. Elytra oblong, narrowed towards the apex, at the latter singly rounded; surface convex, free from tubercles, finely punctured and clothed with a mixture of dark-brown and tawny-brown pile, in which are two short, zigzag, blue-grey fasciæ, one before, the other after the middle; each side in the middle with a tawny spot encircled by a whitish ring. Body beneath ashy tawny. Legs blackish, short, stout; femora clavate.

S. Paulo, Upper Amazons.

Group *Leiopodinæ*.

Genus LEPTURGES.

25. *Lepturges ovalis*, n. sp.

L. ovalis, paulo convexus, griseo-brunneus; elytris crebre punctatis, griseo lineatis, apice oblique sinuato truncatis, angulis productis; femoribus valde clavatis. Long. 2½–3¼ lin.

Head clothed with dingy tawny-brown pubescence. Antennæ dull red, sparingly clothed with short bristles. Thorax widening from the front towards the base; lateral spines short, acute, and situated very near the hind angles; disk with a transverse depression near the hind margin; colour brown, clothed with dingy-grey pubescence. Elytra oval, slightly convex; apex obliquely sinuate-truncate, both angles produced, sutural one very slightly; surface rather closely and coarsely punctured, light brown; each elytron with about eight narrow lines of grey pubescence, interrupted in some places. Body beneath and legs brownish red; femora abruptly clavate.

Santarem. The species will come next to *L. griseostriatus*; but it is shorter and more oval and convex than any other *Lepturges* hitherto described.

26. *Lepturges scutellatus*, n. sp.

L. subovatus, paulo convexus; thorace fusco-nigro, griseo vario, spinis lateralibus validis, rectis, paulo ante basin sitis; elytris ovatis, apice breviter oblique truncatis, fulvo-brunneis, nigro maculatis, macula magna basali communi fusco-nigra fulvo-cinereo marginata. Long. 2¼ lin.

Head clothed with tawny-brown pubescence; epistome and labrum testaceous; palpi black. Antennæ reddish, tips of joints dusky. Thorax widened and rounded from the fore to the hind part; lateral spines stout and uncurved, placed a short distance from the hind angles, and the thorax greatly narrowed behind them; surface blackish, varied with silky grey marks. Elytra ovate, slightly convex, narrowed near the apex, and briefly and obliquely truncated; surface punctured, tawny brown, varied with blackish spots of various sizes, and having over the scutellar region a large black triangular spot broadly margined with tawny ashy. Body beneath dusky tawny, clothed with fine ashy pile. Legs dull red; thighs dusky and distinctly clavate.

S. Paulo, Upper Amazons. The place of this species will be in the second division of the genus, near *L. dorcadioides*.

Genus SPORETUS.

3. *Sporetus decipiens*, n. sp.

S. elongatus, *Colobotheæ* speciei simillimus, setosus, olivaceo-niger; capite cinereo trivittato; thorace vitta lata laterali cinerea, medio nigro lineolata; elytris thorace basi duplo latioribus, elongatis, sinuato-truncatis, maculis cinereis in lineas transversas flexuosas irregulariter ordinatis, apice albo marginatis. Long. 4¾ lin. ♂.

Head narrow, black; forehead with three ashy stripes, besides a streak underneath each eye; vertex with an ashy central line, and a broad lateral stripe, the latter continuous with both the lateral stripe of the forehead and the cheek stripe. Antennæ black, fourth joint ringed with ashy. Thorax very slightly widened from the front to beyond the middle, armed at that point with a minute tubercle, and then narrowed again to the base; surface black, sides each with a broad ashy stripe, in the centre of which is a short black line. Elytra twice the width of the thorax at its base, elongated, narrowed near the apex, the latter broadly sinuate-truncate (angles not produced); surface punctured, olivaceous black, marked with a number of small dingy-ashy spots, most of which are confluent, and tend to form three transverse flexuous lines. Body beneath plumbeous black; sides, from the prothorax to the apex of the abdomen, ashy. Legs black, basal joint of tarsi grey.

♂. Apical ventral segment truncated, sharply notched in the middle; dorsal segment slightly emarginated in the middle.

Pará. The species resembles greatly in form and coloration certain species of *Colobothea*. The absence of a lateral carina to the elytra readily distinguishes it from that genus.

Eurypanus Colobotheides, White (Cat. Long. Col. Brit. Mus. ii. p. 372), belongs also to our genus *Sporetus*.

INDEX TO GENERA.

IX. *Contributions to an Insect Fauna of the Amazon Valley* (Coleoptera, Prionides). By H. W. BATES, F.Z.S., Pres. Ent. Soc.

[Read 15th March, 1869.]

THE following pages contain a description of the genera and species of Longicorn *Coleoptera*, Tribe *Prionides*, obtained by me in the region of the Amazons, and are a continuation of a series of papers commenced in the Annals and Magazine of Natural History, in July, 1861. Those papers completed the tribe *Lamiides*, leaving for subsequent publication the tribes *Prionides* and *Cerambycides*. My reasons for postponing the continuation of the work, on the completion of the first part, were the want of a general classification of the Longicorns founded on a study of the whole family, and a conviction of the inconvenience to science of partial classifications applicable only to a single fauna. Such a classification I was compelled to invent for the *Lamiides* group; which, although it seemed to suit well the material I had before me, I afterwards found impossible to reconcile with the arrangements proposed by other writers, probably equally well-suited to other faunas. This was especially the case with the classification adopted by Mr. Pascoe for the Longicorns of the Malay Archipelago, and the inconvenience to which I have alluded was felt in this way, that it was impossible, with two such distinct arrangements, to institute those comparisons which all Naturalists find so interesting, between the faunas of these two equatorial regions. The work which all Coleopterists interested in this family have been so long expecting, the eighth volume of Lacordaire's "Genera" has at length appeared, containing a new and well-considered classification of the family, and there is no longer need to delay the description of my collections. In so difficult a group it would be presumptuous to alter this classification, without a laborious study of material, as large as that which has been at the command of Professor Lacordaire; and to do so in a partial manner would hinder rather than forward the progress of our knowledge of the group; I shall, therefore, adopt it implicitly in the following descriptions, although I believe, in some points, it is far from natural in its arrangement.

According to Lacordaire, the Longicorns are divisible
into three tribes ; the *Lepturides*, formerly considered by
most authors as a fourth tribe, being sunk to the rank of
a subordinate group under the *Cerambycides*. Having
traced the successive modifications of the forms allied to
Lepturides, in the order given by Lacordaire, I have
recognized the justice of this arrangement, and the error
of the opinion expressed in the introduction to the
Lamiides of the Amazons, on the same subject. The
Prionides are distinguished from the two other tribes by
the pronotum being distinct from the flanks of the prothorax, and by the anterior coxæ lying in transversely-
elongated sockets. In common with the *Cerambycides*,
they differ from the *Lamiides*, by the palpi never terminating in points, and the anterior tibiæ being simple
instead of grooved on their inner sides.

The number of *Prionides* obtained by me from the
Amazons is only twenty-six ; a small proportion of the
whole number found in Tropical America, namely 166.
A great many, however, described from other quarters,
will probably be found to be varieties or opposite sexes
of other species ; in confirmation of which opinion I may
point to the seven false species of one genus only, *Pyrodes*,
which I have reduced to synonyms in the following
descriptions. But the equatorial plains seem to be less
rich in the group than the borders of the tropics, or the
mountainous regions. The species are mostly nocturnal
in their habits, and of great rarity. When found
in sitû, it is generally on the trunks, or under the
bark, of the largest forest trees. They fly abroad at
night, and are sometimes overtaken by a sudden storm,
and cast into lakes or rivers, whence the swell carries
them to the sandy beaches ; several of the species here
recorded have been found under these circumstances.
I have not thought it necessary to insert in the *Prionides*
the sections and "tribus" of Lacordaire; the genera
follow in the order of his classification.

I. Prionides aberrantes.

Genus Parandra.

Latr. Hist. Nat. des Crust. et Ins. xi. p. 252.

1. *Parandra gracillima*, n. sp.

P. elongata, angustata, mandibula dentibus molaribus
basalibus contiguis, apice tridentatis, orbita oculorum

valde elevato acuto, thorace regulariter et forte angustato ab apice usque ad basin, elytris fortiter punctatis.

Long. ♂ (mandib. incl.) 9 lin.

A distinct species remarkable for the gradual tapering
of the thorax from apex to base; so that near the base
it is no broader than the length. The mandibles agree
in shape with the group to which *P. mandibularis* of Perty
belongs, that is, they have in the ♂ a very large basal or
molar tooth, the opposing teeth meeting in the centre,
near the apex is an acute tooth, and the apex itself, being
notched, forms two others; above, each mandible has a
sharply-defined triangular depression, and the surface is
rather coarsely punctured, and black. The submentum
is not separated from the gula or throat by an impressed
line; it is very broad, blackish, opaque, and is covered
with very large and shallow circular pits, the anterior
edge has not a raised border or impressed line, and the
anterior angles are broad, and very obtuse. The orbit
behind the eyes is very abruptly elevated, its upper edge
being above the level of the eyes. The head and thorax
are finely punctured, the elytra coarsely punctured, and
the whole surface less shining than in the allied species.

I took one example only of this species (the only *Parandra* found on the Amazons) at Ega, under the bark
of a dead tree.

The genus *Parandra* has been excluded from the family
of Longicorns, by some modern authors, and restored to
its place recently by Lacordaire. It may perhaps be
objected to the arrangement of the latter, that he includes
it in an artificial group termed "*Prionides aberrants*,"
with a number of forms such as *Hypocephalus, Sceleocan-
tha*, &c., with which it has nothing in common, except
the fact of being aberrant. *Parandra* would seem rather
to be an extreme development of the *Mallodon* type of
Prionides; its chief peculiarity, namely, linear tarsi, with
an onychium furnished with two bristles between the
tarsal claws, being lessened in importance by the fact o
a typical Prionid of the *Mallodon* group, *Hystatus*,
(Thoms.), possessing a distinct onychium. I have moreover noticed that the onychium is absent, or extremely
reduced and destitute of bristles, in at least one species,
the North American *P. brunnea*, F. Another character of the genus, the distinct fourth joint of the tarsi,

doubtless arises from the absence of lobes in the third joint, for in all *Prionides* where these lobes are reduced in amplitude, the fourth joint is more or less visible. No importance is to be attached to the form of the ligula, this point being excessively variable in the *Prionides*.

II. PRIONIDES VERI.

Cohort 1. *Subterranei.*

Genus PSALIDOGNATHUS.

G. R. Gray, in Griffith's An. King. Ins. ii. 115.

1. *Psalidognathus Incas.*

P. Incas, Thoms. Arc. Nat. p. 42.

P. Limenius, Erichs. Archiv. für Nat. 1847, i. 139, ♀.

♀. Ps. cupreo-violaceus; a femina Ps. *Friendii* differt, 1° antennis articulo 3io rugoso-punctato, 2° elytris magis subtiliter vermiculato-rugosis, et magis distincte tricostatis, 3° prosterno fortiter scabroso.

One example, a female, obtained at Tabatinga, on the frontier of Brazil and Peru.

Cohort 2. *Sylvani.*

Genus ENOPLOCERUS.

Serville, Ann. Soc. Ent. Fr. 1832, p. 146.

1. *Enoplocerus armillatus.*

Lin. Syst. Nat. ii. 622; Oliv. Ent. 66, pl. v. f. 17, ♂.

E. maximus, elongato-oblongus, brunneus, cinereotomentosus, elytris cinnamoneis nudis, antennis pedibusque nigris nitidis, ♂ scabrosis, ♀ laevibus.

Long. unc. ♂ 3-4¼, ♀ 3¼.

I obtained three examples only of this species; on the Upper Amazons, on the trunks of dead trees.

Genus ORTHOMEGAS.

Serville, Ann. Soc. Ent. Fr. 1832, p. 149.

1. *Orthomegas cinnamoneus.*

Lin. Syst. Nat. ii. 628; Drury, Ill. i. 89. t. 40. f. 2, ♀.

O. oblongo-linearis, cinnamoneus, aureo-fulvo sericeus, laevis; thorace lateribus pone medium dente magno obliquo, antice denticulis duobus vel tribus armato.

Long. 22-30 lin. (♂, ♀).

Found occasionally, in repose on leaves in the forest, throughout the Amazons region.

Genus MACRODONTIA.

Serville, Ann. Soc. Ent. Fr. 1832, p. 140.

1. *Macrodontia cervicornis.*

Lin. Mus. Lud. Ulr. p. 65; Oliv. Ent. 66, pl. 2, f. 8.

M. magna, depressa, rufo nigroque varia, elytris flavorufis, lineis plagisque nigris variegatis.

Long. unc. 2¼-4. (♂, ♀.)

A rare insect on the Amazons; on dead trees, banks of the Tapajos, and at Ega.

2. *Macrodontia crenata.*

Oliv. Ent. 66. p. 27, pl. 12. f. 45, ♀.

♂. A ♀ differt spina anteriore thoracis brevissima, et mandibulis multo longioribus. Oblonga, depressa, fusco-castanea, elytris cinnamoneis. Caput supra concavum, grosse punctatum, mandibulis capite sesqui longioribus, triquetris, cum antennarum basi (articuli reliqui desunt) nigris. Thorax transversus, quadratus, basi valde angustatus, lateribus inter spinas rectis, crenulatis, angulo antico spina minuta acuta oblique antrorsum spectante armato, spina postica majore sed brevi, angulis posticis distinctis acutis; supra creberrime punctatus, medio plaga longitudinali lineisque aliquot elevatis nitidis, sparsim punctatis. Elytra marginibus pone humeros valde explanatis, margine foliaceo usque ad apicem extenso sed sensim

angustato, apice late rotundato, angulo suturali
spinoso; supra opaca, subtiliter alutacea, cinnamonea,
abaque lineis elevatis.   Corpus subtus nitidum, im-
punctatum, castaneum.   Pedes nigro-castanei, nitidi.
Long. ♂ (mandib. incl.) 28 lin.

I am not aware that the male of this very rare species
has been heretofore described.   It differs greatly from
the female in the punctuation of the thorax, and in the
size of the antero-lateral spine.   A similar sexual difference exists in *M. flavipennis* (Chevrolat), the female of
which is named *serridens* in Chevrolat's collection, and
in *M. Dejeanii* (Gory); the male in all these species
having a finely punctured opaque thorax, with a glossy
space in the middle, and the female being uniformly
scabrous-punctate, slightly shining.

I met with one example only of *M. crenata* on the
Amazons, near Ega.

Genus Titanus.

Serv. Ann. Soc. Ent. Fr. 1832, p. 133.

1.   *Titanus giganteus*.

Linn. Mant. p. 531; Drury, Ill. iii. p. 73, pl. 49. f. 1, ♀.

T. (♂, ♀) fusco-castaneus, thorace lateribus trispinosis,
supra punctato-rugoso, medio late impunctato, tibiis
♂ intus multispinosis, ♀ laevibus, antennis utroque
sexu dimidium corporis haud excedentibus, segmento ultimo ventrali ♂ in medio late exciso, ♀
integro.

Long. 4½-6 unc.

In addition to the sexual differences mentioned in this
short diagnosis, may be mentioned the much greater
width of the tarsi in the ♂ than in the ♀.   The tarsi of
the ♂ are of remarkable width, and the second joint is
transverse quadrangular, instead of triangular as in the
♀ and in Longicorns generally.   On the Amazons this
colossal Longicorn was found only near Manaos, on the
Rio Negro; where it is occasionally picked up on the
shores of the river after a stormy night, the insect being
cast into the water whilst flying across.*

* In the system of Lacordaire the group *Ancistrotides* follows the *Tita-
nides*, but in a note on a subsequent page (Genera, viii. p. 163) he justly
doubts whether they would not be better placed near the group *Tragoso-
mides*.   The following new species of *Acanthinodera* (group *Ancistrotides*)
tends in favour of this emendation.

Genus CTENOSCELIS.

Serv. Ann. Soc. Ent. Fr. 1832, p. 134.

1. *Ctenoscelis ater.*

Oliv. Ent. 66, p. 11, pl. 7, f. 24, ♂.

Ct. piceo-niger, tarsis posticis lobis angustissimis et longe spinosis ; elytris apice apud suturam sinuatis, angulo suturali spinoso.

Long. 3¼-3¾ unc. (♂ . ♀).

♂ . Thorax minutissime et creberrime punctatus, disco utrinque plagis angustis tribus grosse scabrosis nitidis ; antennæ longitudine corporis, intus denticulatæ ; tarsi lobis intermediis et posticis spinosis.

♀ . Thorax omnino grosse scabrosus ; antennæ dimidium corporis attingentes, punctatæ ; tarsi lobis omnibus spinosis.

I obtained many specimens of this fine species at Ega, on the Upper Amazons, on the trunks of large felled trees.

2. *Ctenoscelis Dyrrhacus.*

Buquet, Ann. Soc. Ent. Fr. 1843, p. 235, pl. 9, fig. 1, ♂.

Ct. piceo-niger, tarsorum lobis ovatis, haud spinosis ; thorace utroque sexu scabroso, medio lævi, nitido ; antennis articulo basali breviori et validiori ; elytris apice nec sinuatis nec dentatis.

Long. 3 unc. 4 lin.—4 unc. 4 lin. (♂ ♀) ; lat. elytr. (♂) 18 lin.

Acanthinodera bihamata, n. sp.

♂ . Oblonga, depressa, rufo-castanea, tibiis tarsisque flavo-castaneis, capite, thorace, scutello et pectore quam *A. Cumingii*, ♂ , minus dense flavo-lanuginosis, puncturis grossis capitis thoracisque patentibus. Antennæ robustæ, leviter serratæ, corpore paulo longiores, articulis subæqualibus, primo punctato, reliquis lævibus. Thorax transversus, lateribus utrinque ante valde bihamatis. Elytra oblongo-quadrata, depressa, marginibus explanatis, apice sub-truncata, angulo suturali spinoso, supra subtiliter alutacea et elevato-reticulata costis quatuor vix perspicuis. Pedes elongati, graciles, valde compressi, sparsim hirsuti ; tarsis elongatis, linearibus, articuli tertii lobis angustissimis acutis.

Long. 11 lin.

Hab.—Mendoza. A Dom. Ed. Steinheil recepta.

♂. Antennæ longitudine corporis, articulis basalibus tribus aspere punctatis, omnibus intus denticulatis ; tarsi omnes lati, lobis rotundatis.

♀. Antennæ dimidium corporis attingentes, tenuiter punctatæ, nec denticulatæ; tarsi lobis compressis.

The thorax is coarsely scabrous on the sides in both sexes, and has a broad shining space in the middle, finely punctured ; the basal joint of the antennæ is shorter and thicker than in any other known species. The female has not before been described.

The species occurred at Santarem, Obydos and Manaos, on the Amazons.

8. *Ctenoscelis Nausithous*.

Buquet, Ann. Soc. Ent. Fr. 1843, p. 286, pl. 9, f. 2, ♂.

Ct. gracilior, thorace angustior; piceo-niger, thorace scabroso, spatio mediano lævi subtiliter punctato ; elytris apice rotundatis, angulis suturalibus dentatis. Long. 8 unc. 4 lin.

♂. Antennæ corpore longiores, articulis extus breviter tuberculatis, tribus basalibus subtiliter punctatis, reliquis lævibus.

♀. Antennæ dimidium corporis superantes, sparsim punctatæ; tarsi lati, lobis posticis acutis.

The shape of this species is much more slender than in the rest of the genus; the thorax of the female is scarcely to be distinguished from that of the same sex in *Ct. Dyrrhacus*, but the elytra differ in being much narrower, and especially more parallel-sided ; the tibiæ also are relatively longer, and more slender.

I met with one example only of this species, a female, at Serpa, near the mouth of the River Madeira. Not being acquainted with the male, I have drawn up the above imperfect diagnosis from the description of M. Buquet, who obtained the insect from Bolivia.

Lacordaire founds his generic characters of *Ctenoscelis*, in great measure, on the sexual differences in the punctuation of the thorax, as shown in *Ct. ater* and *acanthopus*; this part of his diagnosis will no longer be applicable, as *Ct. Dyrrhacus* and *Nausithous* show no such differences.

Genus IALYSSUS.

J. Thomson, Syst. Ceramb. p. 296.

1. *Ialyssus tuberculatus.*

Oliv. Ent. 66, p. 20, pl. 6, f. 22.

I. oblongus, fusco-castaneus, elytris (basi scabrosa excepta) rufo-ochraceis, lævibus, opacis, angulis suturalibus spinosis; antennarum articulis primo et tertio longitudine æqualibus.

♂. Thorax subtiliter creberrime punctato-rugosus, opacus, disco utrinque plaga triangulari lineisque duabus, et basi linea transversali, elevato-scabrosis, nitidis.

♀. Thorax omnino scabrosus.

Long. 2 unc. 4 lin.

I found one example only, a male, of this rare species; at Ega, washed up on a sandy beach, after a storm, on the river Teffé.

Genus MALLODONHOPLUS.

J. Thomson, Classif. des Ceramb. p. 320; Lacord. Genera, viii. p. 117.

Distinguished from the genus *Mallodon* by the femora and tibiæ, at least of the anterior legs, being armed with rows of short denticulations, and by the scabrous punctuation of the thorax. The portion of Lacordaire's definition relating to the mandibles, will have to be modified, to include the following species, which is undoubtedly congeneric with the type species, *M. nobilis.*

1. *Mallodonhoplus crassidens,* n. sp.

M. oblongus, paullulum convexus, piceo-niger, capite grossissimé punctato, labro antice liguláque fulvohirsutis, mandibulis capite brevioribus (♂) extus à basi valde rotundato-dilatatis vel tumidis, apice acutissimis, supra scabrosis, intus concavis vix hirsutis et margine interiori medio dilatato quadridentato; antennis dimidium corporis superantibus, articulo primo grosse punctato; thorace quadrato, supra scabroso, in medio elevationibus indistinctis duabus lævioribus; elytris vix nitidis, passim crebre minus profunde punctatis, angulis suturalibus spinosis; sternis omnibus grosse punctatis, pedibus anticis denticulatis, tarsis piceo-rufis; ventris segmentis singulis valde convexis.

Long. ♂ (mand. incl.) 2 unc. 4 lin.

Very similar in shape and sculpture to *M. nobilis*, Thoms., from Venezuela, the thorax being almost exactly of the same outline and surface; it differs in being entirely black, in the absence of fulvous hairs from the epistome, and in the great thickness of the mandibles. These organs are much shorter than the head, and are abruptly dilated externally, the apex of each ending in a long acute point, and the inner edge about the middle being advanced, and armed with four short broad teeth, nearly as in *Mallodon spinibarbis*. The elytra have a marked convexity from base to apex.

I met with only one example of this species, at Ega, cast up on a sandy beach, after a storm.

Genus MALLODON.

Serville; Lacord. Genera, viii. p. 125.

1. *Mallodon spinibarbis.*
Lin. Mus. Lud. Ulr. p. 67.

M. piceo-niger, elytris castaneis, vel omnino fusco-niger, capite grosse punctato, genis sub mandibulis prolongatis acutis; thorace plagis politis septem, interstitiis in ♂ subtiliter crebre punctatis, in ♀ scabrosis; elytris subtilissime sparsim punctatis; processu prosternali plano, ♂ crebre punctato, ♀ glabro.

A widely-distributed insect, found sometimes in great numbers, under the loose bark of felled trees; it is very unstable in the outline of the thorax in both sexes, but it may always be distinguished from the nearest allied species by the angles of the cheeks below the mandibles forming a simple point, instead of being bifid. I have specimens from Mexico, Cayenne, and Rio Janeiro, as well as from the Amazons.

2. *Mallodon bajulus.*

M. bajulus, Erichs. Consp. Ins. Col. Peru. p. 136.

M. occipitale, Thomson, Physis, I. p. 93, ?.

♀. "Oblongus, depressus, fusco-niger, nitidus, genarum processu bidentato: prothoracis disco polito, medio serie punctorum obsoletorum longitudinali notato, lateribus punctato-rugosis, margine obtuse crenulato, angulis posterioribus denticulo acutiusculo terminatis, elytris punctulatis. Long. 1″ 5‴." (Erichs.).

♂. Mandibulis brevibus, suprà et infra grosse punctatis; capite grosse confluenter punctato; thorace plagis septem elevatis politis, omnibus (exteriori utrinque excepta) basi conjunctis, interstitiis crebre grossius punctatis; elytris distincte punctatis; processu prosternali convexo, vix punctato.

Hab.—St. Paulo, Amazons.

I think it extremely probable that *M. occipitale*, Thoms., from Venezuela, belongs to the same species: specimens before me referable to this species differ only in the finer punctuation of the elytra. *Mallodon bajulus*, of Erichson, has been generally cited as the ♀ of *Chiasmetes Linæ* of Guérin, but Erichson's diagnosis lends no support to such an inference; the terms "depressus—genarum processu bidentato" being quite inapplicable to *Chiasmetes*; on the contrary, they fit the present species of *Mallodon* which I found in the same tract of country where Erichson's insect was discovered by the traveller Tschudi. M. Thomson makes no mention, in his monograph of *Mallodon*, of the form of the anterior angles of the cheeks, without which it is scarcely possible to give satisfactory descriptions of the species.

Genus STICTOSOMUS.

Serville; Lacord. Genera, viii. p. 144.

A remarkable genus, distinguished by its oblong-linear depressed form, long acute mandibles curving downwards and a little backwards towards the apex, and by the extremely long claw-joint of the tarsi; the third joint of the antennæ is nearly as long as the four following taken together, and is thickened and cylindrical.

1. *Stictosomus semicostatus.*
Serv. Ann. Soc. Ent. Fr. 1832, p. 153.

St. oblongus depressus, niger, passim grosse punctatus, elytris costis quatuor distinctis.

Long. 1 unc. 10 lin. ♂.

One example, found near Montes Aureos, in the interior, East of Pará.

Genus POLYOZA.
Serville; Lacord. Genera, viii. p. 152.

1. *Polyoza lineata*, n. sp.

P. oblongo-linearis, rufo-fulva, elytris utrinque suturâ et carinis quatuor elevatis, interstitiis nigris: antennis (♂) dimidium corporis vix superantibus, articulis 3—8 basi laminas elongatas emittentibus.

Long. 11 lin. (♂).

Resembles in form *P. Lacordairei*; the head, palpi, mandibles, and eyes offer scarcely any difference: the thorax has on each side three teeth, the foremost one very small, and the hindmost pointing towards the shoulders of the elytra; its surface is sculptured in a raised reticulate pattern. The elytra offer a raised suture and four carinæ, the interstices of which are of a dark colour, the whole surface is finely rugose-punctate, and the sutural angle acute, but not spinose. The antennæ are very much shorter than in *P. Lacordairei*, and the long foliaceous appendages to the joints placed more closely together. The underface and legs are finely punctate-granulate and hairy, and of a paler hue.

One example, taken in a dead tree at Ega.

Cohort 8. *Pœcilosomi.*

Genus MALLASPIS.

Serville, Ann. Soc. Ent. Fr. 1832, p. 188.

The characters of this genus, in Lacordaire's " Genera,"
derived from the antennæ, especially those of the ♀,
are no longer applicable, since species have lately been
discovered * which differ from the types in this respect.
In fact, there is now no character to distinguish the genus
from *Pyrodes*, except the pilose scutellum.

* These are as follows.

Mallaspis Belti, n. sp.

Species distinctissima, antennis in utroque sexu articulis linearibus,
colore ♀ a ♂ valde diverso; mesosterno et metastomo sutura distincte divisis.

♂ Long. 18-20 lin.

Oblonga, convexa, postice attenuata, æneo-castanea, capite thoraceque
æneo-fuscis, antennis tibiis et tarsis rufo-castaneis. Caput magnum,
elongatissimum, subremote medice punctatum, lateribus parallelis,
fronte, late sulcato. Thorax pone medium valde dentatus, antice in
linea subrecta, angustatus, minute irregulariter denticulatus, pone
dentem sinuatus, angulis posticis nullis, supra æqualis, modice crebre
punctatus, subnitidus, lateribus rugulosis. Elytra parte basali convexa, postice declivia, apice obliqua truncata, angulo suturali pro-
minenti acuto, humeris antice protrusis, supra prope basin vermicu-
lato-punctata, postice subtiliter punctato-coriacea. Corpus subtus
æneum, griseo-tomentosum; pedes rufo-castanei, femoribus obscurioribus. Antennæ corpore paulo longiores, rufæ, articulis a 3º
omnibus linearibus, 8º paullo latiori sed lineari, ultimis duobus basi
intus spinosis. Scutellum aureo-tomentosum.

♀ 15 lin. Forma ♂ similis; postice minus angustatus, capite breviori;
thorace et elytris basi fortiter rugoso-punctatis, colore omnino satu-
rate-æneo vel obscure-cyaneo, antennis pedibusque cyaneis; corpore
subtus glabro. Antennæ corporis dimidio longitudine æquales, articulis linearibus, cyaneis. Processus prosterni canaliculatus.

Hab.—Chontales, Nicaragua. Dom. Belt invenit.

Mallaspis Salvini, n. sp.

M. Belti proxime affinis; eadem differentia sexuum formaque prothoracis et antennarum.

♂ Long. 15-16 lin. Oblonga, depressa, postice attenuata, supra tota
late ænea, subtus aureo-refulgens. Caput thorax et elytra ut in *M.
Belti*, sed illud grossius punctatum nitidum, hæc prope apices minus
attenuata, obtusius rotundata. Antennæ corpore longiores, articulis
duobus ultimis basi tuberculatis, rufæ. Pedes rufi, femoribus æneo-
tinctis.

♀ 15 lin. *M. Belti* ♀ forma et colore simillima; caput multo grossius
rugoso-punctatum; antennæ omnino violaceæ; processus prosterni
haud canaliculatus.

Hab.—Costa Rica. A Dom. Salvin receptus.

1. *Mallaspis scutellaris.*

Oliv. Ent. 66, p. 14, pl. 2, a. b.

M. obscure ænea, elytris basi excepta cinnamoneis, thorace lateribus antico rotundatis et multidenticulatis, antennis articulis 4-7 basi et apice 8-11 totis rufis, ♂ corpore multo longioribus articulis compressis denticulatis, ♀ brevioribus articulis dilatato-compressis.

Extremely raro; found only at Nauta, on the Upper Amazons.

Genus PYRODES.

Serville ; Lacord. Genera, viii. p. 177.

1. *Pyrodes pulcherrimus.*

P. *pulcherrimus*, Perty, Del. An. Art. Bras. p. 86, t. 17, f. 4, ♀.

P. *fastuosus*, Erichs. Consp. Ins. Col. Peru. p. 189, ♀.

P. *heterocerus*, Erichs. Consp. Ins. Col. Peru. p. 189, ♂.

P. *antennatus*, White, Cat. Long. Brit. Mus. p. 51, pl. 2, f. 6, ♂; Lucas, Voy. de Castelnan, Ins. pl. 10, f. 8, ♂.

P. *petalocerus*, White, Cat. Long. Brit. Mus. p. 50, ♂.

♂. Medio valde convexus, colore variabilis, fusco-ferrugineus vel æneo-fuscus, vel antice ferrugineo-cupreus, postice ferrugineus; scutello apice prolongato; antennis articulo tertio magno lato compresso, azureo, reliquis rufis; capite thorace et scutello crebre æqualiter punctatis; elytris vermiculato-coriaceis; femoribus cyaneis.

♀. Azureus vel cyaneus, elytris rugulosis nitidis, fascia lata ante medium flavescenti-alba; antennis cyaneis, articulo tertio simplici lineari. Variat thorace maculis duabus rufis.

This beautiful and singular species was not infrequent at Ega. The two sexes, as here described, have always been considered distinct species; but the fact that all of one form are males, and all of the other females, and that they are invariably found together, on the trunks of trees, induces me to consider them as pairs, although I never ound them *in copulâ.*

2. *Pyrodes Smithianus.*

White, Proc. Zool. Soc. 1850, p. 12.

P. pulcherrimo (♀) formâ similis, thorace latiori, lateribus antice rotundato-dilatatis. Cupreo-æneus, elytris
aureo-viridescentibus; antennis brevibus tenuibus;
thorace elytris latiori, cum capite et scutello crebre
distincte punctatis; scutello elongato; elytris sutura
et costis duabus utrinque elevatis, crebre rugosis.

Long. 15 lin. (♀).

I found one specimen of this species on the foliage of
a low tree, at Caripi, near Pará. Unfortunately it was
not reserved for my private collection, and I have drawn
up the above diagnosis from the type specimen in the
British Museum, which was found by Mr. J. P. G.
Smith, also at Caripi.

3. *Pyrodes formosus*, n. sp.

P. pulcherrimo (♀) formâ simillimus, colore omnino
læte saturato-cœruleus, plagis duabus elytrorum
violaceis exceptis; caput thorax et scutellum sub-
opaci, confertim punctati; antennæ ut in *P. pulcherrimo* (♀) lineares, vel prope apicem incrassatæ,
cyaneæ; elytra fortiter, prope apices vix minus
forte, vermiculato-rugulosa, angulis suturalibus haud
productis, nitida; corpus subtus, et pedes, saturate
cœrulei.

Long. 15-18 lin. ♀.

Two female examples only of this beautiful species
were found, near St. Paulo, Upper Amazons. In the
British Museum there is a specimen from Cuenca, Equador, entirely of a beautiful greenish-blue colour, and
rather more coarsely sculptured.

4. *Pyrodes gratiosus*, n. sp.

P. bifasciato, Linn., affinis, minor, colore ♀ valde diverso, et antennis distincte 11-articulatis.

♂. Oblongus, læte æneo-viridis, nitidissimus, elytris
testaceo-translucentibus. Caput et mandibula sparsim punctata, illo antice aureo, sulco frontali fundo
lævi. Thorax quadratus, lateribus denticulatis,.

medio fortiter dentatis, angulis posticis dentiformibus, suprà grosse punctatus, disco impressione magna triloba. Scutellum læve. Elytra oblongo-quadrata, tertia parte basali grosse punctato-rugosa, dein subito crebre subtiliter rugosa ; viridi-ænea, basi et medio testaceo translucentia. Corpus subtus glaberrimum, prosterno gibbo, lævi. Antennæ corpore paulo longiores, robustæ, filiformes, nigræ, articulis 4 basalibus cupreis. Pedes rufi, geniculis fuscis.

Long. 6 lin.

♀. Oblongus, saturate cyaneus, capite thoraceque grossius punctatis, vix nitidis, elytris basi tantum micantibus, miniato-rubris, fascia sub-basali curvata, purpureo-nigra. Sulcus frontalis fundo lævis. Thorax ut in ♂ quadratus, lateribus medio dente forte armatis, angulis posticis dentiformibus, punctatus scabrosus, medio impressione magna triloba. Scutellum grosse punctatum, opacum. Elytra oblonga, basi convexa, prope basin fortiter punctato-rugosa, dein subito subtiliter rugosa. Antennæ corpore paulo breviores, minus robustæ, filiformes, cyaneæ, basi cupreo-violaceæ, articulis 3-7 supra sulcatis. Corpus subtus cyaneum, nitidissimum. Pedes cyanei, geniculis et tibiis apice violaceis, tarsis rufis.

Long. 10-12 lin.

This beautiful species is readily distinguished in the female from *P. bifasciatus* by the elytra having only one dark fascia, and that near the base, commencing below the shoulder, and curving towards the suture, which it does not reach ; the rest of the elytral surface is of a clear red-lead colour, or dark vermillion, almost opaque, except near the base, where it has a metallic lustre. The slender 11-jointed antennæ distinguish it at once from the female of *P. nigricornis*, Guér., besides the colour. I obtained three female specimens of precisely similar colours, two of which are in my own collection. The males of these closely-allied species are less easily to be distinguished ; I obtained only one example with the three females ; but have seen a second, similar in size, in Mr. Fry's collection, also from Pará. The antennæ are notably more slender than in the ♂ of *P. nigricornis*, and the terminal joint is much less elongate, being very little longer than the penultimate, whilst it is half as long

again in *P. nigricornis.* The species seems peculiar to
the neighbourhood of Pará. *P. bifasciatus* is found at
Surinam and Demerara, and all the female specimens I
have examined have the two terminal joints of the antennæ blended into one elongate joint, with a trace of
the articulation.

5. *Pyrodes nodicornis*, n. sp.

♀. *P. bifasciato* simillimus, gracilior, antennis 10-
articulatis, thoracis angulis posticis haud prominentibus. Oblongus, nigro-cyaneus, supra vix nitidus, subtus politissimus. Caput et thorax grosse crebre
punctati, hoc quadrato angusto, in medio paulo dilatato et valde dentato, angulis posticis haud prominentibus, supra impressione profunda triloba.
Elytra postice haud attenuata, medio vix rotundata,
supra præcipue versus basin et suturam grosse rugoso-
punctata, læte miniata, fascia lata communi prope
basin (margines haud attingente) alteraque apicali
(cum precedente vitta lata suturali conjuncta) cya-
neo-violacea. Pedes cyanei, violaceo-micantes.
Antennæ corpore paulo breviores, graciles, articulo
decimo precedentibus latiori, colore cupreo-violaceæ,
articulo tertio haud sulcato.

Long. 11 lin., lat. 5 lin. ♀.

One example, taken at St. Paulo, Amazons. Very
similar in form and colour to *P. bifasciatus;* differs in
the tenth antennal joint being short and ovate, without
trace of constriction in the middle, and also in the
absence of projecting hind angles to the thorax.

6. *Pyrodes nigricornis.*

P. nigricornis, Guérin, Verhandl. zool.-bot. Verein zu
Wien, 1855, p. 598, ♂.

P. rubrozonatus, Lucas, Voy. de Castelnau, Entom.
p. 180, pl. xi. f. 2 (1857) ♂.

♂. Breviter oblongus, variat vel fulvo-testaceus
æneo tinctus, antennis (basi excepta) violaceo-
nigris, vel aureo-viridis splendens, elytris semifascia
rufa, vel cupreo-violaceus, fascia elytrali integra, vel

pallidus, pedibus rufis, vel ut ante coloratus, pedibus
rufis femoribus tibiisque posticis plus minusve viola-
ceo-metallicis. *P. gratioso* differt antennis magis robustis, articulis brevioribus, elytris grossius punctato-
scabrosis, thoracis angulis posticis dentiformibus.
Latitudo thoracis variat.

Long. 8 lin. ♂.

♀. Breviter oblongus, cyaneus, elytris violaceis, basi
splendide cupreis, pone medium fascia interrupta
rufa; thoracis angulis posticis dentiformibus; antennis robustis, corpore multo brevioribus, 11-articu-
latis, articulo tertio haud sulcato : tarsis rufis.

I took many specimens of this species at St. Paulo,
Amazons, on the leaves of trees in the forest. One pair
were taken *in copulâ*. I have examined Guérin's type of
P. nigricornis, and found it to agree with the palest of
my specimens. It was probably taken in the same
locality as mine, by Osculati, who spent some time at
the village of St. Paulo on his voyage down the Amazons.

· 7. *Pyrodes dispar*, n. sp.

P. precedentibus affinis, differt pedibus grosse et profunde punctatis.

♂. Oblongus, læte viridi-æneus; thorace inæquali,
grosse punctato, scabroso, lateribus medio spinosis,
angulis posticis dentiformibus. Elytra passim crebre
et grossius punctato-scabrosa. Pedes rufi, grosse
et profunde punctati, femoribus tibiisque posticis
cyaneis. Antennæ corpore breviores, robustæ, filiformes, cupreæ, articulo ultimo cum precedente quasi
concreto.

Long. 8 lin.

♀. Breviter oblongus, affinibus multo latior ; viola-
ceus, supra passim crebre rugosus, sulci frontalis
fundo haud lævi, thoracis disco haud foveato, angulis
posticis dentiformibus. Elytra plaga basali, altera
laterali pone medium, et margine exteriori inter has,
rubro-cupreis. Corpus subtus cupreo-nitidum, abdomine crebre punctato, pedibus cyaneis, grosse et
profunde punctatis. Antennæ dimidium corporis
vix excedentes, cyaneæ, 10-articulatæ, articulo tertio
haud sulcato, decimo precedente latiori et duplo
longiori.

The nearly coalescent two terminal joints of the antennæ will distinguish the male of this from the three preceding species, as well as the much more deoply and roughly punctured femora and tibiæ; the other characters, and the colour, I think, may be more variable. The female is distinguished from all by its great width of body, and coarsely punctured femora and tibiæ. It differs greatly from *P. nodicornis* in general form of body, and in the prominent dentiform hind angles of the thorax. The two terminal joints of the antennæ are blended into one in both species, without trace of separation, but the joint thus formed is short and ovate in *P. nodicornis*, and very elongate in *P. dispar.*

I have seen one pair only of this species, taken by Mr. J. Hauxwell, at Pobas, on the Amazons.

In this group of *Pyrodes*, the scutellum seems to afford no reliable specific characters, as it varies in shape and sculpture in specimens undoubtedly belonging to the same species; the form is nearly as in the common *P. speciosus*, but its apex is more prolonged. In all the species where the hind angles of the thorax are produced, the lower margin is also dentiform, giving an appearance of two teeth at the angle, one above the other. The margin of the thorax anterior to the lateral spine, in all the species, is irregularly and variably denticulate. The general form and colours of the *bifasciatus* group are so variable, that if future discoveries prove the terminal joints of the antennæ to be variable, the whole will constitute one variable species, remarkable for its inconstancy in structural characters.*

* The genus *Pyrodes*, after withdrawing *P. pictus* (Perty) which, having the sides of the scutellum pubescent is better placed in *Mallaspis*, and *P. columbinus* (Guér.) which belongs to the genus *Esmeralda*, contains the following species: 1. *P. pulcherrimus* (ut supra). 2. *P. formosus.* 3. *P. tenuicornis*, White, ♂ (♀ =marginatus, White, Catal. p. 48, ♂ =angusticollis, Lucas, Voy. de Casteln., p. 179, pl. 11. f. 1; the description leaves no doubt whatever of this synonym, and the locality given by Lucas must be erroneous). 4. *P. Smithianus*, White. 5. *P. speciosus*, Oliv. 6. *P. æneus*, Buq., Ann. Soc. Ent. Fr. 1860, p. 613. 7. *P. bifasciatus*, L., Oliv. 8. *P. nigricornis*, Guér. (rubromaculus, Lucas). 9. *P. gratiosus.* 10. *P. nodicornis.* 11. *P. dispar.* There remains only one undescribed species in all the extensive collections of Longicorns which I have examined in London; this is a fine one from New Granada, represented by a single female example, in Mr. Fry's possession, derived from the Dejean-Laferté Collection.

Genus ESMERALDA.

J. Thomson, Classif. des Ceramb. p. 308 ; Lacord.
Genera, viii. p. 178.

Distinguished from *Pyrodes* by the great width and
length of the scutellum, which, in the male, is nearly
half the length of the elytra ; and by the metasternum
being greatly advanced between the middle coxæ, and
nearly hiding the grooved mesosternum, which lies
obliquely on its anterior face ; the prosternum is also of
great width, and its point does not interlock with the
mesosternum. The antennæ in the ♂ are very robust,
compressed, and subserrate ; the tibiæ also are compressed into thin blades in both sexes, and the tarsi are
excessively short.

The only species of this charming group hitherto described is *E. suavis*, Thoms. But I have no doubt whatever of this being the ♂ of *Pyrodes columbinus*, of Guérin
(said by White, erroneously as I think,* to be the
Cerambyx auratus of Linnæus). I captured the male and
female of the following species together, but not *in
copulâ*, on the trunk of a slender tree, and as the differences between them are not at all greater than in many
species of *Pyrodes*, the conclusion that they are sexes of
one and the same species is not to be resisted.

1.   *Esmeralda lœtifica*, n. sp.

♂.   Oblonga, depressa, viridi-ænea, nitidissima, capite
antice et infra thoraceque toto testaceo-rufis aureo-
tinctis, femoribus 4 anticis et processu metasternali
rufis, elytris violaceis, subtilissime rugoso-punctatis, bicostatis, triente basali excepta sparsim punctatis.

Long. 6 lin.

♀.   Late oblonga, subdepressa, læte cyanea, scutello
et corpore subtus violaceis.

Long. 9½ lin.

Differs from *E. columbina*, Guér. (♂, *suavis*, Thoms.)
in both sexes, by the basal third of the elytra being
glossy, and marked with very few punctures ; the scutellum has a few very fine punctures on each side. The

---

* The phrase of Linnæus "elytra rubro-viridi-aurata" is not at all applicable to any specimen of *E. columbina* which I have seen.

thorax in the ♂ has its lateral margins free from crenulations, and is of a tawny-orange colour above and beneath, except a narrow mark on the hind margin on each side; above, it has a few strong punctures on the sides, and in the deep central fovea. In the ♀ the thorax is crenulate on the sides, and has numerous large punctures on the sides and in the central fovea, most of the elevated portions being impunctate. The posterior part of the elytra, in both sexes, is rendered rough and opaque by the extreme closeness and minuteness of its punctuation. In the ♂, the middle coxæ, the greater part of the anterior and middle femora, and the broad metasternal process, are fulvous-red.

I met with one pair only of this species, on the trunk of a slender tree, in a small clearing in the forest, near St. Paulo, on the Amazons. They were actively moving about in the heat of the mid-day sun.

The following genus seems to be a degraded form of the *Pyrodes* type, wanting the grooved mesosternum, the sulcate head, and many other minor characters of the group. Its projecting metasternum indicates a nearer affinity with *Esmeralda* than with the *Pœcilosomina*, in which group it is placed by Lacordaire. No other member of the *Pœcilosominæ* or *Solenopterinæ* is found on the Amazons.*

* The following very interesting species has recently been discovered by Mr. Belt at Chontales, Nicaragua; it connects the two North-American genera *Holonotus* and *Sphenostethus.*

Holonotus nigroæneus, n. sp.

Elongatus, convexus, scaphiformis, *Sphenostetho serripenni* similis, sed scutello haud elongato, aterrimus nitidus, supra præcipue elytris æneo-tinctus. Caput parvum, supra late sulcatum, grosse sparsim punctatum. Antennæ (♀) dimidio corporis breviores, articulis valde compressis latis, vix punctatis. Thorax a basi usque ad apicem attenuatus et declivis, marginibus lateralibus integris, disco sparsim subtiliter, lateribus grosse et rugose punctatus, margine postico elytris angustiore, utrinque sinuato, lobo mediano truncato. Scutellum latum, triangulare, apice depressum, læve. Elytra a basi usque ad apicem regulariter attenuata, humeris obliquis rotundatis, apicibus rotundato-truncatis serratis; supra coriacea, submitida, passim modice punctata. Subtus glaber, pectoris lateribus punctatis, vix pilosis; mesosternum crassum, apicem prosterni incumbens. Pedes nigerrimi, nitidi, tarsis brevibus, latis.

Long. 1 unc. 2 lin., lat. 5 lin. ♀.

Hab.—Chontales, Nicaragua; in Mus. nostr.

Genus Nicias.

Nicias, J. Thomson, Arch. Entom. i. p. 136.
Hamadryades, Thomson, *lib. cit.* p. 22 (*olim*).

1. *Nicias alurnoïdes.*
J. Thomson, Arch. Ent. i. p. 23, pl. 9, f. 3.

N. niger, nitidus, elytris stramineis, quarta parte apicali maculaque utrinque mediana transversa nigris ;
abdomine rufo-testaceo.

Long. 7 lin. ♀ .

I found two examples of this pretty and singular
Prionid, on different days, on the foliage of trees at
Ega ; one of them is now in my own collection, the other
I believe was sent to Paris. The beautiful figures of
M. Thomson render further description unnecessary.

Dr Le conte
with the authors comp

XVI. *Contributions to an Insect Fauna of the Amazon Valley* (Coleoptera, Cerambycidæ). By H. W. Bates, F.Z.S., late Pres. Ent. Soc.

[Read 4th July, 1870.]

The present memoir is a continuation of a former one on the *Prionides* (Trans. Ent. Soc. 1869, p. 37), and the classification, with trifling modifications, is that established by Lacordaire in the eighth and ninth volumes of his "Genera des Coléoptères."

Fam. CERAMBYCIDÆ.

Section A. Eyes coarsely facetted.

Sub-fam. ŒMINÆ.

Antennæ without spines, anterior coxæ with their sockets lengthened externally, intermediate sockets open.*

Genus ATENIZUS.

Bates, Entom. Monthly Mag. iv. 28 (1867).

(Charac. emend.). Corpus parvum, sublineare, depressum, pubescens. Caput rotundatum, thorace latius, genis brevissimis, fronte convexa, vertice tuberculo magno instructo; oculis magnis emarginatis, lobo inferiori ante tuberculos antennarum producto, superiori brevi; palpis articulo terminali conico, maxillaribus (?) elongatis, ♂ pendentibus. Antennæ filiformes, hirsutæ, articulo basali apice infra dilatato. Thorax ovatus, depressus, inermis. Elytra linearia, apice rotundata. Pedes breves, lineares, tarsis posticis elongatis. Coxæ anticæ et intermediæ subconicæ, contiguæ, exsertæ, acetabula antica extus angulata, intermedia aperta; laminæ sternales inter coxas obsoletæ.

This is one of the genera which M. Lacordaire was unable to place in the rigid system of classification

---

* That is, the mesothoracic epimera are inserted between the meso- and meta-thorax so far as to reach the orbit of the sockets.

adopted by him. On a careful examination, I have no
doubt it belongs to his group *Œmides*, and that its place
would be probably in Section I. of that group. The
angulation of the anterior sockets is not so strongly
pronounced as in *Œme* and the other genera of the
group; but this may be attributable to the narrow form
of the prothorax. The sternal processes between the
anterior and middle coxæ appear to be wholly wanting.
The abdominal segments are normal, and not distorted
as in the *Obrioninæ*.

1. *Atenizus laticeps*, Bates, *l. c.*

" Sublinear, reddish-testaceous ; antennæ from the
third joint brown, bases of joints pale testaceous. Body
and limbs finely setose ; head and thorax sparingly punctured ; elytra regularly and closely punctured."

Long. 2½–4½ lin. ♂ ♀ .

Hab.—Pará and Santarem, Amazons ; on dry twigs.

Genus NIOPHIS.

Bates, Entom. Monthly Mag. iv. 27 (1867).

(Charac. emend.). Corpus parvum, elongatum, lineare,
depressum, pubescens. Caput postico haud angustatum,
genis brevissimis, fronte concava ; oculis magnis præcipue lobo inferiori, supra longe separatis ; palpis apice
truncatis ; tuberculis antenniferis haud elevatis, vertice
plano. Antennæ (♂) corpore multo longiores, longe
pubescentes, articulis 3–5 longitudine subæqualibus.
Thorax elongatus, inermis, à medio usque ad basin angustatus. Elytra postice attenuata, utrinque longe spinosa.
Pedes elongati, femoribus valde elongatis, compressis,
gradatim clavatis ; tarsis articulo basali elongato. Coxæ
anticæ exsertæ, conicæ, extus modice angulatæ, processu
sternali angustissimo ; acetabula intermedia extus aperta,
processu sternali latiusculo plano. Abdomen (♂) segmento basali cæteris haud longiori.

This genus is evidently allied both to *Atenizus* and
Œme, and on this account, although unable to ascertain
the texture of the ligula, I have no hesitation in placing
it in the present group. The antennal joints are clothed
all round with a long pubescence. The buccal aperture
is close to the lower margin of the eyes, there being no
muzzle. The eyes are emarginate, with well-developed
upper lobe.

1. *Niophis coptorrhina,* Bates, *l. c.*

Tawny reddish, opaque, clothed with fine erect hairs; apices of antennal joints darker, tips of thighs black; thorax with two broad and shallow longitudinal dorsal channels; elytra finely punctured, the apex of each with an acute spine.

Long. 4½ lin. ♂.

Hab.—Santarem, River Tapajos.

Genus ŒME.

Newman, Entom. i. 8; Lacord. Gen. Col. viii. 222.

In this genus, the thorax (unarmed) is abruptly narrowed near the base, and the prosternal process is reduced to an extremely narrow vertical partition.

1. *Œme picticornis,* n. sp.

Elongata, linearis, depressa, pallido-fulva, antennarum articulis (a tertio) et tarsis nigris; capite et antennarum articulo basali crebre et grosse punctato; thorace subquadrato, basi subito constricto, dorso subtiliter creberrime punctato, sericeo; elytris pube erecta tectis, crebre punctulatis, apice conjunctim rotundatis; (abdomen docet).

Long. 8 lin. ♂.

Apparently allied to *Œ. annulicornis,* Buq., which, however, is described as having the head smooth, and the thorax "on ovale très allongé." In the present species, the thorax, except the constricted hind portion, forms a square, almost exactly as broad as long, with the angles rounded. Besides the black tips of the antennal joints and the tarsi, the tergum of the mesothorax, uncovered by the base of the thorax, has a distinct black spot. The antennæ are regularly ciliate beneath, and have only a short pubescence above.

One example, taken at Ega, evidently a male.

Genus PHRYNOCRIS.

Bates, Entom. Monthly Mag. iv. 20; Lacord. Gen. Col. viii. 226.

1. *Phrynocris notabilis,* Bates, *l. c.*

Body elongate, subdepressed. Head and thorax coarsely and scantily tomentose, the rest of the body clothed with

short hairs. Thorax subquadrate, armed on each side
with a spine, surface uneven, covered with small scattered
tubercles, reddish-tawny, with the depressed parts black.
Elytra reddish-tawny, ornamented with three strongly
undulated black belts, the apex also black; surface
shining, punctured and roughened with three or four
rows of small tubercles. Legs reddish, tips of thighs and
tibiæ black.

Long. 10 lin. ♂.

Hab.—Ega.

Genus ZATHECUS.

Bates, Entom. Monthly Mag. iv. 20; Lacord. Gen. viii.
230.

In addition to the characters given in the places
quoted, may be here mentioned the sockets of the anterior
coxæ angulate externally, and those of the intermediate
coxæ open. These characters show that the genus is
related to *Œme*. The markings of the elytra, however,
are very similar to those of *Ibidion*, and allied genera.
The thorax is subquadrate, narrowed behind and unarmed,
slightly uneven above, without transverse impressions,
and opaque; the thighs are elongate, and strongly and
abruptly clavate.

1. *Zathecus graphites*, Bates, *l. c.*

Elongate, linear, depressed. Testaceous, head and
thorax clothed with a silky tomentum; vertex dusky,
basal joints of antennæ blackish beneath. Thorax uneven,
black, with a curved testaceous belt across the anterior
part. Elytra near the base and suture marked with a
black patch, followed behind by two curved black streaks;
the testaceous apical half with a brownish cloud in the
middle; whole surface roughened with small scattered
tubercles, and irregular but not large punctures. Legs
and under-surface testaceous, sides of prothorax and
breast and basal part of hind thighs blackish.

Long. 8 lin. ♂.

Hab.—Ega.

Genus MALACOPTERUS.

Serville, Ann. Soc. Ent. Fr. 1833, p. 565; Lacord. Gen. viii. 227.

1. *Malacopterus lineatus*, Guérin, Icon. Règne Anim. p. 222.

Elongatus, depressus, pallidus; elytris utrinque brunneo bivittatis, thorace medio carinato, margine postico producto-lobato; antennis fortissimis (♂) apicem versus attenuatis.

Long. 10 lin.

Hab.—Pará.

Sub-fam. ACHRYSONINÆ.

This sub-family differs from the *Œninæ* only in the anterior haunches being less angulate externally, with the sockets having a corresponding narrower and shorter opening on their outer side.

Genus ACHRYSON.

Serville, Ann. Soc. Ent. Fr. 1833, p. 572; Lacord. Gen. viii. 232.

1. *Achryson surinamum.*

Cerambyx surinamus, Lin. Syst. Nat. ii. 632.

A widely-distributed and well-known insect, cylindrical in form, of pale reddish-testaceous colour, with a black circumflex mark on the posterior disc of each elytron, and a few smaller spots on the anterior part of the same.

Common throughout the Amazons; the earlier states are passed in the interior of certain trees having wood of a light texture, and the insect is often found in the neighbourhood of houses.

2. *Achryson nanum*, n. sp.

Parvum, lineare, rufo-testaceum, unicolor, corpore toto longe piloso; thorace quam in *A. surinamo* longiori et magis cylindrico, creberrime subtiliter rugoso, tuberculis acutis subseriatis asperato, linea longitudinale et foveolis duabus disci anterioris impresso; elytris asperato-punctatis, apicibus in dente lato sub-obtuso productis.

Long. 8½ lin. ♂ (?).

Hab.—Tapajos.

3. *Achryson pictum*, n. sp.

Minus lineare, thorace subovato, postice angustato, rufum, sparsim breviter pubescens, thorace disco plagis confluentibus nigris; elytris apice aculeatis, pallide brunneis, maculis magnis nigris, scilicet, una circa scutellum, altera obliqua elongata humerali, plaga triangulari discali pone medium, et una apicali; antennis pedibus et episternis nigris.

Long. 7½ lin.

Of shorter and less cylindrical form than *A. surinamum*: thorax shorter, more rounded on the sides, and attenuate from the middle to the base. Clothed with a moderate tawny pubescence, sub-erect on the elytra and legs; colour red, varied with black patches, elytra yellower and shining, thorax opaque; antennæ, legs, and side-pieces of the sterna black. Head very coarsely rugose, thorax minutely rugose, and with scattered elevated granules; elytra punctured, more coarsely and densely so near the base. The black marks on the thorax are on the disc, and consist of a lateral vitta expanding on the front margin, and two central vittæ extending only from base to middle, and there united by a cross belt; but these marks are sometimes more or less blended. On the elytra the base is spotted with black, and there is a squarish black spot in the scutellar region, an oblique stripe from the shoulder, a triangular discal patch behind the middle, and a spot at the apex, including the apical spine.

Hab.—Pará; also found at Cayenne.

4. *Achryson hirsutulum*, n. sp.

Parvum, lineare, thorace medio paulo rotundato, elytris apice acutis, castaneum, fulvo-hirsutum; thorace et elytris pilis crassis decumbentibus, illo lineis tribus dorsalibus, his vittis irregularibus nudis; elytris coriaceis opacis, apice politis, basin versus sparse granulatis.

Long. 4¼ lin.

Allied to *A. ornatipennis* (Perroud) from Guadeloupe; but differing in the sculpture and apical armature of the elytra, besides the less regular arrangement of vittæ on the latter. According to Perroud's description, the elytra are "très faiblement tronquées à leur extrémité," whereas *A. hirsutulum* has the apex of each prolonged into an acute tooth, distinct enough, but not spiniform, as in *A. surinamum.* The pubescence is very coarse and decumbent on the body, but the elytra have besides erect setæ springing from the few acute granules on their surface.

Hab.—Tapajos.

Sub-fam. TORNEUTINÆ.

Large robust insects with exserted and robust mandibles in the males, and a broad apex to the abdomen in both sexes.

Genus COCCODERUS.

Buquet, Rev. Zool. 1840, p. 293; Lacord. Gen. viii. 243.

1. *Coccoderus amazonicus*, n. sp.

Elongatus, parallelogrammicus, rufo-testaceus, elytris (basi excepta) pallidis, maculis eburneis utrinque tribus: capite grosse punctato, genis infra oculos lobo subhamato productis, mandibulis magnis curvatis intus fortiter dentatis; thorace grosse rugoso-punctato, tuberculis atris nitidis, duobus dorsalibus, alteris duobus marginalibus, prope margines anticum et posticum arcto constricto; elytris glabris nitidis, macula eburnea basali, altera discoidali paulo ante medium, alteraque post medium: antennis omnino inermibus.

Long. 13½ lin. ♂.

Although a true *Coccoderus*, this species differs from the definition of the genus given by Lacordaire, in not having the 3-5 joints of the antennæ spinose at the apex. It seems to approach nearest *C. bisignatus*, of Buquet, which, however, has only one ivory-like spot on each elytron. It differs from *C. sexmaculatus* of the same author, in the coarsely sculptured thorax.

Hab.—Tapajos; one example.

Sub-fam. CERAMBYCINÆ.

Genus HAMMATICHERUS.

Serville, Ann. Soc. Ent. Fr. 1834, p. 15; Lacord.
Gen. viii. 255.

1. *Hammaticherus Batus.*

Cerambyx batus, Lin. Mus. Lud. Ulr. Reg. p. 69;
C. Batus, Lin. Syst. Nat. ii. 625.

Omnino fusco-niger, tarsis palpisque solum fulvis; thoraco rugis profundis non interruptis circiter decem transversis breviter cinereo-tomentosis; elytris pube brevissima cinerea vix punctulatis, apice recto truncato, utroque angulo longe spinoso; corpore subtus et pedibus cinereo-tomentosis. Antennæ ♂ corpore triplo longiores, ♀ corpore paulo longiores; utroque sexu articulis 3-6 apice mucrone valido recurvo armatis.

Long. 1 un. 4 lin.—1 un. 7 lin. ♂ ♀.

The Linnæan name is sometimes applied to an allied but distinct species, from South Brazil, which has chesnut-coloured elytra, narrowly edged with black, and golden pubescence on the thorax.* The excellent original description of Linnæus, in which both thorax and elytra are described as fuscous, leaves no doubt which form he described; and besides, at the early date when his description was written, the Entomology of South Brazil was almost unknown in Europe, although large numbers of insects had been received from Surinam.

Hab.—Obydos, Guiana side of Lower Amazons.

The species was rare in the Amazons; found on the boughs of felled trees.

2. *Hammaticherus plicatus.*

Cerambix plicatus, Olivier. Entom. No. 67, p. 40,
pl. xviii. p. 136.

Corpus nigrum, cinereo-argenteo-sericeum; thorace rugis profundis transversis paulo undulatis circiter de-

---

* This species may be thus defined:—

Hammaticherus castaneus.

H. Bato maxime affinis, corpore piceo, subtus cinereo-tomentoso, capite nigro, thorace aureo-tomentoso, rugis profundis circiter 10 transversis, elytris castaneis, cinereo-pubescentibus, vix nitidis, marginibus omnibus nigro-fuscis, apice truncatis utrinque bispinosis.

Long. 1 un. 9 lin. ♂.

Hab.—Brazilia.

com; elytris rufo-castaneis opacis sericeis, nigro marginatis. Antennæ ♂ corpore sesqui longiores, articulo
basali apice infra tuberculo acuto armato, articulis 3-10
apice spinosis, ♀ similes sed paulo breviores.

Long. 1 un. 3 lin. ♂ ♀.

Hab.—Amazons; generally distributed.

3. *Hammaticherus glabricollis*, n. sp.

Brevior, niger, nitidus, subtus (cum pedibus) cinereo
leviter tomentosus; capite glabro, grosse punctato;
thorace rugis latioribus circiter 10 subinterruptis fundo
sparsim punctatis, omnino glabro; elytris apice angustioribus, truncatis, bispinosis, spina suturali multo breviori,
supra crebre punctulatis punctis majoribus interspersis,
fulvo-castaneis nigro-marginatis. Antennæ ♂ corpore sesqui longiores, articulo 4to precedente dimidio
breviori, articulis 5-10 apice infra productis, acutis haud
spinosis.

Long. 8 lin. ♂.

Hab.—Ega; one example only.

4. *Hammaticherus macrus*, n. sp.

Magnus, thorace parvo, spinis lateralibus obtusis; elytris amplis, medio leviter dilatatis, apicem versus rotundatis, prope suturam oblique truncatis et bidentatis;
omnino cinnamomeus fulvo-tomentosus; oculis supra
distantibus; tuberculis antenniforis supra dentatis, antennis (♀) corpore multo brevioribus, articulis 3-10 apice
infra mucronatis vel dentato-productis; thorace rugis
medianis interruptis; elytris subopacis, subtiliter punctulatis.

Long. 2 un. ♀.

Of much less cylindrical form than the other species;
head narrower than the thorax, and the latter only half the
width of the elytra. The elytra are far from being parallel
sided, and are somewhat dilated about the middle of their
length, and broadly rounded towards the apex; in con-
sequence of this form, the apical truncature is confined
to a small portion of the apical margin near the suture,
and the exterior spine is placed about the middle of the
apex; the sutural spine is very small. The colour of the

entire insect is that of cinnamon, a little more ruddy (and rather shining) on the antennæ and legs. Only those rugæ of the thorax are regular which lie near the anterior margin, the rest are much interrupted, and the interstices are here and there thickened; the lateral spines are reduced to smallish conical tubercles. The antennæ are much shorter than the body, the apices of all the joints from 3-10 are produced and acute, but only the third and fourth are really spinous.

The species seems allied to *H. bellator* of Serville, which I have not seen; but it differs in colour and in several points of structure. The anterior haunches and their sockets are much angulate externally, as according to Lacordaire they are in *H. bellator.*

Hab.—Villa Nova (now Villa Bella), Amazons; one example.

Genus CRIODION.

Serville, Ann. Soc. Ent. Fr. 1833, p. 571 ; Lacord. Gen. viii. 270.

1. *Criodion torticolle,* n. sp.

Magnum, parallelogrammicum, depressum, castaneum, fulvo-griseo dense subtiliter tomentosum; capite vix punctato; antennis (♀) tomentosis, infra ciliatis, supra basin versus setis raris vestitis, articulis apice nullomodo angulatis; thorace quadrato, supra valde inæquali, sulcis brevibus flexuosis torto, lateribus foveolis nonnullis profundis nigris; elytris coriaceis, apice rotundatis, sutura spinosis; femoribus et tibiis intermediis et posticis apice valde spinosis.

Long. 2 un. 4 lin. ♀ .

Closely allied to the type of the genus, *C. tomentosum,* Serv., differing chiefly in the very irregular surface of the thorax, which resembles a cerebral surface in its convoluted elevations and fissures. The antennal joints 5-8 have not their apical inner angles produced, and the antennæ are much less setose altogether than in most of the allied species.

Hab.—Pará.

2. *Criodion rhinoceros*, n. sp.

Magnum, parallelogrammicum, vix depressum, fuscum, fulvo - griseo - tomentosum ; mandibulis suprà medio utrinque cornu valido acuto armatis; thorace transversim quadrato, supra inæquali, plagis elevatis nonnullis politis sulcisque rectis et curvatis; elytris subtilissime coriaceis, vermiculato-rugosis, apice truncatis et utrinque bispinosis; pedibus robustis, femoribus intermediis et posticis apice bidentatis.

Long. 2 un. ♂.

Notwithstanding the very remarkable armature of the mandibles, this species is evidently a true *Criodion*, all other parts of structure agreeing with the typical species of the genus. The horn-like processes arise from the upper edge of the organs about the middle, are nearly as long as the mandibles themselves, and incline towards each other, crossing at the apices; together with the broad corrugate cheeks, they give to the head of the insect, viewed in profile, a curious resemblance to that of a Rhinoceros. The thorax is relatively much broader than in other species, and the irregular surface is marked in the middle with grooves forming a large trilobed figure, with the lobes directed towards the head.

Hab.—River Tapajos. I beat an example out of a tree in the forests near the mouth of the Tapajos, in 1852.*

Genus SPHALLENUM, nov. gen.

This genus is formed for the reception of certain species allied to *Criodion*, which differ from that group in having the sockets of the intermediate haunches

* The following is an undescribed species of *Criodion*, differing in the nearly smooth thorax from the more typical forms :—

Criodion hirsutum.

Elongatum, angustatum, fuscum, fulvo-griseo-hirsutum, pilis thoracis et elytrorum decumbentibus. Caput rugosum, vertice inter oculos tuberculo elongato. Thorax quadratum, lateribus paulo rotundatis et rugoso-tuberculatis, dorso sublævi, punctis grossis et tuberculis lævibus tribus notato. Elytra eroberrime punctulata, apice rotundato, sutura solum spinosa, supra pilis sparsis decumbentibus vestita. Femora intermedia et postica apice unispinosa, tibiis apice extus valde spinosis. Acetabula intermedia extus paulo hiantia.

Long. 1 un. 7 lin.; lat. elytr. 4½ lin.
Hab.—Bahia (a Dom. Reade captum).

closed exteriorly. The closure is not produced by the elongation of the outer branch of the mesosternum to meet the corresponding part of the metasternum, but by a small prominence or tubercle at the anterior edge of the latter. The form of body is more cylindrical, and the derm more naked than in *Oriodion*, and there is a striking difference in the antenniferous tubercles, which are contiguous to each other, and form, in fact, a short transverse bicuspid ridge between the roots of the antennæ. The intermediate tibiæ have a spine externally at their apices, which character distinguishes the genus from *Xestia*, where the tibiæ are unarmed.

I believe *Cer. setosus*, of Germar, belongs to this genus.

1. *Sphallenum puncticolle*, n. sp.

Elongatum, subcylindricum, nigro-fuscum, sparse setosum, antennis thorace scutello lateribusque pectoris fulvo-griseo-tomentosis; thorace punctis magnis discretis impresso; elytris castaneis, subtiliter punctulatis, apice utrinque bispinosia, femoribus medio rufo-castaneis.

Long. 1 un. 2 lin.—1 un. 8 lin. ♂ ♀.

Differs from the following species in the separated punctures of the thorax, and in the dense and fine tomentose clothing of the same member. I should have taken it to be the *Oriodion castanopterum* of Erichson, if there had been any allusion in that author's description to the tomentose thorax. It is also allied to *Sph. setosus*, of Germar; but differs in wanting the erect yellow hairs on the elytra, mentioned in that author's description, and in the red femora. The elytra have only very minute, almost microscopic bristles in the punctures.

Hab.—Upper and Lower Amazons; generally found in repose on the leaves of trees in the forest.

2. *Sphallenum femorale*, n. sp.

Oriodion castanopterum, Erichson, in Schomburgk's Reise, iii. 572 (?).

Elongatum, subcylindricum, nigro-fuscum, sparse setosum, antennis scutello lateribusque pectoris fulvo-griseo-

tomontosis; thorace nudo, grosse et confuse rugoso-
punctato; elytris castaneis, subtiliter punctulatis, apice
utrinque bispinosis; femoribus medio læte rufis.

Long. 1 un. 2 lin.—1 un. 6 lin. ♂ ♀.

Agrees with Erichson's description of *Criodion castanopterum* in all points, except the broad clear red ring
round the middle of all the femora. It is possible,
therefore, that Erichson's species may form a third and
distinct one of this group.

Hab.—Upper and Lower Amazons; in the same situations as *S. puncticolle.*

3.   *Sphallenum tuberosum*, n. sp.

Minus elongatum et vix convexum, nigro-fuscum, glabrum, nitidum, antennis scutello lateribusque pectoris
leviter tomentosis; capite thoraceque impunctatis, hoc
tuberibus magnis circa 13 notato, toto lævi, polito;
elytris vix punctulatis, apice utrinque bispinosis; pedibus
piceo-rufis, femoribus medio et apice tibiisque basi fuscis
exceptis.

Long. 1 un. 2 lin.

Hab.—Tapajos.*

Genus Xestia.

Serville, Ann. Soc. Ent. Fr. 1834, p. 16; Lacord.
Gen. viii. 271.

Restricted to those species which have the intermediate
sockets quite closed, and the middle and posterior femora
and tibiæ without spines at the apex.

---

* *Criodion erythropus* (Lucas, in Voy. de Casteln. p. 187, pl. xi. f. 6),
from South Brazil, will, according to the views of Lacordaire (Gen. viii.
271, note), which I have here adopted, require to be separated from
Criodion, on similar grounds to those on which *Sphallenum* is instituted.
The genus may be termed *Butherium*, with the following characters:—

BUTHERIUM, nov. gen.

Corpus oblongum, nudum. Caput tuberculis antenniferis basi late
separatis; antennis articulo 4to (utroque sexu) haud 5to breviori. Femora apice simplicia, tibiis intermediis et posticis apice extus spinosis.
Acetabula intermedia anguste hiantia.

Type. *B. erythropus*, Lucas, *loc. cit.*

Xestia nigropicea, n. sp.

X. spinipenni (Serv.) proxime affinis; differt colore piceo-nigro polito; capite ut in *X. spinipenne* pone oculos constricto; antennis (♂) corpore paulo brevioribus, articulo primo apice intus producto-angulato, 5-11 valde serratis, ultimo precedenti triente longiori et fere diviso apud divisionem angulatim producto; thorace grossissime sparsim punctato, dorso punctis in rugis transversis sitis, plaga mediana lævi; elytris longe bispinosis, supra piceo-nigris, unicoloribus nitidis, haud coriaceis, subtilissime sparsim punctulatis; femoribus pectore abdominoque (partim) rufo-piceis.

Long. 11 lin. ♂.

Hab.—Pará.

2. Xestia brevipennis, n. sp.

X. spinipenni (Serv.) affinissima, corpore (præcipue elytris) distincte breviore robustiore; nigro-picea, elytris obscure castaneis, distincte coriaceis, subsericeo-opacis, passim punctulatis; capite cum tuberibus antenniferis grosse punctatis; thorace latiori, antice minus angustato, lateribus rectioribus, antice subito constrictis, supra grossissime irregulariter rugoso-punctatis; pedibus rufo-piceis, femoribus apice fuscis; antennis ut in *X. spinipenne* (♂) corpore multo brevioribus, articulo basali apice rotundato.

Long. 1 un. ♂ ♀.

Hab.—Ega.

3. Xestia glabripennis, n. sp.

Subcylindrica, castanea, polita; capite parvo, oculis haud prominentibus; thorace transversim strigoso, disco postice lævi; elytris flavo-castaneis, vix punctulatis, glaberrimis, apice bispinosis; femoribus clavatis; pectoris lateribus tenuiter fulvo-sericeis, mesosterno tuberculato.

Long. 8 lin. ♀.

Hab.—Tapajos.

Distinguished from *X. spinipennis*, Serv., by the small size of the head, and the peculiarly flattened eyes, besides its glabrous integument. The sculpture of the thorax is also entirely different, consisting of a number of distinct and rather fine transverse furrows, which cover the whole surface, leaving only a small space on the hinder part of the disc smooth.

4. *Xestia ochrotænia*, n. sp.

Oblongo-linearis, vix convexa, nigra, elytris castaneis, vitta utrinque ochracea ab angulo humerali usquo ad apicem extensa, antice intus solum angustata.

Long. 1 un. 2 lin. ♀.

Belongs to a group of species of less cylindrical form than *X. spinipennis* and its allies, and having much less robust antonnæ without perceptible difference in length between the fourth and fifth joints. They agree, however, in the closure of the intermediate sockets, and in the spineless apices of the hinder femora and tibiæ, and are, moreover, connected with the typical forms by species showing all the intermediate gradations.

X. ochrotænia is closely allied to *X. lateralis*, Erichs.; judging from the description, there is no difference between them, except the mode in which the yellow vitta is narrowed to the humeral angle. Erichson's words are "vitta laterali antice extus abrupte, intus sensim attenuata." In *X. ochrotænia* the vitta shows the inner gradual narrowing, but the outer edge is perfectly straight. The head and thorax are coarsely punctate-rugose, or scabrous; the elytra are finely coriaceous and punctulate, the apex is rounded, and there is a small spine only at the sutural angle. The ochreous vitta forms a well-defined moderately broad stripe, of equal width throughout, except the narrowing near the base, and not quite touching either the base or the apex; it is moderately distant from the lateral margin, and curves slightly towards the sutural angle. The sides of the elytra near the base have a depressed space rather more distinctly sculptured than the rest of the surface.

Hab.—Upper Amazons.

Genus MELATHEMMA, gen. nov.

Xestiæ affinis, sed antennis gracilibus, filiformibus, elongatis. Oculi magni, lobis inferioribus tubera antennifera superantibus, his valde obtusis sulcatis; collo haud constricto. Antennæ graciles, filiformes, corpore (♂) multo longiores, glabræ, sparsim setosæ; articulo basali brevi, oblongo, tertio elongato, 4to et 6to precedenti brevioribus subæqualibus, cæteris æquilongis, 11mo duplo longiori excepto. Thorax inermis, subquadratus. Elytra subcylindrica, apice inermia. Prosternum arcuatum, acetabula extus angustissime emarginata; mesosternum planum, acetabula intermedia anguste aperta. Abdomen glabrum, apicem versus attenuatum. Pedes breves, inermes, femora compressa subclavata, tarsi articulo primo 2ndo 3ioque conjunctim breviori. Corpus subcylindricum, politum, sparse hirsutum.

This genus is formed for the reception of a species which is closely allied to *Xestia* in its principal characters, but differs greatly from it in facies, and in the long slender filiform and non-tomentose antennæ, which, in the male (the only sex I know), are longer by one-half than the body, and have a short oblong (not conical) basal joint. The head is not constricted behind the eyes, the antenniferous tubercles are very obtuse, the upper edges being rounded; and they are separated from each other at their bases by a narrow portion of the forehead. The thorax is scarcely broader than the head, and of subquadrate outline, glabrous, with fine transverse striæ. The elytra are quite unarmed at the apex.

1. *Melathemma polita*, n. sp.

Subcylindrica, nigra, polita, sparsim griseo-hirsuta; elytris vittulis duabus ochreis vel omnino nigris punctatis; scutello tomento griseo fimbriato; thorace transversim subconfluanter rugoso, disco tri-tuberculato, tuberculo mediano elongato, lateralibus rotundis.

Long. 9½ lin. ♂.

Of the two male examples which I obtained of this species, one is wholly of a glossy deep black colour, and the other has on each elytron two short ochreous vittæ, one very short and linear near the middle of the disc,

and the other much longer on the posterior part of the
elytron. The antennæ are of a shining black, or pitchy-
black throughout, fringed beneath with longish hairs in
their basal part, and rather more densely clothed with
hairs in their apical portion. The elytra are naked and
glossy, except near the base, where there are numerous
very long, gray, erect hairs. The body beneath is very
glossy, except the sides of the meso- and meta-sternum,
which are finely tomentose. The elytra are rather
thickly punctured throughout.

Hab.—Ega.

Sub-fam. HESPEROPHANINÆ.

Genus HESPEROPHANES.

Mulsant, Col. de France, Longic. p. 66; Lacord.
Gen. viii. 275.

1. *Hesperophanes amazonicus.*

Obrium Amazonicum, White, Cat. Longic. Brit.
Mus. p. 240.

Oblongo-linearis, fusco-castaneus, passim griseo-pu-
bescens; capite exserto thoraceque subcylindrico rugoso-
punctatis; elytris punctatis, linea indistincta elevata;
antennis articulo 3io triente 4to longiori.

Long. 5½-8 lin. ♂ ♀.

I do not know on what grounds Mr. White placed this
species in the genus *Obrium,* to which it bears very little
resemblance. All the characters are those of the typical
Hesperophanes, with the exception that the head is more
exserted, with a more convex neck, and the thorax more
elongate. The thorax is, however, essentially of the
same form as in *Hesperophanes,* being dilated and
rounded at the sides anteriorly. The whole insect is of
a reddish-brown colour, and covered with rather coarse
erect grayish pubescence; the elytra are uniformly punctured, with a faint raised line from shoulder to apex;
the head and thorax are coarsely rugose-punctate, or
scabrous. The abdominal segments are normal in both
sexes; the apical ventral plate being truncate in the
♂, and rounded in ♀. The antennæ are of the length
of the body in the ♂, and two-thirds the length in the
♀, with the third joint about one-third longer than the
fourth, and much shorter than the fifth.

Hab.—Santarem; taken flying into houses at night.

Genus ANOPLOMERUS.

Thomson, Classif. des Ceramb. p. 249; Lacord. Gen.
viii. 279.

Anoplomerus gracilis, n. sp.

Cylindricus, rufo-testaceous; thorace rotundato-ovato,
linea abbreviata discoidali elevata nigra; elytris utrinque
maculis eburneis duabus geminatis fusco-cinctis, una
paulo ante medium, altera inter medium et apicem,
maculaque fusca ad angulum suturalem; pedibus præcipue
femoribus elongatis.

Long. 6 lin. ♂.

Apparently closely allied to *A. globulicollis* (Buquet),
but very much smaller. Head opaque, sometimes with
a black spot on the occiput; thorax ovate, with sides
equally rounded, surface opaque, owing to the minute
sculpture, centre with a short elevated line covered by a
black spot. Scutellum black. Elytra linear, of same
width as the thorax; apex narrowly sinuate-truncate,
with each angle of the truncature briefly spinous; surface
granulate-punctate, with an erect dark bristle arising
from each puncture, the sculpture much weaker near the
apex; each elytron has two geminate elevated ivory
spots, one at one-third, the other at two-thirds the length,
and both encircled by a dusky ring; each spot is divided
into two by a line of coarse punctures, and the inner
portion is shorter than the outer; the latter, also, is more
elevated, forming part of an elevated line extending
down the elytron. There is a dusky spot within the
sutural apex, which is connected by means of an indistinct dusky line with the dark ring of the posterior
ivory spot. The legs are elongate, especially the femora,
the posterior pair extending much beyond the apex of
the elytra; the knees are black.

Hab.—River Tapajos; also Cayenne.

2.   *Anoplomerus brachypus*, n. sp.

Elongatus, testaceo-rufus; thorace oblongo-ovato,
grosse punctato, maculis quatuor nigris transversis
alteraque utrinque ad marginem anticum; elytris apice
unispinosis, maculis eburneis elongatis utrinque tribus,

una basali, alteris duabus paulo post medium; pedibus brevibus, robustis.

Long. 8 lin. ♂.

Of less cylindrical form than the preceding, the elytra tapering towards the apex, and each prolonged there into an elongate black spine; the thorax is oblong, rounded in the middle, and very closely covered with large punctures, or foveæ, giving a reticulate appearance; lying across the middle are four black spots, beside one on each side on the anterior margin. Elytra coarsely punctured, and with minute punctures on the interstices between the larger ones; setose, the apical third nearly smooth and shining; the basal eburneous spot is large and oblong, bordered with black behind; the two posterior spots consist of a smaller inner one, and a much larger outer one, the smaller a little in advance of the other, and separated distinctly from it; they are edged with black before and behind. The legs are short and stout, the hind femora not reaching, by a long way, the apex of the elytra; the knees are black.

This very distinct species occurred only at Pará.

Genus OPADES.

Lacordaire, Gen. viii. 288.

1. *Opades vittipennis*, n. sp.

Elongatus, cinnamomeo-fuscus, pube subtili sericea vestitus; elytris oblongis, vix convexis, suturâ et vittis utrinque tribus obscurioribus notatis.

Long. 1 un. 3 lin. ♂.

Differs from *O. costipennis*, according to the descriptions of Buquet and Lacordaire, in its broader and less cylindrical form, and in the colour of its fine dense pubescence, which in *O. costipennis* is " greenish-gray," and in our species is of a dingy brown, or cinnamon-brown hue. Both species have two elevated and almost spiniform black tubercles on the disc of the thorax. The dark vittæ of the elytra lie along the interstices of the costæ, and are distinctly seen only in certain lights.

Hab.—Ega.

Genus CHLORIDA.

Serville, Ann. Soc. Ent. Fr. 1834, p. 31; Lacord. Gen.
viii. 289.

1.   *Chlorida festiva.*

Cerambyx festivus, Lin. Syst. Nat. ii. 623.

This common and well-known tropical American insect
is generally distributed throughout the Amazons region.
I found it frequently at night, especially at sugar smeared
on palings to attract moths.

2.   *Chlorida curta.*

Thomson, Archives Entomologiques, i. 288.

Similar to *Chl. festiva;* but different in the markings
of the thorax, and in the distinct sharply-elevated costæ
of the elytra, especially the lateral one, which extends
from the humeral callus to near the apex, where it joins
the two inner ribs.   The antennæ are black, with the basal
joint red.   The upper surface of the head is black.   The
thorax is dark red, with a very broad vitta on each side,
and a central spot or stripe, very much wider on the fore
margin than on the hind, black; the surface of the thorax
is uneven and coarsely sculptured, as in *Chl. festiva.*
The elytra are somewhat shorter relatively than in *Chl.
festiva,* and besides the strong elevation of the ribs, offer
a differential character in the thick punctuation of all the
basal portion.   Body beneath and legs red.

Long. 10 lin. ♀ .

The species offers a very remarkable feature, unnoticed
by its original describer, in the apex of the abdomen
(in the ♀ at least) being greatly dilated and swollen;
the edge of the last ventral segment is straightly truncate, but the pygidium, or last dorsal segment, is rounded,
slightly sinuate in the middle and on each side.   This
feature forms the chief character of Lacordaire's "Groupe
Torneutides;" and it is a further instance of the instability of diagnostic characters in the *Longicornia,* that an
isolated member of a distinct group should show it in so
high a degree of development.

Hab.—Ega.

8. *Chlorida fasciata*, n. sp.

Angustata, capite thoracoque supra fusco-olivaceis, grosse punctatis; elytris viridibus, basi fasciaque dentata obliqua ante medium flavis.

Long. 8 lin. ♂.

Narrower than *Chl. festiva*. Head and thorax above dark olive-brown, coarsely punctured. Antennæ black, with joints one and two, and the base of the third, pitchy-red. Elytra glaucous-green, with a spot in the middle of the base, and an oblique belt of spots, beginning in a long line from the shoulder and terminating on the suture before the middle, pale yellow; the costæ are three in number on each elytron, the two inner alone united before the apex. Body beneath, and legs, red; prothorax with a dusky belt before the coxæ.

Allied to *Chl. denticulata*, Buq., differing in the situation of the yellow marks of the elytra.

Hab.—St. Paulo, Amazons.

Sub-fam. EBURIINÆ.

Genus STYLICEPS.
Lacordaire, Gen. viii. 291.

1. *Styliceps sericatus*.

Ocragenia sericata, Pascoe, Trans. Ent. Soc., 2 ser., v. 16 (1858).

Ocragenia amazonica, Thoms. Classif. des Ceramb. p. 210 (1860).

Styliceps sericans, Lacord. Gen. viii. 202, note (1869).

"Læte rufo-fulvus, vix nitidus, prothoracis tuberculis disci, elytrorum apice summo, femorumque spinis apicalibus nigris; pectore abdomine elytroque singulo vittis duabus longitudinalibus aureo-sericeis." (Lacord.)

Long. 1 un.—1 un. 2 lin. ♂ ♀.

Distinguished from the genus *Ceragenia*, to which it bears a great general resemblance, by the sectional character of the coarse granulation of the eyes. The thorax is glossy red, with deep transverse furrows in front and behind, and the intermediate space covered with rounded smooth tubercles, two of which, in the middle, are black. Both sexes have the elevated tubercle on the crown which has suggested the name of the genus.

Hab.—Upper Amazons; also Cayenne.

Genus EBURIA.

Serville, Ann. Soc. Ent. Fr. 1834, p. 8; Lacord. Gen. viii. 293.

1. *Eburia longicollis*, n. sp.

Elongata, angustata, fulvo-ochracea; thorace angustato, lateribus acute spinosis (antice haud tuberculatis), disco tuberculis duobus elevatis conicis acutis nigris, supra haud profunde punctato-rugoso, rugis undulatis transversis; elytris fortiter punctatis, vitta prope suturam minute rugoso-punctata opaca, tertia parte apicali lævi, subtiliter flavo-pubescentibus cum setis longioribus nonnullis ejusdem coloris, maculis eburneis duabus elongatis geminatis, una basali, altera apud medium, antice et postice nigro-marginatis, apice bispinosis; pedibus elongatis, femoribus linearibus, apice nigris, intermediis et posticis longe unispinosis.

Long. 10 lin. ♀ .

Resembles the species of *Eburodacrys* in form, and in the elongate femora, but has no trace of the groove along the third and fourth antennal joints, which is the chief character that distinguishes *Eburodacrys* from *Eburia*. The sculpture of the thorax consists of large shallow punctures, forming on the disc short very irregular transverse furrows. The elytra have the basal two-thirds thickly covered with circular punctures or foveoles, but near the suture these are replaced by a minute sculpture, rendering that part opaque, the apical third is smooth, or with very slight punctuation; the pubescence is fine, and of a golden yellow, with a few scattered longish bristles of the same colour; the apex only is glossy; the ivory spots are somewhat elongate, and the pairs of which each consist do not differ notably in relative length.

The species is evidently allied to the true *E. 4-maculata* of Linnæus, which, however, according to the description in the "Systema Naturæ," is destitute of the lateral thoracic spines.

Hab.—Ega.

2. *Eburia costulata.*

Elongata, depressa, fusco-cinnamomea, flavo-griseo tomentosa; thorace haud distincte punctato, transverso, depresso, sex-tuberculato, tuberculis duobus utrinque lateralibus duobusque disci, omnibus nigris et subæque conicis;
antennis rufescentibus, infra usque ad apicem densissime
ciliatis; elytris punctatis et utrinque bicostulatis, apice
longe unispinosis, maculis eburneis duabus geminatis,
una basali minus elongata et æquali, altera pone medium
majori et inæquali macula externa multo majori; pedibus
rufo-flavis, femoribus apice fuscis, intermediis et posticis
bispinosis, spinis interioribus paulo longioribus.

Long. 11 lin. ♂.

Distinguished by its depressed form, and the two
distinct costæ of the elytra, which pass through the
ivory spots, but do not reach the apex; the latter with
only one elongate spine. The colour is a light tawny-
brown, with the antennæ and legs rather yellower; the
antennæ are remarkable for the long and dense fringe of
hairs which extends nearly to the apex; the rest of the
antennæ has a shorter pile. The thorax is depressed,
without punctures apparent through the rather close
ashy tomentum; the two lateral tubercles, that of the
middle and that near the anterior angle, and the two
tubercles on the disc, are all black and nearly equally
prominent. The twin spots composing the basal spot of
the elytra are similar in form, the exterior a little the
longer; but the middle spots are very unequal; they are
level on their front edge, but behind, the exterior one
passes the other by one-third its length, and they are
edged with black at both ends.

Hab.—Ega.

3. *Eburia unicolor,* n. sp.

Elongata, subcylindrica, rufescens, pube tenui fulva
vel aureo-fulva vestita; antennis articulo basali antice
sulcato; vertice tuberculo obtuso erecto; thorace transverso, aureo-tomentoso, supra et infra foveolis grossis
insculptis, dorso tuberculis obtusis duobus, lateribus
utrinque bituberculatis, tuberculis omnibus concoloribus;
elytris passim punctulatis, absque maculis eburneis, lateribus anguste nigro-marginatis; pedibus rufis, femoribus
intermediis et posticis apice bispinosis. Antennis ♂
articulo 1mo penultimo sesqui longiori.

Long. 1 un.—1 un. 4 lin. ♂ ♀.

Distinguished from all other *Eburia* as yet described
by the total absence of ivory-like spots from the elytra.
A small oblong smooth callus, which exists in the middle
of the basal margin, may be taken as the sole vestige of
these characteristic spots, but this is rufous, like the
rest of the elytra. The insect is, nevertheless, a true
Eburia, and is, in fact, very closely allied to the common
E. octoguttata (Germ.) of South Brazil, having the same
coarse punctuation or pitting of the surface of the
thorax, and a similar but rather more elevated tubercle
on the crown of the head, like the genus *Styliceps*.* The

* *Eburia octoguttata* exists in some collections under the name of *E.
didyma* of Olivier. This must be wrong, as Olivier's insect, according to
his description, has no lateral spines to the thorax, like *E. 4-maculata* of
Linnæus, and is probably a West Indian species. A fine undescribed
species of the *octoguttata* group is the following:—

Eburia maculicornis, n. sp.

Robusta, elongata, postice attenuata, supra minus convexa. Caput
nigricans, fulvo-tomentosum, vertice tuberculo elevato obtuso. Antennæ
(♂) corpore duplo longiores, dimidio basali infra griseo-ciliatæ; con-
dylis rufis, articulo basali antice sulcato, nigro, apice extus macula rufa,
2ndo nigro, cæteris testaceo-rufis, apice nigris. Thorax transversus, nigricans, fulvo-tomentosus, supra grosse punctatus, tuberculis duobus minus
elevatis, lateribus medio unispinosis. Elytra postice gradatim attenuata,
apice truncata et bispinosa; supra dorso deplanata, apud latera declivia,
fere lævia, fulvo breviter pubescentia, maculis parvis flavis vix elevatis,
haud eburneis, utrinque quatuor, apud basin duabus discretis, externa
minori, alteris duabus paulo pone medium etiam discretis, interna minori.
Pedes rufo-testacei, validi, femora compressa, apice nigra, intermedia et
postica bispinosa.

Long. 1 un. 3 lin. ♂.
Hab.—Brazilia merid.

Another undescribed species differs from all others known to me in the
peculiarly short and thick basal joint of the antennæ. I name it after
the skilful Entomological traveller, Mr. Rogers, who has recently discovered it in South Brazil.

Eburia Rogersi, n. sp.

Elongato-oblonga, capite thoraceque vix elytris angustioribus; fulvo-
rufa, elytris pallidioribus. Caput genis infra productis subspinosis; tuberibus antenniferis apice extus acutis, productis. Antennæ (♂) corpore
duplo longiores, infra longe ciliatæ, articulo basali brevi, crasso, basi extus
dilatato, subauriculato, antice concavo, articulo 3tio supra subcanaliculato.
Thorax transversus, grosse et dense punctatus, supra tuberculis nigris duobus elevatis, lateribus spina mediana rufa. Elytra opaca, passim punctata, breviter setosa, bicostulata, apice bispinosa, spinis nigris, externa
multo longiori; supra macula flava eburnea elongata basali, band elevata,
extus et postice nigro-marginata, alteris duabus pone medium multo
longioribus, linearibus et bene separatis, externa duplo interna longiore
et hanc antice superante, haud nitidis, antice et postice nigro-marginatis. Pedes unicolores; femora intermedia et postica spinis duabus
brevibus nigris, interna majori.

Long. 1 un. 2 lin. ♂.
Hab.—Santa Fé, Minas Geraes. A Dom. Rogers capta.

colour of tho derm is tawny-rufous, and this is covered
by a fine and close tawny pubescence, which is of a silky
golden-yellow hue in fresh specimens. On the head and
thorax this pubescence is tomentose, but on the elytra
and the under-surface of the body, it consists of very
fine short hairs. There are no long erect bristles, as in
many other species, but the antennæ have the usual
fringe underneath the basal joints. The narrow black
lateral margin to the elytra occupies the groove formed
by the upturned lateral edge. It exists also in *E. octoguttata*, but is here rendered more conspicuous by the
light tawny-reddish hue of the surface.

Hab.—Pebas, Upper Amazons; also Venezuela, where
it was taken by Mr. Goering, in the neighbourhood of
Lake Valencia.

Genus EBURODACRYS.

Thomson, Classif. des Ceramb. p. 288; Lacord. Gen.
viii. 296.

Distinguished from *Eburia* by the more abruptly clavate form of the anterior femora, and especially by the
grooved third and fourth joints of the antennæ. The
middle and hind femora are more elongate and slender,
and always terminated by a single elongate spine.*

* Lacordaire gives the glabrous surface of the body as one of the
distinguishing characters of *Eburodacrys*, but *E. sexmaculata, E. strac-
guttata* (Thoms.), *E. longipilis* and others here described, are clothed
above with very long hairs. The following fine large species also is thickly
clothed with long erect hairs.

Eburodacrys caeca (Dej. Cat.), n. sp.

Hujus generis species maxima, fulvo-rufa, pilis elongatis fulvis erectis
vestita. Caput grosse punctatum. Antennæ articulo basali crasso (♀),
grosse punctato, antice concavo ; articulis 3io et 4to sulcis haud profundis-
Thorax subquadratus, grosse densissime rugoso-punctatus, medio linea
glabra, lateribus spina valida nigra antice linea nigra connexa, dorso
tuberculis validis conicis nigris duobus instructus. Elytra dense punctata,
postice læviora, nitida, apice oblique truncata et bispinosa, utrinque
maculis eburneis elongato-ovalibus geminatis duabus, una basali macula
externa dimidio minore, altera pone medium macula externa duplo
majore, antice et postice internam superante, omnibus maculis nigris,
lanceolato-terminatis. Pedes minus elongati, femora apice nigra, intermedia et postica longe unispinosa.

Long. 1 un. 3 lin. ♀.

Hab.—Cayenne.

1. *Eburodacrys megaspilota.*

White, Cat. Longic. Brit. Mus. p. 95, pl. iii. f. 4.

Elongata, testaceo-rufa; thorace angustiori, supra leviter transversim rugoso, medio spatio elevato lævi, spinis dorsalibus duabus validis, lateralibus duabus magnis, nigris; elytris apice transversim truncatis, spina suturali minima, marginali longissima, supra grosse punctatis, parce setosis, apice sublævibus, macula magna rotundata eburnea basali, alteris duabus magis elongatis pone medium antice contiguis postice divergentibus, interiori oblonga, exteriori duplo longiori leviter curvata; pedibus valde elongatis, gracilibus, femoribus apice nigris, intermediis et posticis unispinosis.

Long. 9 lin. ♂.

Mr. White suggested that this species might form a new subgenus near *Holacanthus* (*Nyssicus*, Pasc., Lacord.); it is, however, a true *Eburodacrys*, and, perhaps, the most typical of the genus.

Hab.—Ega.

2. *Eburodacrys longipilis*, n. sp.

Elongata, subcylindrica, testaceo-rufa, pilis longissimis sparsis passim hirsuta; thorace grosse transversim punctato-rugoso, spina laterali acuta nigra, antice cum tuberculo anteriore linea nigra indistincta connexa, dorso tuberculis nigris duobus, interdum in linea nigra postice continuatis; elytris dense punctatis, postice sublævibus, macula elongato-ovata eburnea basali, alteris duabus pone medium magis elongatis, exteriori paulo longiori, antice conjunctis postice divergentibus, apice transverse truncatis et extus unispinosis; pedibus minus elongatis, femoribus nigris unispinosis.

Long. 7-8 lin. ♂ ♀.

Allied to *E. puella*, Newman, but apparently distinct. In two of my specimens (from Cayenne) there are two indistinct black lines on the thorax, posterior to the black dorsal tubercles, but in the third (from Ega) these are absent. The two median ivory spots of the elytra commence exactly together at their anterior extremity; they have there a triangular black spot common to both;

at their hind extremity, each has a longer triangular black spot. There is a short fulvous pubescence on the elytra, besides the longer hairs.

Hab.—Ega; also found at Cayenne.

3. *Eburodacrys hirsutula*, n. sp.

E. longipili valde affinis, differt maculis eburneis elytrorum posticis paulo magis separatis, interiori antice exteriorem superanti; testaceo-rufa, pilis longissimis sparsis passim hirsuta; thorace grosse transversim punctato-rugoso, spina laterali acuta nigra, dorso tuberculis duobus nigris; elytris dense punctatis, postice sublævibus, macula elongato-ovata eburnea basali, alteris duabus pone medium haud longioribus, exteriore paulo magis retrorsa; pedibus elongatis, femoribus apice haud nigris, intermediis et posticis unispinosis.

Long. 6¼ lin. ♂.

The elytra, as in *E. longipilis*, have a short fulvous pubescence, besides the longer hairs; the apices are unispinose, with a black streak proceeding from the spine. The thorax has no black lateral streak, and the legs are entirely unicolorous. The posterior spots of the elytra are not longer than the basal one, but are a little more pointed.

Hab.—Santarem, Amazons.

4. *Eburodacrys rufispinis*, n. sp.

Elongata, sublinearis, fulvo-testacea; thorace subcylindrico, spinis lateralibus parvis vix conspicuis fulvis, tuberculo laterali antico nigro, supra transversim rugoso et tuberculis obtusis rotundatis nigris, medio plaga elongata elevata; elytris glabris, grosse punctatis, apice sublævibus, macula eburnea oblonga basali, alteris duabus contiguis pone medium, exteriori paulo longiori; spinis apicalibus, geniculis spinisque femorum nigris.

Long. 7-8 lin. ♂ ♀.

Also closely allied to *E. longipilis;* differs in its glabrous surface, having but very few long hairs, except on the antennæ and legs, and wanting entirely the short pubescence. The lateral spines of the thorax are very small and acute, which gives the thorax a more cylindrical appearance. The ivory spots are margined before and behind, as usual, with black spots; the basal spot is oval; the posterior ones are close together, very little more elongate than the basal one, and the exterior is distinctly posterior to its companion in front, but is much longer and broader behind. *

Hab.—Ega.

5.   *Eburodacrys sexmaculata.*

Cerambix 6-maculatus, Oliv. Entom. No. 67, p. 47, pl. xv. f. 108; *Stenocorus 6-maculatus*, Fabr. Ent. Syst. I. ii. 295.

Elongata, testaceo-fulva, pilis longis sparsis hirsuta et breviter pubescens; thorace spina laterali brevissima nigra cum linea nigra connexa, supra grosse punctato-rugoso, bituberculato lineisque duabus abbreviatis dorsalibus nigris; elytris dense punctatis, apice sublævibus, utrinque maculis ovatis tribus bene separatis flavo-eburneis; spinis et geniculis nigris.

Long. 7½-9 lin. ♂ ♀ .

Var. 1.   Thorace supra tuberculis solum nigris et lateribus spina maculaque nigris haud nigro-lineatis. *Hab.*—Pará, Amazon. sup.

Var. 2.   Elytrorum maculis duabus posticis eburneis magis minusve postice distantibus, interdum pro parte parallelis. *Hab.*—Amazon. sup., Venezuela.

Var. 3.   Geniculis concoloribus, spinis solum nigris. *Hab.*—Pará.

Var. 4.   Spina laterali thoracis obsoleta. *Hab.*—Ega.

In a large series of this species before me, there are no two specimens exactly alike. The posterior spots

* A species closely allied to *E. rufispinis* is—
Eburodacrys variapilla, n. sp.

Ab *E. rufispini* differt corpore pilis raris hirsuto, maculis duabus eburneis posticis late separatis divergentibus, exteriore magis retrorsa. Testaceo-rufa, vertice macula nigra; thorace spina laterali parva et tuberculo antico rufis, supra valde transversim rugoso, tuberculis duobus obtusis nigris, modio spatio elevato; elytris fortiter confluenter punctatis, apice levibus, macula oblonga eburnea basali, alteris duabus pone medium bene separatis, maculis elongatis nigro-terminatis, exteriore haud longiore, prope medium interioris incipiente; spinis geniculisque nigris.

Long. 8¼ lin. ♂ .
Hab.—Cayenne.

especially vary much in relative position. In specimens which agree with the type of Olivier, the third spot is far from reaching the level of the apex of the second. I find this character only in specimens from Cayenne, Pará, and South Brazil; but the South Brazilian differ in other points, and perhaps merit specific separation. In other examples, the third spot at its base is nearly or quite level with the apex of the second. This form occurs with the type at Pará, and is the prevalent form on the Upper Amazons. Lastly, examples occur in which the third spot is so much advanced, that it is parallel with the second for about one-fourth their respective lengths. Such examples are furnished by Venezuela and the Upper Amazons. These approach in the position of the spots *E. longipilis* and the allied species, but *E. sexmaculata* is a larger and more robust insect; and besides, I have not yet seen specimens in which the posterior spots are quite contiguous.*

Sub-fam. SPHÆRIINÆ.

Genus NYSSICUS.

Pascoe, Trans. Ent. Soc., 2 ser., v. 17; Lacord. Gen. viii. 314.

1. *Nyssicus quadrinus*, n. sp.

Minus elongatus, depressus, testaceo-fulvus, nudus; capite crebre punctato; thorace lateribus breviter obtuse tuberculato, supra inæquali plagiatim punctulato, linea mediana elevata, macula nigra apud marginem anticum, altera ad marginem posticum; elytris apice unispinosis, angulo suturali nullo, supra sparsim setosis punctulatis

---

* The following may be added to the now numerous list of species of this genus:—

Eterodacrys arcifera, n. sp.

Elongata, gracilis, depressa, fulvo-testacea; thorace elongato, transversim punctato-rugoso, linea dorsali lævi, spina minuta laterali; elytris apice truncatis, spina laterali elongata obliqua nigra, suturali minuta fulva, supra punctatis nitidis, sparsim longe pilosis, linea eburnea recta basali, alteris duabus longioribus pone medium, interiori recta, exteriori duplo longiori, arcuata.; femoribus elongatis gradatim incrassatis, intermediis et posticis spina longa nigra armatis.

Long. 8 lin. ♂.

Hab.—Santa Fé, Minas Geraes. A Dom. Bogars capta.

nitidis, maculis eburneis utrinque duabus ovatis, una
(antice nigro-marginata) paulo ante medium, altera (paulo
exteriori) pone medium, macula humerali nigra; geniculis fuscis.

Long. 7 lin. ♂.

Hab.—Tapajos.

Genus SPHÆRION.

Serville, Ann. Soc. Ent. Fr. 1834, p. 68; Lacord. Gen.
viii. 315.

1. *Sphærion callidioides*, n. sp.

Depressum, ferrugineum, capite antice, antennis pedibusque nigris; elytris nigris vel ferrugineis, pube subtile
cinerea indutis et sparsim nigro-setosis, punctulatis, apice
unispinosis, inter spinam et suturam breviter sinuatis,
angulo suturale acuto; capite dense punctato; thorace
transverso, lateribus tuberculo lato conico alteroque antico, supra quinque-tuberculato, sparsim punctato; femoribus clavatis; antennis articulis 3-5 unispinosis.

Long. 6¼-7¼ lin. ♂.

Undoubtedly congeneric with the typical form *S.
cyanipenne*, Serv. The species somewhat resembles in
general form and range of colours *Callidium (Phyma-
todes) variabile*, but is broader. It seems to be closely
allied to *Sph. rusticum* (Burmeister), from Uruguay; but
I cannot feel sure of the identity of the two from the
description.

Hab.—Santarem; Tapajos.

Found flying at night in and around houses.

2. *Sphærion ducale*, n. sp.

Magnum, subdepressum, rufo-castaneum, antennis (articulo basali excepto), tibiis, maculis basalibus et marginalibus lineaque suturali nigris; capite inter antennas
subplano; thorace magno, lateribus medio tumido rotundato, supra tuberculis duobus obtusis alterisque linearibus
lævibus, interstitiis rugoso-punctatis, lateribus punctatis
opacis; scutello fulvo-sericeo; elytris apice unispinosis,
angulo suturale obtuso, supra punctatis, pube subtilis-
sima cinerea indutis, macula basali, marginibus (apicalibus
exceptis), sutura usque pone medium et ibidem fascia

interrupta, nigris; femoribus robustis, clavatis; corpore subtus pube sericea cinereo-fulva dense vestito, prothorace opaco lanuginoso punctato.

Long. 1 un. 1 lin. ♂.

Allied to *Sph. procerum*, Erichs. (in Schomb. Reis. Brit. Guiana), differing chiefly in colour, and in the markings of the elytra.

Hab.—Tapajos. I found it in repose on a leaf in the forests.

Genus PERIBŒUM.

Thomson, Syst. Ceramb. p. 245; Lacord. Gen. viii. 318.

According to Lacordaire, this genus is distinguished among the *Sphæriinæ*, by the pedunculate femora, *i.e.*, slender at the base and clavate beyond the middle, by their unarmed apices, and by the integument being glabrous or not clothed with fine pubescence as in *Sphærion* proper. One of the species here described invalidates this definition, as it possesses the abruptly clavate form of the femora characteristic of *Peribœum*, and their bispinose apices, which is distinctive of the genus *Nephalius*.

1. *Peribœum pubescens.*

Cerambix pubescens, Olivier, Ent. No. 67, p. 33, pl. xviii. f. 135.

Minus elongatum, postice attenuatum, rufo-castaneum, nitidum, longe et sparsim griseo-hirsutum; capite et dimidio apicale elytrorum nigris, interdum capite solum nigro; thorace tuberculo valido laterali aliisque quatuor et carina mediana dorsalibus; elytris sparsim punctulatis, punctis piliferis asperatis, apice truncatis et unispinosis.

Long. 4-8½ lin. ♂ ♀.

Common throughout the Amazon region. Varies from clear reddish-chesnut, with the head alone black, to dark chesnut, with the head, thorax, apical half of the elytra, and abdomen, black. The head, basal joint of the antennæ, and underside of the prothorax, are clothed with grayish tomentum.

274 Mr. H. W. Bates *on Cerambycidæ*

2. *Peribœum ebeninum.*

P. pubescenti valde affinis; differt corpore toto aterrimo, politissimo, capite, articulo basali antennarum, et prothorace subtus opacis griseo-tomentosis, sternis lateraliter sericeis.

Long. 9 lin. ♂.

Hab.—Pebas, Upper Amazons.

3. *Peribœum lissonotum*, n. sp.

Angustatum, castaneo-rubrum, nitidum, antennis pedibusque nigris; thorace elongato subcylindrico, lateribus rotundatis, supra lævi; elytris apice sinuato-truncatis, bispinosis, spina suturali brevi; pedibus brevibus, femoribus prope apicem clavatis, intermediis et posticis bispinosis.

Long. 5¼ lin. ♂.

A species belonging to *Nephalius* (Lacord.), by the form of the thorax and the bispinose femora, but to *Peribœum* by its abruptly clavate femora, and especially by its evidently close relationship to *P. pubescens.* The colour is reddish-chesnut, with the antennæ and legs pitchy-black, the derm shining, but trunk and limbs clothed equally with very long and straight hairs. The head is strongly punctured, and naked; the thorax is elongate, rounded, and quite unarmed on the sides, polished on the disc, and with a very strong bi-arcuate transverse furrow near the hind margin; the sides are strongly punctured in patches, and the underside is evenly punctured. Scutellum naked. The elytra are not much wider than the thorax, and relatively not very elongate; the surface is coarsely punctured, except near the apex; the extreme tip, and the spines, are black.*

Hab.—River Tapajos.

* Other species exist which supply still further connecting links between *Peribœum* and *Nephalius*, as defined by Lacordaire; among them is—

Nephalius fragilis, n. sp.

Elongatus, subdepressus, nitidus, pilis sparsis erectis hirsutus, rufotestaceus, antennis, elytris, femoribus basi, tibiis et tarsis pallidioribus. Caput plagiatim punctatum. Thorax elongatus, lateribus medio paulo rotundato-dilatatis, supra antice et postice constrictus et transversim sulcatulus, supra disco omnino lævi. Elytra postice attenuata, apice truncata et bispinosa, spina suturali brevi, ambabus nigris; supra lævia, versus basin solum sparse punctato. Femora distincte clavata, apice bispinosa, spina interiori longiori.

Long. 5¼ lin.

Hab.—Rio Janeiro (E coll. Dom. Rov. Hamlet Clark).

Genus Aposphærion, nov. gen.

Thorax elongatus, angustatus, cylindricus, omnino lævis. Antennæ articulis 3-11 sulcatis, carinatis, et (11mo excepto) apice spinosis. Pedes breves; femora abrupte clavata, inermia.

Differs from all the other genera of *Sphæriinæ* in the form of the thorax, which is greatly elongate, and almost perfectly cylindrical, without a trace of lateral spine or dorsal inæqualities or punctures; it has only a single transverse curved impression near the base. The elytra are nearly twice the width of the thorax at the base, and taper regularly thence to the apex, where they are truncate and bispinose, the sutural spine much the smaller; the surface is nearly impunctate, except at the base, where, for a small space, they are very strongly punctured, and there are only a few long hairs. The chief peculiarity of the genus, however, is in the antennæ, without which I should have hesitated to separate it from *Peribæum*. This consists in the grooves and carinæ, which in the allied genera are confined to the third, fifth, or sixth joints, but are here extended to the apex; all these joints being spinose at the tips, except, of course, the eleventh. The legs are short, and the femora abruptly clavate, and quite unarmed at the tips. The palpi are extremely short, with the terminal joints triangular. The intermediate sockets are closed; the anterior haunches are globular, with the narrow prosternum sunk between them.

Notwithstanding the cylindrical form of the thorax, the genus has not at all the facies of *Ibidion*, a circumstance which arises from the thorax not having the arched appearance characteristic of the *Ibidion* group.

1. *Aposphærion longicolle*, n. sp.

Angustatum, castaneum, politum; capite antice sparse punctato; thorace cylindrico, lævi, prope marginem posticum arcuatim sulcato; elytris postice attenuatis, apice bispinosis, supra lævibus, prope basin aspere punctatis, postice prope suturam punctis nonnullis setiferis.

Long. 4½-7½ lin. ♂ ♀.

Hab.—Obydos, Lower Amazons; on branches of dead trees.

Genus PANTONYSSUS, nov. gen.

Allied to *Nephalius*, as defined by Lacordaire, but differing essentially in the middle and hind femora being linear, or nearly linear, with a single long spine externally at their apices, as in *Eburodacrys*. In *Nephalius* (with which I think *Castiale*, Pascoe, ought to be incorporated, as it offers precisely the same generic characters), the posterior femora are more or less fusiform, or gradually enlarged from the base, and the spines at their apices are always two in number; moreover, when there is an inequality in the length of these femoral spines, it is the interior one which is the longer; quite the opposite of what is seen in *Pantonyssus*. The head, antennæ, and sterna offer no differences. The antenniferous tubercles are united, and form an even elevation across the forehead; the third to sixth or seventh antennal joints are spined, and the third and fourth are grooved and carinate; the basal joint is concave in front.

1. *Pantonyssus Erichsoni.*

Sphærion Erichsonii, White, Cat. Longic. Brit. Mus. p. 108.

Elongatus, angustatus, minus convexus, pilis elongatis erectis griseis hirsutus; capite antico, antennis, pedibus, apiceque elytrorum nigris; thorace elongato, inerme, supra obsolete quinque-tuberculato et transversim rugoso; elytris apice truncatis et bispinosis, spina suturali brevissima, supra irregulariter punctatis, apice sublævibus, punctis nonnullis asperatis; femoribus valde elongatis, linearibus.

Long. 7-7½ lin. ♂ .
Hab.—Parí. *

* A second species of this genus is the following:—

Pantonyssus nigriceps, n. sp.

Elongatus, minus convexus, sparsim flavo-hirsutus, flavo-testaceus, capite articuloque basale antennarum nigris. Caput plagiatim punctatum. Thorax oblongus, lateribus rotundatis, supra obsolete 5-tuberculatus, interstitiis et lateribus sparsim grosse punctatis. Elytra postice attenuata, apice truncata et bispinosa, spina suturali parva, marginali nigra, supra sparsim haud profunde punctata, nitida. Femora linearia, leviter fusiformia, spinis apicalibus fuscis.
Long. 6½ lin. ♂ .
Hab.—Rio Janeiro (a Dom. Rev. Hamlet Clark lectus, prope Tejuco).

Genus ATHARSUS.

Bates, Entom. Monthly Mag. iv. 25 (1867); Lacord.
Gen. viii. 323.

Lacordaire suggests that Burmeister's *Sphærion rusticum* may be closely allied to *Atharsus;* but I have no
doubt, from Burmeister's description, that his insect is
a true *Sphærion,* and near my *Sph. callidioides. Atharsus*
differs in having a slight trace of the antennal grooves
only on the third joint, and in the absence of spines at
the apex of the elytra. The third to fifth joints of the
antennæ have a short spine at the apex. The great
relative length of the maxillary palpi, and the depressed
form of body, with its clothing of excessively fine ashy
pile, show a close relationship to *Sphærion* proper. The
femora are very gradually clavate, and unarmed at the
tips. There is only a feeble trace of carina on the hind
tibiæ.

1. *Atharsus nigricauda,* Bates, *loc. cit.*

Brevis, depressus, rufo-testaceus, vix nitidus, pube
subtili cinerea indutus, haud pilosus, elytris, capite
antice, antennis, quinta parte apicali elytrorum, et pedibus nigris; antennis sparsim subtus ciliatis.

Long. 5 lin. ♂.
Hab.—Tapajos.

Genus TEREPNISSA.

Bates, Entom. Monthly Mag. iv. 25 (1807); Lacord.
Gen. viii. 324.

Elongata, sublinearis, opaca, sparsim hirsuta. Caput
antice brevissimum, palpis maxillaribus valde elongatis:
antennis setaceis, corpore dimidio longioribus, articulis
3-5 unicarinatis, apice unispinosis. Thorax rotundato-
ovatus, lateribus medio angulatis, supra convexus, paulo
inæqualis. Elytra linearia, ante apicem rotundata, deinde
breviter truncata, angulo externo spinoso, suturali haud
producto. Prosternum inter coxas angustissimum, coxis
orbiculatis; mesosternum latum, acetabula clausa. Pedes
elongati, femoribus clavatis, pedunculatis, apice inermibus; tibiis compressis, vix perspicue intus sulcatis; tarsis
brevissimis.

The genus undoubtedly belongs to the *Sphærinæ* by the majority of its characters, although the facies is entirely different, owing to the sub-orbicular form of the thorax; the tibiæ are not perceptibly sulcate.

1.   *Terpnissa listropterina*, Bates, *l. c.*

Nigricans, pube tenui cinerea induta, et sparsim nigro-setosa; antennis (articulis quatuor basalibus exceptis) flavis; capite postice thoraceque rufis, rufo-sericeis; elytris punctulatis, cinereis, basi, lateribus, vittaque mediana abbreviata, nigris.

Long. 5¼ lin. ♂.

Hab.—Tapajos.

Resembles certain species of *Listroptera* in the ashy clothing of the underside, and the markings of the elytra.

Genus MALLOCERA.

Serville, Ann. Soc. Ent. Fr. 1833, p. 567; Lacord. Gen. viii. 320.

1.   *Mallocera amazonica*, n. sp.

Elongata, nigra, pube variante argenteo-sericea vestita, elytris passim breviter nigro-setosis, capite subtus et pectore longe hirsutis; antennis subtus ciliatis, articulis 3-6 carinatis et apice unispinosis; thorace elongato, lateribus medio tuberculo magno, dorso quadri-tuberculato et medio linea impressa; elytris apice oblique truncatis et unispinosis, argenteo-sericeis, fasciis duabus latis indistinctis nigris, una prope basin transversali dentata, altera ad medium obliqua, apice certo situ nigricante; pedibus valde elongatis, robustis, femoribus medio paulo incrassatis, apice inermibus.

Long. 10 lin. ♂.

Much resembling *M. glauca*, Serv., the type of the genus, but differing in the setose elytra, and in the apex of the same having one spine only instead of two. The setæ of the elytra are rather short, black, and rigid, and cover the whole surface with the greatest regularity; the punctures from which they spring are not visible, owing to the dense changeable silky pile with which the surface is clothed. This silvery or pale golden pile seems spread over the whole elytra, but black markings always appear, which vary according to the position in which the insect is held; their most constant form, however, appears to be that of an undulating belt near the base, and an oblique belt (from the suture rearwards towards the margin) about the middle. The under-surface of the body is clothed with a similar silky pile, but the throat and the centre of the breast have a very dense beard of long pale soft hairs.

Hab.—Ega and Pebas, Upper Amazons; two males.

Genus APPULA.

Thomson, Syst. Ceramb. p. 245; Lacord. Gen. viii. 322.

1. *Appula nigripes,* n. sp.

A. laterali et *undulante* (White) differt corpore magis cylindrico elytrisque multo brevioribus; pube nigra argenteo-sericea vestita; thorace cylindrico, antico et postice constricto, medio paulo rotundato, supra æquali, linea dorsali abbreviata glabra; elytris ante apicem rotundatis, recte truncatis, angulo exteriori spinoso, suturali producto, supra sparsim hirsutis et punctatis, medio maculis et fasciis nigris, apice certo situ nigricantibus; pedibus validis, setosis, femoribus paullo incrassatis.

Long. 8 lin.

Resembles much *Mallocera glauca* and *amazonica,* in the silky changeable pile with which it is clothed, and the vague black markings of the elytra, but differs in the long single exterior spine of the middle and hind femora, which in this group is a tolerably stable generic character. The thorax, too, has no trace of tubercles, either on the sides or disc, and in this respect the insect recedes more from the *Mallocera* type than do *Appula lateralis* and *undulans.* The elytra, instead of the short dense black bristles, have a more scanty clothing of fine long erect hairs.

Hab.—Tapajos.

Sub-fam. Piezocerinæ.

Genus HARUSPEX.

Thomson, Syst. Ceramb. p. 221; Lacord. Gen. viii. 326.

1. *Haruspex lineolatus*, n. sp.

H. brevipedi (White)* simillimus, sed antennis elongatis filiformibus, haud serratis. Rufescens, subtus nitidus, supra opacus, breviter sparsim setosus; capite dense punctato; thorace subcylindrico, angustato, lateribus vix rotundato, postice constricto, supra æquali, eleganter reticulato-punctato; elytris truncatis et bispinosis, supra denae subrugose punctatis, medio vitta irregulari nigricante, lineolas duas flavas includente, una prope basin longiore, altera pone medium multo breviore, intus paulo dilatata, lineolis supra lineam discoidalem elevatam sitis; pedibus brevibus, femoribus clavatis.

Long. 5½ lin. ♂.

Although the antennæ have elongate linear joints, unlike the majority of the *Piezocerinæ*, which have the antennæ flattened and serrate, they agree in being grooved and carinate to near the apex, and they are finely and sparsely hirsute above and beneath. The surface of the thorax is closely covered with shallow pits, and has a honeycombed appearance, without any impunctate interval; this character, together with the narrow form of the thorax, only half the width of the elytra, readily distinguishes the species from the common Brazilian *H. brevipes*.

Hab.—Santarem, Amazons.

2. *Haruspex modestus.*

Phyton modestum, White, Cat. Longic. Brit. Mus. p. 289.

Angustatus, cylindricus, testaceo-rufus, opacus, prothorace pectoreque subtus nigris; antennis corpore multo longioribus, filiformibus, vix pilosis; thorace oblongo-ovato, postice valde constricto, supra quadri-tuberculato, subtiliter rugoso et punctato, postice utrinque linea obliqua nigra; elytris apice sinuato-truncatis et breviter bidentatis, supra grossissime punctatis, linea longitudinali elevata, pone medium utrinque fascia obliqua et post hanc macula communi triangulari fuscis.

* *Ozodes brevipes*, White, Cat. Longic. Brit. Mus. p. 218.

Long. 2-3¾ lin. ♂.
Belongs undoubtedly to the genus *Haruspex*, from the grooved and carinate antennal joints and tibiæ.
Hab.—Tapajos.

3. *Haruspex maculicornis*, n. sp.

Cylindricus, fulvo-testaceus; thorace rotundato, vittis discoidalibus duabus; elytris macula triangulari humerali plagaque magna communi postica violaceo-fuscis, apice breviter emarginatis.
Long. 4 lin. ♂.
Of cylindrical form, the head and thorax narrower than the elytra. The head is opaque, yellowish; the antennæ, a little longer than the body, are filiform, not at all serrate, pubescent, grooved and carinate to the tenth joint, the four basal joints chesnut-red, the rest testaceous yellow, with the tips of the joints brown. The thorax forms a somewhat regular oval, and is not much constricted behind, with the groove not strictly marked; the surface is densely and confluently punctate, testaceous yellow or tawny, with a broad vitta on each side of the disc purplish-brown, not touching either front or hind margin. The elytra are rounded, and much narrowed close to the tips, the latter simply notched; the surface is covered with large deep circular punctures; the longitudinal elevated line of the disc becomes wavy near the middle, and then disappears; the colour is testaceous, with a large triangular humeral spot, and a spot occupying the whole apical half, violet-brown, this spot being advanced and rounded on the suture. Beneath, entirely testaceous-yellow, shining; the legs the same, with the tips of the thighs and base of the tibiæ brown.
Hab.—Parú.

4. *Haruspex ornatus*, n. sp.

Brevis, cylindricus, fulvo-testaceus; thorace rotundato, crebre rugoso-punctato, postice profunde flexuoso sulcato, supra utrinque vitta lata abbreviata fusco-violacea; elytris apice oblique truncatis, extus breviter late dentatis, supra crebre grosse punctatis, linea elevata mediana, maculis vel fasciis obliquis abbreviatis fusco-violaceis utrinque tribus, una subhumerali, secunda longiori pone medium, tertiaque transversa subapicali; antennis fili-formibus, breviter pilosis.

Long. 4½ lin. ♂.

Rather shorter in form and more regularly cylindrical than the other species, wholly fulvo-testaceous in body and limbs, except two purplish-brown vittæ on the thorax, and three belts or spots of the same colour on each elytron. The punctuation of the elytra is so large and dense, even to the apex, that the whole surface seems honeycombed.

Hab.—Santarem, Amazons.

5. *Haruspex pusillus*, n. sp.

Parvus, linearis, aparse setosus, ferrugineo-testaceus, elytris vitta suturali indistincta et pone medium inter-rupta flavescenti; capite reticulato-punctato, antennis subfiliformibus, serratis, corpore (♂) paulo longioribus, (♀) paulo brevioribus, articulo basali scabroso, reliquis usque ad apicem sulcato-carinatis; thorace subcupuli-formi, basi fortissime constricto, supra parum profunde reticulato-scabroso, opaco; elytris minus convexis, apice breviter sinuato-truncatis, extus longe dentatis, angulo suturali breviter spinoso, supra grosse punctatis opacis, suturam et apicem versus minus dense, subnitidis: pe-dibus validis, femoribus modice clavatis, tibiis omnibus apice extus spinoso-productis.

Long. 2¼-2½ lin. ♂ ♀.

A curious little species, approaching *Piezocera* in hav-ing the apices of the tibiæ externally produced and acute, but without the sharp exterior edge of these members, which is a distinguishing character of *Piezocera*. The thorax does not differ essentially in form from that of *H. brevipes*, but it has a peculiar appearance, owing to the sides not being rounded, and the usual constriction near the hind margin being unusually strong, with a cor-respondingly deep sulcus; this is, however, much more marked in the ♂ than in the ♀.

Hab.—Santarem, Amazons.*

* There may be added to the above :—

Haruspex simplicior, n. sp.

Elongatus, longe pilosus, testaceo-ferrugineus, elytris et corpore subtus nitidis. Caput, articulus basalis antennarum, et thorax, reticulato-scabrosi,

Genus PYRGOTES.

Bates, Entom. Monthly Mag. iv. 27 (1867).

Corpus elongatum, angustum, capite thoraceque valde angustatis, lævibus, nitidis. Antennæ articulis 3-11 valde explanatis, a basi abrupte dilatatis, apice utroque angulo æque producto. Thorax angustus, cylindricus, postice constrictus, post medium tuberculo laterali. Pedes validi, tibiæ mox pone basin compresso-dilatatæ. Reliqua ut in *Piezocera* (Serv., Lacord.).

Lacordaire, judging from the description alone, concluded that the present genus was identical with *Piezocera*, and even that the species might be the same as *Piezocera bivittata*, Serv. The description of the antennæ of *Pyrgotes* ought to have prevented him from falling into this error. In fact, the form of the antennal joints is entirely different; both the apical angles of the third to the tenth being equally produced and pointed. The genus, in fact, is much more distinct from *Piezocera* than are *Haruspex* and *Gorybia*. Between *Pyrgotes æneus* and *Piezocera bivittata* there is no resemblance of form, and but little of sculpture or colour.

1. *Pyrgotes æneus*, Bates, *l. c.*

Lævis, nitidus, sparsim breviter fulvo-setosus, castaneus, elytris (marginibus angustis exceptis) læte viridi-æneis; thorace nitido, impunctato, nigro-setoso, medio nigricanti sericeo; elytris sparsim punctulatis, apice oblique truncatis, angulis rotundatis.

Long. 3¼ lin.

Hab.—Ega.

opacl. Antennæ articulis a 3io usque ad 10mum serratis, deplanatis, et cum 11mo sulcato-carinatis, sparse ciliatis. Thorax oblongo-ovatus, postice constrictus. Elytra sinuato-truncata, angulo externo longe dentato, suturali acuto, supra grosse sed non dense punctata, nitida. Femora distincte clavata, nitida: tibiæ apice haud productæ nec compressæ, carinatæ.

Hab.—Rio Janeiro (E coll. Dom. Rev. H. Clark).

Genus HEMILISSA.

Pascoe, Trans. Ent. Soc., 2 ser., iv. 238; Lacord. Gen. viii. 327.

1. *Hemilissa sulcicollis*, n. sp.

Elongata, subcylindrica, nigro-castanea, abdomine rufescenti, setis longis hirsuta, nitida; capite grosse sparsim punctato, nitido, tuberibus antenniferis intus vix perspicue productis; antennis corpore longioribus, articulis haud dilatatis sed distincte serratis, basali scabroso, a 3io ad 11mum carinatis; thorace oblongo, grossissime scabroso, sulco medio lato profundo, fundo politissimo; elytris truncatis, extus longe spinosis, angulo suturali recto, supra basin versus asperato-punctatis, punctis seriatis, parte apicali sublævi polita; femoribus abrupte clavatis.

Long. 4½ lin. ♂.

Resembles in form and colouring *Sphærion* (*Peribœum*) *pubescens*; but clearly allied to *Hemilissa gummosa*, the type of the present genus.

Hab.—Ega.

2. *Hemilissa cornuta*, n. sp.

Elongata, breviter pilosa, fusco-violacea, supra opaca, infra nitida, abdomine rufescenti; capite opaco, grosse punctato, tuberibus antenniferis intus valde productis cornutis; antennis opacis, compressis, serratis, articulo basali scabroso; thorace oblongo, postice constricto, sine sulco, medio dilatato rotundato, opaco, reticulato-scabroso et asperato, medio supra nigricanti; elytris apice truncatis, extus unispinosis, supra lineatim punctatis, postice punctis minoribus et magis confusis; pedibus validis, femoribus clavatis scabrosis.

Long. 7 lin. (♀ ?).

A handsome and remarkable species; differing from *H. gummosa* in the total absence of gloss from the elytra, but presenting in a still more marked degree the generic character of pointed inner angles of the antenniferous tubers.

Hab.—River Tapajos.

One example, found on a leaf in the forests of the Cupari.

Sub-fam. IsIDIINÆ.

Genus HEXOPLON.

Thomson, Syst. Ceramb. p. 219.

Antennæ with all the joints linear; femora elongate, linear, or very little incrassate, the intermediate with a long spine on the *inner* side of the apex, the posterior with a long spine on the *outer* side.

Lacordaire unites *Hexoplon* with *Gnomidolon*.

1. *Hexoplon flaveolum*, n. sp.

Angustatum, sublineare, flavo-testaceum, capite plus minusve infuscato vel nigro; elytris fasciis duabus testaceo-albis, prima suturam haud attingente ante medium, secunda integra pone medium, apicem versus interdum infuscatis, apice truncatis et extus unispinosis; toto insecto longe sparsim setoso et nitido; thorace impunctato; elytris seriatim punctatis, apicem versus laevibus, supra medio vix depressis.

Long. 4½–5 lin.

Hab.—Tapajos and Ega.

2. *Hexoplon quincunx.*

Thomson, Physis, i. 162.

Lineare, elytris postice ut in genere *Ctenostoma* (*Cicindelidarum*) valde convexis, longe sparsim griseo-setosum; capite antice flavo, postice sericeo-nigro; thorace testaceo-fulvo, dorso plaga magna postice trifida sericeo-nigra; elytris dimidio basali flavido, figuram magnam X-formem includente, post hanc partem brunneo-rufescentibus, deinde juxta apicem albis, parte antica seriatim punctata, postica subtilissime rugulosa, opaca; pedibus flavo fuscoque variegatis.

Long. 4–5½ lin. ♂ ♀.

Hab.—Ega; found abundantly, concealed in folded leaves of trees in the day-time.

Genus GNOMIDOLON.

Thomson, Syst. Ceramb. p. 219; Lacord. Gen. viii. 380.

Antennæ with all the joints linear; femora moderately elongate, and slightly thickened towards the middle, a long spine only at the apex of the hind femora, on the *outer* side.

1.   *Gnomidolon Clymene.*

Thomson, Physis, i. 161.

Angustum, lineare, testaceo-rufum, longe pilosum, nitidum; capite sparse punctato; antennis unicoloribus; thorace medio modice convexo, polito, interdum plaga dorsali nigra, vel toto nigro-castaneo polito; elytris medio vel dimidio basali nigro-castaneo, macula triangulari marginali et paulo post hanc vitta obliqua albis, margine laterali (prope basin excepto) et apice testaceo-albis; elytris medio depressis.

Long. 3½–4½ lin. ♂ ♀.

Common.   The space between the white triangular spot and the oblique fascia of the elytra is not wider than the white belt; it is sometimes of a darker hue than the rest of the elytra, and looks then like a distinct dark fascia.

Hab.—Ega.

2.   *Gnomidolon rubricolor,* n. sp.

G. Clymeni valde affinis, differt colore ferrugineo obscuriori et spatio inter fascias elytrorum multo latiori; ferrugineum, tibiis tarsisque obscurioribus, sparsim hirsutum, politum; elytris apud medium fasciis duabus obliquis albo-testaceis, prima latiori suturam haud attingente, secunda angustiori integra, apice testaceo-albis.

Long. 4½ lin. ♂.

The two white marks of the elytra constitute two fasciæ, the first not being triangular.   The distance between the two is twice the width of the posterior fascia.

Hab.—Tapajos.

3. *Gnomidolon conjugatum.*

Ibidion conjugatum, White, Cat. Longic. Brit. Mus.
p. 231.

Elongatum, lineare, sparse hirsutum, politum, nigrum, antennis femoribusque flavo-ferrugineis, tibiis articulisque primis antennarum nigro-fuscis; elytris fasciis duabus abbreviatis obliquis albis, margine connexis, apice albis.

Long. 2¾-3¾ lin.

Among the glossy black species with white belts, the present is distinguishable by both belts halting far from the suture; the anterior is twice the width of the posterior, and is of the form of a triangular spot; the dark space between the two is also elongate-triangular.

Hab.—River Tapajos.

4. *Gnomidolon eganum,* n. sp.

G. conjugato affinissimum, differt fascia secunda elytrorum integra et prima haud triangulari.

Long. 4 lin.

Hab.—Ega.

5. *Gnomidolon picipes,* n. sp.

G. conjugato affine, sed minus, et pedibus piceo-rufis facile distinguendum; lineare, nigro-piceum, politum, sparse hirsutum; antennis rufo-piceis, basi obscurioribus; thorace lævi; elytris striato-punctatis, apud medium fasciis duabus testaceo-albis, prima latiori suturam haud attingente, secunda angustiori integra, marginibus lateralibus apiceque testaceo-albis; pectore rufo; pedibus rufo-piceis, femoribus apice longe unispinosis.

Long. 3 lin.

Hab.—Ega.

6. *Gnomidolon humerale,* n. sp.

Lineare, sparse hirsutum, politum, nigrum, antennis piceo-rufis, basi obscurioribus, femoribus rufo-ferrugineis;

thorace lævi; elytris apud medium fasciis duabus albo-testaceis, prima latiori suturam haud attingente, maculaque elongata humerali rufo-ferruginea, apice albo-testaceis.

Long. 4 lin.

Closely allied to *G. conjugatum*, but having a rufous spot on the elytra, extending from the humeral angle to the first white belt; the second narrow white belt does not quite reach the suture; the femora and apical half of the tibiæ are clear rufous.

Hab.—Pará.

7. *Gnomidolon melanosomum*, n. sp.

G. picipedi affine, multo majus, thorace medio inæquali subtuberculato facile distinguendum; nigrum, ebeninum, politum, griseo-pilosum, antennis articulis 5 primis rufo-piceis, reliquis pallidis, pedibus rufo-piceis; elytris apud medium fasciis duabus obliquis albo-testaceis, prima vix latiori suturam haud attingente, secunda integra, apice albo-testaceis, hoc truncato et extus unispinoso; corpore subtus lateraliter argenteo-sericeo.

Long. 6 lin. ♂.

Hab.—Ega.

8. *Gnomidolon subeburneum.*

Ibidion subeburneum, White, Cat. Longic. Brit. Mus.
p. 234.

Lineare, ferrugineo-fulvum, politum, griseo-setosum; capite grosse sparsim punctato; antennis unicoloribus; thorace lævi, postice plaga magna fusca polita, vel toto fusco; elytris medio valde depressis, seriatim punctatis, maculis duabus paulo ante medium, fascia obliqua integra pone medium, et apice late albo-testaceis; femoribus posticis spina valde elongata.

Long. 4-5½ lin. ♂ ♀.

Hab.—Tapajos and Ega; common in folded leaves of trees, in repose in the day-time.

The Ega specimens are bright rusty-tawny; those from the Tapajos are much darker, and are those in which the thorax is wholly blackish-brown above, the apices of the femora and the tibiæ are also of dark hue. The species also occurs at Cayenne, and is there still darker in its colours, the upper-side being black, with rufous lines on the elytra, and the legs partly reddish. In all, the markings are the same; the two whitish spots, which take the place of the anterior fascia, being elongate and nearly parallel, the outer one a little posterior, and near the lateral margin.

9. *Gnomidolon biarcuatum.*

Ibidion biarcuatum, White, Cat. Longic. Brit. Mus. p. 228.

Elongatum, sparse pilosum, politum; capite nigro; thorace lævi, nigro antice et postice fulvo, vel fulvo macula postica antice trifida nigra; elytris depressis, melleo-flavis, apice late albo-testaceis, arcu longo nigro, utrinque a humero usque ad marginem pone medium, suturam haud attingente, maculas pallidas marginales includente, et postice fascia obliqua pallida marginata; elytris lateribus impunctatis; pedibus rufo-fulvis.

Long. 6-7 lin.

Hab.—Tapajos and Ega.

In Tupajos examples only, the thorax is black, with reddish margins.

10. *Gnomidolon simplex.*

Ibidion simplex, White, Cat. Longic. Brit. Mus. p. 230.

Angustissimum, lineare, politum, nigro-piceum, sparse pilosum; elytris testaceo-fulvis, linea arcuata fusca, a medio baseos prope ad marginem lateralem pone medium, suturam haud attingente, maculam marginalem pallidam includente, apice indeterminate albo-testaceis; antennis (articulis 3 basalibus nigro-piceis exceptis) pallidis; pedibus elongatis, fulvo-testaceis, femoribus posticis dimidio apicali nigris.

Long. $3\frac{1}{2}$-$4\frac{1}{2}$ lin. ♂ ♀.

Hab.—Forest of Altar do Chaõ, Tapajos; common.

11. *Gnomidolon dubium*, n. sp.

G. simplici valde affine, differt statura majore, linea
fusca elytrorum postice apud discum terminante; tibiis
intermediis tarsisque nigris; elytris obscurioribus.
Long. 5 lin. ♂.
Hab.—Tapajos.

Genus Octoplon.

Thomson, Syst. Ceramb. p. 218.

Hinder thighs slightly and gradually thickened, neither
linear nor clavate, both the intermediate and posterior
with two short equal or subequal spines at the tip.
Thorax tuberculate, clothed with silvery tomentum.

I think this genus may be better limited to the second
section, as characterized by Lacordaire, Gen. viii. 331,
note.

1. *Octoplon Orpa.*

Ibidion Orpa, White, Cat. Longic. Brit. Mus. p. 227.

" Thoracis dorso monticuloso, antice nigerrimo, lævi,
postice argenteo sericeo; elytris flavescentibus, fascia
subapicali ferruginea, apice albis, mucronatis, lineola
[obliqua] media nigra et arcu nigro basali." (White.)

Long. 5¼ lin.

In the male, the third antennal joint is gradually
thickened, and the fourth very much shorter than either
the third or fifth. The thorax has five distinct large
flattish tubercles. The elytra are polished, and without
punctures, except the widely placed ones bearing long
setæ, and which run in lines; the basal two-thirds are
fulvous, then follows a broad reddish belt, and the apical
part is testaceous yellowish-white; the basal fulvous portion is marked with a blackish curved line enclosing
laterally a paler spot, behind this there is an oblique
dark line, between which and the reddish belt is a paler
fascia. The femora are elongate, the hind ones rather
longer than the elytra, and armed with two short, distinct, nearly equal spines; the legs are pale testaceous-
red, the tips of the hindmost thighs dusky.

Hab.—River Tapajos.

2. *Octoplon polyzonum,* n. sp.

Linear, head and thorax black, the latter very uneven, with the front and hind parts and sides very densely clothed with white tomentum, leaving an opaque portion in the middle, and two posterior tubercles black. Elytra pale tawny-brown, with five dark brown belts, the first very oblique near the shoulder, the second also oblique, much dilated towards the sutural end near the scutellum, the third straight, linear, and entire behind the middle, the fourth a little posterior to the third, much broader, rather paler and slightly oblique, and the fifth transverse, near the tip; behind the fifth belt, the elytra are testaceous-white, and between the third and fourth pale tawny; the surface is shining and impunctate, except a few setiferous punctures arranged in rows. Legs testaceous-tawny, hind thighs rather longer than the elytra, gradually but rather considerably thickened, the tips with two equal projecting broad spines.

Long. 5¼ lin. (♀ ?, third joint of antennæ not thickened).

Hab.—River Tapajos. The fourth antennal joint is much shorter than the preceding and following.

3. *Octoplon tetrops,* n. sp.

Angustatum, thorace antice attenuato et constricto, capite valde exserto, collo distincto, rufo-testaceum ; capite subtiliter ruguloso et sericeo, oculis divisis ; thorace supra inæquali, longitudinaliter elevato, subnitido, plagiatim sparse tomentoso ; elytris postice attenuatis, supra seriatim sparse punctatis, fasciis duabus medianis approximatis, prima prope suturam interrupta, et apice lato albo-testaceis ; pedibus rufis, tarsis pallidioribus, femoribus sublinearibus, posticis apicem elytrorum attingentibus, breviter bidentatis ; antennis et tibiis posticis carinatis.

Long. 4¼ lin. ♀ .

The singular aberration in the form of the head amply distinguishes this species, as well as illustrates in a striking manner the instability of form of the most important organs in a genus of Longicorns; for the species, notwithstanding the division of the eyes into two on each side, and the dilatation of the cheeks behind the eyes, offers all the other characters of the *Ibidiinæ*. The thorax is more narrowed anteriorly than in most other species, and is constricted there, as well as near the hind margin. The antenniferous tubercles are obtuse, but the specimen is a ♀.

Hab.—Tapajos.

4.   *Octoplon unoculum*, n. sp.

Cylindricum, longe pilosum, nigrum, nitidum; thorace flavo-griseo-pubescente, tuberculo discoidali nigro nitido; elytris postice valde convexis, ante medium fascia interrupta alba, dimidio apicali flavo-griseo-pubescente, antice albo-marginato, sparse punctatis, apice truncatis, extus breviter spinosis, spina alba; femoribus piceo-rufis, posticis elytris longioribus, apice distincte bispinosis; antennis piceo-rufis, articulis 2 basalibus nigris, reliquis apice fuscis.

Long. 5¼ lin. ♀.

An elegant species, of cylindrical from, distinguished by the single glossy black tubercle on the disc of the pubescent griseous thorax, and also by the convex posterior part of the elytra. The basal half of the elytra is glossy black, with very few punctures, and ornamented at about the third of the elytral length by a whitish belt, broadest on the lateral margin, and disappearing before reaching the suture; the apical half is covered with a dense griseous pubescence like that of the thorax, the anterior margin of this is advanced and rounded on the suture, and is there edged with whitish, which forms an indistinct, oblique, and entire fascia. The antennæ and tibiæ are carinate, and the hinder femora are nearly linear with distinct apical subequal spines.

Hab.—River Tapajos.

5.   *Octoplon striatocolle.*

Ibidion striatocolle, White, Cat. Longic. Brit. Mus. p. 224.

Elongatum, minus convexum, breviter sparse setosum, nigrum, nitidum; thorace inæquali, transverse forte rugato,

plagiatim argenteo-tomentoso; elytris utrinque ante
medium macula discoidali et longe post medium fascia
transversa marginem lateralem haud attingente fulvis,
sparsim punctatis; femoribus paulo incrassatis, posticis
elytris ♂ paulo longioribus, ♀ multo brevioribus, apice
breviter bispinosis vel dentatis.

Long. 5-6½ lin. ♂ ♀.

Distinguished by the numerous sharp transverse ridges
across the disc and hinder part of the thorax. The basal
joint of the antennæ is remarkably short, and arcuate-
clavate in form. In some specimens the femoral spines
are very short, and do not project beyond the articulation of the tibiæ; the femora also approach the clavate
form; the species stands, therefore, on the confines of
the two groups *Octoplon* and *Ibidion* proper.

Hab.—Pará; also Cayenne, whence I have a specimen
collected by M. Bar.

6. *Octoplon callispilum*, n. sp.

Elongatum, piceo-nigrum, nitidum, sparsim pilosum;
thorace valde inæquali, quadri-tuberculato, et dorso spatio
elevato, plagiatim argenteo-sericeo, impunctato, nitido;
elytris basin versus plaga laterali magna intus rotundata
et fascia lata pone medium antice ad suturam valde angulata testaceo-flavis, sparsim punctatis, apice longe unispinosis; femoribus paulo incrassatis, distincte bispinosis;
antennis nigris, apicem versus sensim rufescentibus.

Long. 7½ lin. ♀.

The thorax is of similar elongate-cylindrical shape to
the allied species, but is rendered unequal both on the
sides and surface, by the sharp tubercles; the centre of
the disc has an elongate flattish elevation, the two anterior
tubercles of the disc are very acute; the pale markings
of the elytra are very large, and leave between them a
black cross-like mark, or rather, the space between the
anterior lateral spot and the posterior broad fascia, forms
a distinct oblique belt of the ground-colour of the elytra,
open to the equally black base by the concolorous suture;
the apex is black. The thighs are not at all clavate, and
are distinctly bispinose, so that the species cannot come
within the definition of *Ibidion*.

Hab.—Pará.

7. *Octoplon calligrammum*, n. sp.

Magnum, elytris haud linearibus, latis, apicem versus rotundato-attenuatis, piceo-nigrum, vix nitidum; thorace inæquali, tuberculato, sericeo-opaco et argenteo-tomentoso; elytris basin versus macula magna laterali subquadrata, et longe post medium fascia lata postice ad suturam indentata fulvis, supra passim punctatis, apice unispinosis; femoribus incrassatis, haud clavatis, apice bidentatis; antennis piceis, apicem versus pallidioribus.

Long. 9 lin. ♀.

A large species, of much less linear figure than usual in this group. The elytra much wider, with the sides rounded, and tapering towards the apex; the thorax is very uneven, and projecting a little in the middle of the sides, but the tubercles are not distinct or acute; the whole surface is silky and sub-opaque. The elytra are thickly punctured throughout, and have short erect setæ; the spots are of very large size and fulvous; the space of ground-colour left between the anterior spot and the posterior fascia, forms a straight belt, owing to the fascia not being advanced on the suture, and the anterior spot being narrowed on the sides instead of widened.

Hab.—Ega.

8. *Octoplon cinctulum*, n. sp.

Minus elongatum, cylindricum, nigro-piceum, griseo-pilosum, vix nitidum, antennis, pectore et pedibus ferrugineis; thorace inæquali, medio paulo dilatato, tuberculo discoidali magno rotundato, sericeo-opaco; elytris punctis setiferis seriatim ordinatis et inter hæc punctulatis, basi rufescentibus, macula lineari transversa versus basin fasciaque angusta integra pone medium testaceo-flavis; femoribus paulo incrassatis, apice bidentatis.

Long. 3¼ lin. ♂.

Of shorter form than its allies, linear or cylindrical. The thorax is silky and opaque, the elytra moderately shining, with very distinct setiferous punctures, and between them punctulate; the yellow marks are a transverse spot across the disc, not far from the base, and a narrow straight fascia considerably after the middle.

Hab.—River Tapajos.

9. *Octoplon polychromum*, n. sp.

Elongatum, minus convexum; thorace valde inæquali, lateribus medio tumido et supra tuberculis quinque magnis, cum capite dense sericeo-tomentoso; elytris dimidio basali rufo, fascia obliqua versus basin (ad suturam late interrupta) testaceo-flava nigro-æneo-marginata, pone medium fascia testaceo-flava obliqua antice nigro-æneo-marginata, spatio apicali nigro-æneo, apice ipso testaceo-flavo, supra nitidis, punctis parvis setiferis; antennis piceis; femoribus rufis, tibiis tarsisque nigris.

Long. 6¼ lin. ♀.

The design and colours of the elytra are much varied. Near the base is an oblique, moderately wide, yellowish fascia, which is far from reaching the suture, and is margined with brassy-black, this latter colour occupying the whole humeral space, leaving a spot of pale rufous only near the scutellum; behind the middle is an oblique and entire fascia, also of moderate but considerable width, margined anteriorly with brassy-black; the whole space behind this is brassy-black, except the white tips; the whole surface is very glossy, and is roughened only by the setiferous punctures, which are very minute.

Hab.—River Tapajos.

10. *Octoplon carissimum.*

Ibidion carissimum, White, Cat. Longic. Brit. Mus. p. 223, pl. v. f. 9.

" Pallide luteum lævissimum; capite, antennarum articulis duobus primis aterrimis; elytris, parte basali pallide lutea, macula alba, et postice fascia transversa, ornata, parte postica aterrima, apice extremo cum spinis albis." (White.)

Long. 4 lin.

Hab.—Pará.

This seems to be the position of this species, which, although taken by me, was not reserved for my own collection.

11. *Octoplon rugicolle*, n. sp.

Lineare, breviter setosum, nitidum; thorace medio elevato et grosse transversim rugato, rufescente, antice cum capite obscuriori, lateribus sericeo-tomentoso; elytris apice unispinosis, angulo suturali producto, supra punctis setiferis sparsis, versus basin aspere tuberculatis, rufescentibus, tertia parte apicali nigra, macula indistincta versus basin alteraque recte transversa suturali communi testaceo-fulvia; pedibus rufis, femoribus incrassatis, apice bidentatis.

Long. 4½ lin. ♀.

Hab.—River Tapajos.

12. *Octoplon thoracicum.*

Ibidion thoracicum, White, Cat. Longic. Brit. Mus. p. 228.

Cylindricum, opacum, sparse setosum, pallide ochreum; thorace opaco, nigro, annulo magno per totam superficiem submarginalem rufo-testaceo et aureo-tomentoso; elytris punctis setiferis asperatis, pallidis, fasciis angustis quinque pallide fuscis, duabus primis prope suturam conjunctis, spatio basali et inter fascias 4tam et 5tam pedibusque rufescentibus.

Long. 4¼ lin. ♂.

The third antennal joint in the ♂ is gradually thickened, and is carinate, without grooves. The femora are much thickened and almost clavate, at the apices distinctly bidentate.

Hab.—River Tapajos.

13. *Octoplon Rutha.*

Ibidion Rutha, White, Cat. Longic. Brit. Mus. p. 227.

Cylindricum, capite rufo-testaceo, cinereo-tomentoso; thorace inæquali, nigro, opaco, cinereo-tomentoso, punctis cinereis consperso, postice testaceo-rufo; elytris unispinosis, supra nitidis, sparsim longe fulvo-setosis, punctis setiferis asperatis, basi fulvo-brunneis, deinde fascia perobliqua a scutello ad medium marginis extensa

testaceo-alba antice nigro-marginata, pone medium fascia
testaceo-alba recta, et prope apicem fascia obscura fusca,
spatio inter fascias primam et secundam nigro, inter
hanc et fasciam tertiam fulvo-brunneo, spatio apicali tes-
taceo-albo; antennis femoribusque fulvo-testaceis, his
nigro-maculatis, femoribus posticis incrassatis, quasi
clavatis, apice bidentatis; antennis articulo 3to carinato
et bisulcato; tibiis posticis vix perspicue carinatis.

Long. 4½ lin.
Hab.—River Tapajos.

14. *Octoplon charile*, n. sp.

Elongatum, lineare, opacum; capite testaceo-rufo, vertice nigro; thorace valde inæquali, tuberculoso, disco
tuberculis minoribus setiferis sparsis, testaceo-rufo, sericeo-tomentoso, margine antico et disco nigris; elytris
sordide testaceo-albis, passim punctulatis, punctisque
setiferis asperatis, plaga magna scutellari alteraque
minore prope apicem castaneis, pone medium fascia obliqua
et paulo infra macula communi suturali nigris; antennis
pallide rufescentibus, articulis 3-5 forte sulcatis et carinatis; femoribus posticis paulo incrassatis, bidentatis,
tibiis haud perspicue carinatis.

Long. 5 lin. ♀.
Hab.—River Tapajos.

Genus IBIDION.

Serville, Ann. Soc. Ent. Fr. 1834, p. 108; Lacord.
Gen. viii. 331.

This genus is here restricted, following Lacordaire, to
those species having the third antennal joint and posterior
tibiæ carinate, and the hinder thighs distinctly or abruptly clavate and simple at their apices.

1. *Ibidion monostigma*, n. sp.

Elongatum, elytris postice gradatim attenuatis, thorace
antice constricto; castaneo-rufum, nitidum; thorace
postice punctulato et sparsim tomentoso, medio dorsi

208 Mr. H. W. Bates on *Cerambycidæ*

transversim elevato et tri-tuberculato; elytris basi inæqualibus, apice unispinosis, punctis setiferis lineatim
seriatis, macula magna oblonga ante medium, paulo
obliqua, ochrea, fusco indeterminate marginata.

Long. 6½ lin. ♂.

Hab.—River Tapajos.

The legs are elongate, the hind femora (in the ♂)
reaching a little beyond the apex of the elytra; they are
distinctly clavate, *i. e.*, the base is slender, and towards
the middle they become rapidly enlarged, so as to form
an elongate club.

2. *Ibidion œdicneme*, n. sp.

Elongatum, supra nigro-piceum; thorace multi-tuberculato, opaco, argenteo-sericeo; elytris unispinosis, nitidis,
punctis setiferis sparsis, macula rotundata utrinque versus
basin, fascia obliqua pone medium alteraque recta apicali
testaceo-albis; corpore subtus pedibusque rufo-piceis;
antennis rufescentibus, basi nigris; femoribus elongatis,
abrupte tumide clavatis.

Long. 6 lin. ♂.

The thorax is very unequal on its sides and surface,
partly caused by the transverse elevation across the
anterior part having seven irregular tubercles; anterior
and posterior to this, the thorax is constricted, and there
are other tubercular elevations behind; the whole surface
is opaque, and clothed with silky tomentum. The white
markings of the elytra are rather broad, and there is a
wide space between the anterior rounded spot and the
posterior fascia. The thighs are abruptly clavate, and
the club tumid or dilated in the middle.

Hab.—St. Paulo, Upper Amazons.

3. *Ibidion rubellum*, n. sp.

I. œdicnemi valde affine, differt semper colore pallide
fulvo-rufo, femoribus minus inflato-clavatis, elytris maculis flavis anticis majoribus antice versus basin extensis.
Elongatum, fulvo-rufum; thorace opaco, argenteo-sericeo
tomentoso, medio transversim elevato et multi-tuberculato; elytris macula ovata prope basin antice angus-

tiori, fascia lata obliqua pone medium, apiceque testaceo-albis; femoribus clavatis.

Long. 4-5¼ lin. ♂ ♀.

In the female, the antennal carinæ are with difficulty perceived, being visible only in certain lights, and un-accompanied by a groove.

Hab.—River Tapajos, Pará, Ega.

4. *Ibidion Leprieuri*, n. sp.

I. œdicnemi affine. Elongatum, capite thoraceque opacis, hoc argenteo-sericeo, antice transversim elevato et tuberculoso, disco postice bi-tuberculato; elytris apice extus unidentatis, supra politis, punctis setiferis paucis, obscure fulvo-brunneis, plus minusve fusco-plagiatis, macula magna ovali fusco-marginata versus basin, vitta lata obliqua pone medium (ad suturam angustata), et apice testaceo-albis; antennis pallide rufescentibus, articulo 3io lateraliter vix perspicue carinato; femoribus fortiter clavatis.

Long. 5¼-6 lin. ♀.

Hab.—Obydos, Lower Amazons; also Cayenne.

Received from Paris, under the MS. name here adopted.

5. *Ibidion dilectum*, n. sp.

Elongatum, testaceo-rufum, fulvo-setosum; capite piceo; thorace dorso quinque-tuberculato, argenteo-sericeo; elytris unispinosis, nitidis, parte basali ultra medium fulva, parte apicali nigra, macula magna rotundata versus basin, vitta obliqua pone medium, apiceque testaceo-albis; antennis (articulo basali piceo excepto) et pedibus pallide testaceo-rufis; pedibus elongatis, femoribus abrupte clavatis.

Long. 4¼ lin. ♂.

Hab.—Ega.

6. *Ibidion digrammum*, n. sp.

Elongatum, rufo-castaneum, nitidum; thorace medio valde angustato, medio dorsi bispinoso, postice bituberculato; elytris linea flava paulo obliqua discoidali ante medium.

Long. 6¼ lin. ♂.

One of the species which resemble the genus *Gnoma* in the curious form of the thorax narrowed in the middle. The whole surface is glossy, and of a reddish-chesnut hue. On the middle of the thorax are two small spines or very acute tubercles, and behind, near the hind margin, are two obtuse rounded tubercles. The yellow lines on the elytra are about one-fifth the length of the wing-cases, and lie a little obliquely, the lower end being near the suture.

Hab.—River Tapajos.

7. *Ibidion sulcicorne.*

White, Cat. Longic. Brit. Mus. p. 232.

Elongatum, robustum, ferrugineum; thorace æquali, nudo, opaco; elytris nitidis, passim punctulatis, apice uni-spinosis, toto disco nigricante, linea longitudinali ante medium suturæ parallela, altera pone medium obliqua pallidis; pedibus robustis, femoribus grosse clavatis; antennis (♂) articulo 3io reliquis latiori et cum 4to et 5to carinato et bisulcato.

Long. 6 lin. ♂.
Hab.—River Tapajos.

8. *Ibidion sphæriinum*, n. sp.

Facies *Peribœi* (*Sphæriinarum*), castaneo-fuscum, nitidum; capite grosse confluenter punctato; thorace elongato, medio paulo latiori, subnitido, tenuiter tomentoso, punctulato et punctis nonnullis magnis setiferis; elytris apice sinuato-truncatis, bispinosis, spina suturali minori, supra politis, punctis setiferis seriatim ordinatis, interstitiis sparsim punctulatis; antennis sparsim setosis, articulis 3io et 4to carinatis; pedibus curtis, femoribus valde clavatis.

Long. 4¼ lin. ♀.

In its shining concolorous dark chesnut hue, without pale markings, and its bispinous elytra, this resembles species of *Peribœum* in the sub-family *Sphæriinæ*, but is readily distinguished by the unarmed antennæ. The elytra taper a little from the base to the extremity, the hind thighs are very much shorter than the apex of the elytra, but the specimen is most likely a female.

Hab.—St. Paulo, Amazons.

9. *Ibidion unicolor.*

White, Cat. Longic. Brit. Mus. p. 233.

Parvum, angustum, lineare, castaneo-rufum, nitidum; capite impunctato; thorace cylindrico, æquali, lævi; elytris apice bispinosis, spina exteriori valde elongata obliqua, supra passim punctulatis, haud setosis; femoribus modice elongatis, clavatis.

Long. 2¾ lin. (♀ ?).

Hab.—River Tapajos.

10. *Ibidion lineolatum,* n. sp.

Elongatum, lineare, angustum, setis perpaucis vestitum, castaneum, politum; capite et thorace sericeis, punctulatis, hoc cylindrico, æquali; elytris apice sinuato-truncatis et bispinosis, spina exteriori majori recta, supra politis, punctulatis (apice lævibus), medio utrinque lineola discoidali elevata alba, suturæ parallela, extus late nigro-marginata; antennis articulis 3-5 paullo incrassatis et infra longissime ciliatis; femoribus abrupte clavatis.

Long. 4¼ lin. (♂ ?).

Hab.—Ega.

Genus COMPSA.

Perty, Del. An. Art. Bras. p. 92; Lacord. Gen. viii. 333.

Characters of *Ibidion*, with the exception that the hinder tibiæ have no trace of carinæ. It seems to me preferable to limit the genus to those species in which the third joint of the antennæ is carinate, leaving those in which there is no trace of carina either on the antennæ or tibiæ in the genus *Heterachthes.*

1. *Compsa basalis.*

Ibidion basale, White, Cat. Longic. Brit. Mus. p. 229.

Elongata, testaceo-ferruginea, fusco-variegata, griseo longe setosa; antennis pallidis, articulis 1-4 nigris, 3-4 (♂) incrassatis; thorace dorso tuberculis quinque magnis rugosis nigris, interstitiis argenteo-tomentosis; elytris unispinosis, supra nitidis, tuberculis setiferis sparsim asperatis, disco longitudinaliter depressa, lateribus indeterminate albo-testaceis, medio vitta irregulari obliqua a humeris ad suturam pone medium, maculisque subapicalibus nigris; femoribus clavatis et cum corpore subtus rufo- et nigro-variegatis.

Long. 4-4¾ lin. ♂ .

Closely allied to *Octoplon Ruthu, charile,* &c. (*ante,* p. 297), in form and in the peculiar coloration.

Hab.—River Tapajos.

2. *Compsa histrionica,* n. sp.

Elongata, pallide fulvo-rufa; thorace valde inæquali, multi-tuberculato et tuberculis setiferis consperso, nigro-variegato, opaco; elytris pallide testaceo-fulvis, basi rufescentibus, fascia curvata prope basin, alteris duabus tenuibus undulatis pone medium, et quarta obliqua posteriori nigris, apice unispinosis; antennis pallidis, articulis 4 basalibus nigris; femoribus clavatis, fusco-testaceis, tibiis tarsisque pallidis.

Long. 5 lin. ♀ .

Belongs to the same natural group as *C. basalis, Octoplon Ruthu,* &c., but from the absence of tibial carinæ, coming within the definition of the genus *Compsa.* The thorax is much shorter in comparison with the elytra than in the allied species. The dark markings of the elytra consist of an oblique belt, commencing at the shoulder, and bending down to the suture, not very far from the scutellum; then, beyond the middle follow two slender, parallel, undulate belts, oblique in an inverse direction to the basal belt, and immediately behind these is a fourth belt, oblique in the same direction as the basal one, namely, from the lateral margin backward to the suture; this last belt gradually widens as it approaches the suture. The specimen is a female; in the male the third and fourth antennal joints are probably thickened.

Hab.—Ega.

3. *Compsa quadriguttata.*

Ibidion quadriguttatum, White, Cat. Longic. Brit. Mus.
p. 226.

Angustata, linearis, fulvo-testacea, capite thoraceque griseo-tomentosis; hoc cylindrico, paulo inæquali, haud tuberculoso, lævi; elytris sinuato-truncatis et bispinosis, nitidis, lævibus, punctulis setiferis vix conspicuis, fulvis, macula obliqua lineari laterali prope basin, alteraque simili prope apicem nigris, testaceo-albo-marginatis.

Long. 4 lin. ♀.
Hab.—River Tapajos.

Genus HETERACHTHES.

Newman, Entom. i. 9.

Compsa (part), Lacord. Gen. viii. 383.

Antennæ and tibiæ free from carinæ; the femora are clavate, and destitute of spines at the tip. Notwithstanding these differences of structure, the species offer no peculiarity of facies to distinguish them from *Ibidion*, *Gnomidolon*, and other sub-divisions.

1. *Heterachthes decipiens*, n. sp.

Elongatus, nigro-piceus, nitidus, setosus, antennis femoribusque rufo-piceis; thorace elongato, angusto, polito, lævi, dorso medio tuberculo magno conico; elytris bispinosis, spina suturali minori, supra (punctulis setiferis exceptis) lævibus, macula laterali ante medium fasciaque obliqua paulo post medium testaceo-flavis.

Long. 4½ lin. ♂ ♀.

This species has a very close resemblance to *Gnomidolon melanosomum* (*ante*, p. 288) and the allied species, but differs in the antennal joints being entirely free from carinæ in both sexes; the third joint is perfectly cylindrical. In the ♂, the third to the sixth joints are a little stouter than in the ♀; in the femora, the club is distinct, but not abruptly formed, and the slight projections at the apex are not sufficiently advanced or pointed to be termed spines or teeth.

Hab.—Ega.

2. *Heterachthes corallinus*, n. sp.

Elongatus, postice attenuatus, rufus, nitidus, elytris fascia lata basali alteraque simili apicali nigro-æneis; antennis articulis a 3io ad 5tum et 6to basi paulo incrassatis; thorace paulo inæquali, polito, lævi; elytris apice utrinque in spinam prolongatis, supra punctis setiferis seriatim ordinatis; femoribus clavatis.

Long. 4 lin. ♂.

A brilliantly-coloured and elegant species; distinguished besides by the elytra not being truncate at the apex, but tapering each into a spine.

Hab.—River Tapajos.

3. *Heterachthes involutus*, n. sp.

Elongatus, minus cylindricus, castaneo-rufus, setosus, nitidus; capite dense punctato; thorace minus elongato, medio paulo dilatato, supra plagiatim punctato, linea dorsali abbreviata elevata, vittis duabus nigro-fuscis; elytris apice breviter truncatis, extus unidentatis, supra lævibus (punctis setiferis exceptis), usque ultra medium fulvis, linea arcuata a margine sub humero incipiente et ad marginem ultra medium terminante suturam vix attingente et parte anteriore incrassata fusco-castanea, triente apicali etiam fusco-castanea; femoribus posticis (♂) apicem elytrorum haud attingentibus.

Long. 4-6½ lin. ♂ ♀.

Hab.—River Tapajos.

4. *Heterachthes longipilis*, n. sp.

Elongatus, longe griseo-pilosus, rufo-castaneus, nitidus; thorace supra quinque-tuberculato, tenuiter plagiatim griseo-tomentoso, haud opaco; elytris ante apicem rotundatis, apice sinuato-truncatis bispinosis, spinis subæqualibus, supra (punctulis setiferis exceptis) lævibus, vitta sub-recta a basi prope ad medium, fascia valde obliqua pone medium, apiceque flavo-testaceis; femoribus utroque sexu apicem elytrorum longe haud attingentibus, clavatis; antennis longe pilosis, articulis a 3io ad 5tum (♂) paulo crassioribus.

Long. 4½-5½ lin. ♂ ♀.

Hab.—River Tapajos.

5. *Heterachthes ægrotus*, n. sp.

Angustatus, linearis, pallide testaceus, longe setosus; thorace supra quinque-tuberculato, argenteo-griseo tomentoso, nitidulo; elytris apice oblique truncatis, angulis haud productis, supra nitidis, passim haud profunde punctatis et punctis setiferis conspersis, macula triangulari communi basali alteraque simili subhumerali; pallide rufescentibus, fascia obliqua pone medium, altera recta transversali prope apicem, apicibusque albo-testaceis; femoribus elongatis, gradatim sed distincte clavatis.

Long. 4 lin. ♂.

Hab.—Ega.

6. *Heterachthes sylphis*, n. sp.

Gracilis, linearis, capite thoraceque angustioribus, rufo-castaneus, nitidus; capite fortiter punctato, tuberibus antenniferis obtusis; thorace angusto, elongato, vix inæquali, nitido, postice subtiliter punctulato; elytris subplanis, apice oblique truncatis, angulis haud productis, supra (punctulis paucis setiferis exceptis) lævibus, fascia lata transversa recta ante medium alteraque simili apicali flavo-testaceis; antennis pallide rufo-testaceis, articulo 3io ♂ incrassato; pedibus elongatis, femoribus clavatis longissimis, posticis apicem elytrorum multo superantibus.

Long. 3¼ lin. ♂.

Differs wholly in facies from the majority of the genus which have short femora.

Hab.—River Tapajos.

7. *Heterachthes deliciolus*, n. sp.

Gracilis, linearis; capite piceo-rufo, subnitido; thorace valde inæquali, supra quinque-tuberoso, nigro-piceo, sericeo-opaco, argenteo-tomentoso; elytris apice sinuato-truncatis, angulis paulo productis, supra (punctulis setiferis exceptis) lævibus, nitidis, castaneis, macula magna utrinque basali, fascia lata pone medium ad suturam antice dilatata, apicibusque testaceo-albis, parte basali pallide fulva; pedibus antennisque fulvis, femoribus clavatis, modice elongatis.

Long. 3¼ lin. (♀ ?).

Hab.—Obydos, Lower Amazon.

Genus CYCNIDOLON.

Thomson, Syst. Ceramb. p. 217; Lacord. Gen. viii. 333.

The third and sometimes the fourth antennal joints in the ♂ are greatly thickened, fusiform or oval, and carinate in both sexes; femora abruptly clavate, and the hind pair bidentate at the apex; tibiæ carinate.

1.   Cycnidolon Batesianum.

Ibidion Batesianum, White, Cat. Longic. Brit. Mus. p. 230, pl. vi. f. 6.

Cylindricum, nigro-fuscum; elytris dimidio apicali cinereo-pubescenti, antice fascia tenui testaceo-alba marginato, dimidio basali nitido, macula triangulari laterali ante medium, apice longe unispinosis; antennis pallide rufo-testaceis, articulo 3io a basi gradatim incrassato, 4to ovato (♂); pedibus testaceo-rufis.

Long. 3¼ lin. ♂.

I do not find a female example of this species among my series.

Hab.—River Tapajos and Ega.

2.   Cycnidolon binodosum, n. sp.

C. Batesiano valde affine; differt (♂) articulo 3io antennarum basi pedunculato abrupte clavato, 4to etiam clavato sed pedunculo breviori; elytrorum fascia obliqua albo-testacea a parte cinerea bene distante.

Long. 2⅓ lin. ♂.

Hab.—Ega.

3.   Cycnidolon approximatum.

Ibidion approximatum, White, Cat. Longic. Brit. Mus. p. 231.

A *C. Batesiano* differt articulo 4to antennarum haud inflato, lineari, macula fasciaque elytrorum apud latera approximatis, a plaga postica cinerea distantibus.

♀.   Articulis 3-4 antennarum linearibus.

Long. 3½ lin. ♂ ♀.

Hab.—River Tapajos, Ega, St. Paulo, Upper Amazons.

Genus PHORMESIUM.

Thomson, Syst. Ceramb. p. 217; Lacord. Gen. viii. 335.

Differs from all the preceding genera of *Ibidionina* in the short antennæ, scarcely longer than the body in the ♂, shorter in the ♀. The third and fourth antennal joints and the tibiæ are carinate; the femora are clavate, shorter than the elytra, and bidentate at the apex. The carinæ of the antennæ are sometimes very faint, and scarcely perceptible. The third antennal joint is fusiform in the ♂.

1. *Phormesium melanodacrys.*

Ibidion melanodacrys, White, Cat. Longic. Brit. Mus.
p. 235.

"Parvulum, flavo-testaceum, elytris singulis guttis duabus nigris marginalibus, primâ pone humerum, secundâ ad medium, antennis articulo tertio incrassato." (White.)

Long. 2½ lin. ♂ ♀.

Hab.—River Tapajos.

All the antennal joints are carinate from the third to the tenth. The black "guttæ" of the elytra are linear, lateral, and oblique; the surface is polished, and almost impunctate.

2. *Phormesium albinum*, n. sp.

Elongatum, angustum, nitidum, longe setosum, flavotestaceum; elytris fasciis duabus obliquis testaceo-albis, una versus basin (suturam haud attingente), altera apud medium; thorace medio tri-tuberculato, polito; elytris sinuato-truncatis, extus spinosis; antennis vix perspicue carinatis.

Long. 3 lin. ♀.

Hab.—River Tapajos.

Genus Aphatum, nov. gen.

Corpus lineare.   Oculi laterales, supra valde distantes; tubera antennifera rotundata, late separata.   Antennæ corpore breviores, filiformes, articulis nec incrassatis nec carinatis, quarto brevissimo.   Thorax elongatus, supra subplanus, antice latior, haud constrictus, postice gradatim attenuatus.   Elytra apice utrinque longe bispinosa. Pedes breves, femora clavata, apice bidentata, tibiæ haud carinatæ.

Allied to *Phormesium*, but differing from it, as from all other genera of *Ibidioninæ*, in the form of the thorax, which is wider in front than behind, and has no trace of constriction, except near the hind margin.   The widely distant eyes, and short antennæ, which are exactly filiform, *i. e.* are as thick at the apex as at the base (except the first joint), also distinguish the genus.

1. *Aphatum rufulum.*

Ibidion rufulum, White, Cat. Longic. Brit. Mus. p. 234.

Parvulum, rufo-testaceum, parce pilosum, nitidum; thorace elytrisque (punctis setiferis exceptis) lævibus.

Long. 2¾ lin.

Hab.—River Tapajos.

Sub-fam. Obeiinæ.

Genus Obrium.

Serville, Ann. Soc. Ent. Fr. 1834, p. 93; Lacord. Gen. viii. 361.

1. *Obrium cordicolle*, n. sp.

Pallide flavo-testaceum, passim pilosum, nitidum; capite subtiliter rugoso-punctato; thorace lateribus antice valde dilatato vel tumido, postice constricto, supra polito, lævi, dorso depresso; elytris apice rotundatis, supra (punctis piliferis exceptis) lævibus, punctis, fascia tenui prope basin, secunda antice arcuata pone medium, tertiaque arcuata sensu inverso prope apicem, pallide fuscis; femoribus clavatis.

Long. 2½ lin.

Closely allied in form and colours to an undescribed species from Texas. The pale brown fasciæ of the elytra form very slender lines, the second fascia is arcuate towards the base, the third towards the apex, so that the two together form a large ring on the apical half of the two elytra. The abdominal segments in the ♀ are distorted in a similar way to the European types of the genus.

Hab.—Santarem, Amazons.

Genus ARÆOTIS.

Bates, Entom. Monthly Mag. iv. 26 (1867); Lacord. Gen. viii. 398.

(Charac. emend.). Corpus lineare, tenue. Caput supra inter antennas planum, ultra oculos paulo prolongatum, attenuatum; oculi distantes; palpi breves, apice truncati; antennæ (♂) corpore dimidio longiores, tenues, filiformes, breviter sparsim setosæ, articulo basali elongato, leviter incrassato, articulis 3–11 longitudine subæqualibus, simplicibus. Thorax angustus, elongatus, medio paulo dilatatus, lateribus subtuberculatus. Elytra linearia, plana, apice rotundata. Pedes elongati, graciles; femora abrupte clavata, postica longe ultra apicem elytrorum extensa. Coxæ anticæ globulosæ, exsertæ. Acetabula intermedia extus aperta, mesosterni epimera angusta. Abdominis segmenta ♂ normalia, primo magis elongato.

This genus, the position of which Lacordaire considered doubtful, appears to belong to the group *Obriinæ*, with the species of which it also agrees in facies.

1. *Aræotis fragilis*, Bates, l. c.

Flavo-testacea, opaca, supra nuda, articulis antennarum apice fuscis; capite thoraceque creberrimo punctulatis; elytris alutaceis et passim punctatis; pedibus breviter setosis.

Long. 2½ lin. ♂.
Hab.—River Tapajos.

Genus Dodecosis.

Bates, Entom. Monthly Mag. iv. 27 (1867); Lacord.
Gen. viii. 398.

(Charac. emend.). Corpus subcylindricum. Caput
antice verticale, breve, tuberibus antenniferis elevatis,
supra acutis; palpi brevissimi, apice truncati, vix securiformes; antennæ corpore duplo longiores, filiformes,
robustæ, distincte 12-articulatæ, longe pilosæ, articulo
1mo brevi claviformi, 3io paulo 4to breviori, a 4to ad
12mum æqualibus, simplicibus. Thorax cylindricus, elytris
multo angustior, inermis, antice et postice transversim
impressus. Elytra linearia, apice breviter truncata.
Pedes modice graciles, femoribus paulo incrassatis. Prosternum inter coxas angustum; coxis exserto-conicis,
extus angulatis; acetabula intermedia extus aperta.

This genus seems not to fit into any of the sub-families
or "groupes" into which Lacordaire has divided the
Cerambycidæ. In its completely 12-jointed and excessively short antennæ, and perpendicular forehead, it
is unlike any other genus known to me. In facies,
however, it resembles somewhat the *Obriinæ*, and it
seems less out of place in this sub-family than in any
other.

1. *Dodecosis sapordina*, Bates, *l. c.*

Fulvo-testacea, sub-opaca, longe tenuiter pilosa; antennis (articulo primo excepto) fusco-nigris, articulis basi
pallidis; fronte convexa; thorace lateribus antice breviter
tuberculato, disco paulo elevato; elytris dense punctulatis, sutura margine laterali carinaque dorsali ante apicem
desinente elevatis.

Long. 8½ lin. ♂.

Hab.—River Tapajos.

One example only.

The insect resembles, in its general shape, and to some
extent in the filiform antennæ, certain slender species of
Sapordinæ.

Section B. Eyes finely facetted.

Sub-fam. LEPTURINÆ.

Genus OPHISTOMIS.

Thomson, Archiv. Entom. i. 319; Lacord. Gen. viii. 451.

1. *Ophistomis bivittatus*, n. sp.

♂. Gracilis, postice valde attenuatus, luteo-flavus, breviter setosus; antennis longitudine corporis, a medio usque ad apicem incrassatis, nigris; capite thoraceque nigrobivittatis, crebre punctatis, illo vertice etiam nigro; elytris humeris obtusis, postice incurvatim attenuatis, apice oblique truncatis et bidentatis, supra passim punctatis, punctis setiferis, setis incumbentibus, sutura, margine basali, macula laterali apud medium, fasciisque duabus posticis (secunda apicali), nigris; femoribus apice, tibiis tarsisque nigris; abdomine rufo, coxis posticis et metasterni episterno nigro-maculatis.

♀ a ♂ valde differt. Robustior, humeris multo latioribus, fulvus; antennis dimidium corporis paulo superantibus, articulis 5 basalibus fulvis, reliquis nigris, incrassatis; capite thoraceque nigro-bivittatis, crebre punctatis, hoc linea dorsali lævi; elytris humeris latis rotundatis, postice recte attenuatis, apice oblique truncatis et bidentatis, supra punctulatis et setosis, utrinque sutura, maculis tribus discoidalibus, alterisque duabus minoribus juxta humeros, nigris; pedibus fulvo-rufis, apicibus femorum tibiarumque posticarum, et tarsis nigris.

Long. 7 lin. ♂ ♀.

Hab.—Ega; many examples.

Like all other species of *Ophistomis*, it is found in the forest, slowly flying about the underwood in fine weather, and settling on the slender stems of climbing and other plants; sometimes the species are seen on flowers.

2. *Ophistomis paraensis*, n. sp.

♂. *O. bivittato* valde affinis, differt colore magis fulvo, et elytris apice haud nigro-fasciatis. Gracilis,

fulvus, punctatus, setosus; capite thoraceque nigro-bivittatis; elytris valde attenuatis, sutura, margine basali,
vitta laterali ante medium, fasciaque pone medium, nigris;
femoribus apice, tibiis, tarsis, maculisque pectoris nigris.

♀. Ignota.

Long. 6½ lin.

Hab.—Pará.

3. *Ophistomis ochropterus*, n. sp.

♂. Gracilis, postice gradatim lateribus haud incurvatim
attenuatus, fulvo-ochraceus, flavo-setosus, punctulatus,
haud nitidus; antennis a medio modice incrassatis, nigris;
epistomate, vertice, maculaque magna antico-dorsali thoracis nigris; elytris marginibus apicibusque nigris;
pedibus nigris, femoribus antice subtus flavo-testaceis;
abdomine apice nigro.

Long. 4½-6½ lin.

Hab.—Ega.

4. *Ophistomis melanostomus*, n. sp.

♀. Robustior, postice minus attenuatus, fulvo-testaceus,
subnitidus, epistomate, vertice, maculaque magna antico-
dorsali thoracis nigris, hoc minus crebre, vertice creberrime punctatis; antennis a medio vix incrassatis, piceo-
rufis, basi et apice fuscis; elytris vittis suturali et
marginali (hac in fascia ante-apicali terminata) apicibus
que nigris; pedibus nigris, femoribus dimidio basali
fulvis; abdomine apice nigro.

Hab.—Ega.

Possibly the ♀ of *Oph. ochropterus*.

5. *Ophistomis rubricollis*, n. sp.

♀. Curtus, robustus, postice attenuatus, niger, thorace læte rufo; antennis apicem versus paulo incrassatis; thorace creberrime punctato, opaco; scutello rufo;
elytris margine basali rufescente, crebre punctatis, subnitidis.

Long. 5 lin.

Hab.—Ega.

6. *Ophistomis semifulvus*, n. sp.

♀. Minus elongatus, postice vix attenuatus, fulvorufus; elytris (margine basali maculaque humerali exceptis) nigris; antennis vix incrassatis; thorace crebre punctato, linea dorsali lævi; apicibus femorum, tibiarum et tarsorum nigris.

Long. 4¼ lin.

Hab.—Ega.

7. *Ophistomis albicollis*.

Euryptera albicollis, Pascoe, Journ. Entom. i. 68.

Elongatus, niger, thorace (vitta dorsali excepta) maculaque elongata elytrorum testaceo-flavis; capite testaceo-flavo, epistomate nigro-plagiato; antennis articulis basi, coxis femoribusque plus minusve testaceis.

♂. Gracilis; antennis longitudine corporis, apicem versus vix incrassatis; elytris medio paulo angustatis, ante apicem paululum iterum dilatatis.

♀. Robustior; antennis dimidium corporis paulo superantibus, medio (haud apicem versus) incrassatis; elytris elongatis, parallelogrammicis, ante apicem paululum rotundatis.

Hab.—St. Paulo, Amazons.

The form of the rostrum, and the terminal ventral segment of the male, demonstrate that this species belongs to *Ophistomis* rather than to *Euryptera*.

Genus EURYPTERA.

Serville, Encycl. Meth. x. 688; Lacord. Gen. viii. 454.

1. *Euryptera atripennis*, n. sp.

Curta, postice paululum dilatata, punctata, pubescens, subnitida, fulvo-rufa, elytris (macula humerali excepta) nigris, tibiis tarsisque posticis fuscis; antennis robustis, medio paulo incrassatis; capite antice satis elongato et angustato, epistomate punctato polito; thorace minus dense punctulato; elytris ante apicem paulo rotundato-dilatatis, apice late vix oblique truncatis, extus spinosis, supra linea elevata prope suturam, griseo-pubescentibus subnitidis; pedibus robustis, tarsis posticis curtis, tibiis multo brevioribus; abdomine (segmento basali rufo excepto) nigro, segmento ultimo ventrali apice truncato, angulis longe spinosis.

Long. 4½ lin. (♀ ?).

In the form of the terminal ventral segment, and the shortness and stoutness of the hind tarsi, this species agrees with *E. latipennis*, Serv.

Hab.—Ega.

Sub-fam. Necydalinæ.

Genus Sphecomorpha.

Newman, Entom. Mag. v. 396; *Sphecogaster*, Lacord. Gen. viii. 471.

1. *Sphecomorpha chalybea*, Newman, *l. c.*

Sphecogaster biplagiatus, Lacord., *lib. cit.* p. 472, n. (?).

"Chalybeo-nigra; antennis nigris, articulis 4to et 5to subtus testaceis; elytris vitta subhumerali maculaque dorsali albidis, hac fascia ænea transversa divisa."

Long. 13 lin.

Hab.—Ega.

This bears the closest resemblance to a large wasp of the genus *Epipone*, and was captured by me as such, flying in the forest at Ega; it was only after examination at the bottom of my net, that I found it was not a wasp. Subsequently I saw three examples at myrtle blossoms, but by an unlucky shaking of a branch, missed them all, as they took to flight instantly.

Lacordaire appears entirely to have overlooked Newman's genus. I am inclined to think the species he describes is the same as Newman's.

Sub-fam. Molorchinæ.

Genus Stenopterellus, nov. gen.

Merionœdæ et *Stenoptero* affinis; differt antennis (♀) longitudine corporis, setaceis. Corpus tenue, de-

prossum. Caput exsertum, angustum, antice paulo prolongatum et attenuatum; oculis paulo prominentibus;
palpis brevibus, articulis terminalibus apice attenuatis
truncatis; antennis pilosis, apicem versus attenuatis,
articulis tenuibus, apice intus paulo productis. Thorax
tuberosus, antice valde angustatus et forte constrictus.
Elytra plana, abdomine quarta parte breviora, apicem
versus attenuata et dehiscentia, apice obtuse acuminata.
Pedes pilosi, postici longiores et robustiores; femora
omnia abrupte clavata; tarsi breves. Coxæ anticæ
conicæ, exsertæ. Abdomen (♀) segmento ventrali primo
magno, integro, secundo paulo minori, postice semicirculariter emarginato et longe ciliato, reliquis profunde
depressis.

1. *Stenoptrellus culicinus*, n. sp.

Niger, nitidus; thorace tuberibus octo inæqualibus,
duobus medianis linearibus; elytris pallide ochrois, passim grosse punctatis, linea elevata laterali a medio usque
ad apicem; abdomine segmentis duobus basalibus rufis.

Long. 3¼ lin. ♀.

Hab.—Ega; on flowers.

Sub-fam. Necydalopsinæ.

Genus Sthelenus.

Buquet, Ann. Soc. Ent. Fr. 1859, p. 621; Lacord. Gen.
viii. 494.

1. *Sthelenus braconinus*, n. sp.

S. ichneumoneo (Buquet) affinis, differt thorace antice
nigro, pedibus anticis et intermediis totis rufis. Linearis,
testaceo-rufus, capite, antennis, plaga antica thoracis,
postpectore, femoribus posticis apice, tibiis tarsisque
nigris; elytris pallidis, fascia paulo post medium fusca.

Long. 5-7 lin. ♂ ♀.

Hab.—St. Paulo, Amazons.

Resembling to deception certain common species of
Bracon, Fam. *Ichneumonidæ.* Found flying at mid-day
on low bushes; the limbs, as in *Oxodes* and the *Olyti*, are
extremely fragile, breaking off almost at a touch.

Sub-fam. RHINOTRAGINÆ.

Genus OXYLYMMA.

Pascoe, Trans. Ent. Soc., 2 ser., v. 21 ; Lacord. Gen.
viii. 500.

1. *Oxylymma lepida.*

Pascoe, *lib. cit.* p. 22, pl. ii. f. 3.

Testaceo-flava, glabra, nitida, vertice, antennarum basi,
fasciis duabus elytrorum, maculaque triangulari humerali,
nigris ; elytris acuminatis, fortiter punctatis.
Long. 5½ lin. ♀ .
Hab.—Ega.

2. *Oxylymma telephorina,* n. sp.

Elongata, passim pallide setosa, flavo-testacea, capite
supra maculisque humeralibus posticisque elytrorum
nigris ; capite rostro valde abbreviato, antice et partibus
oris flavis, supra cum collo nigro, grosse sparsim punctato ;
antennis testaceo-flavis ; thorace ovato, postice constricto,
supra antice grosse (♀ minus) punctato ; elytris apice
obtuse rotundatis, angulo suturali paulo producto, supra
dense punctatis, macula humerali per marginem continuata, alteraque transversa prope apicem (interdum
obsoleta), nigris ; postpectore nigro.
Long. 4½-5 lin. ♂ ♀ .
Differing from *O. lepida* in the anterior part of the
head not being prolonged into a muzzle, but moderately
short. Notwithstanding this important difference, it can
scarcely be placed in a separate genus, as almost every
other character of the insect agrees with *Oxylymma;*
it has the same peculiar form of the head, eyes, insertion
of the antennæ, thorax, and legs.
Hab.—Ega.

Genus ERYTHROPLATYS.

White, Cat. Longic. Brit. Mus. p. 201 ; Lacord. Gen.
viii. 511.

Lacordaire doubts whether this genus can belong to the
Rhinotraginæ; but in all essential points of structure it

has the closest affinity with *Rhinotragus*. The middle
sockets are widely open externally, and the mesosternum
has elevated and acute lateral margins, and is vertical
anteriorly. The metasternum is large and inflated.

1. *Erythroplatys corallifer*.
White, Cat. Longic. Brit. Mus. p. 202, pl. v. f. 2.

The figure here quoted will give an accurate idea of
this singular insect, which by reason of its widely-dilated
elytra, coarse sculpture, and bright red and black colours,
becomes the mimetic analogue of *Cephalodonta spinipes*,
Fabr., of the family *Hispidæ*. It is not found, however,
in company with that insect. At least, the few examples I
met with were found on the flowers of a low tree, named
Pitomba, in the neighbourhood of Santarem, whilst the
Cephalodonta was seen only on the foliage of a climbing
plant, generally in great numbers.

Genus Rhinotragus.
Germar, Ins. Sp. Nov. p. 513; Lacord. Gen. viii. 500.

1. *Rhinotragus trilineatus*.
White, Cat. Longic. Brit. Mus. p. 200.

" R. flavus ; antennis, capite, thoracis maculis duabus
dorsalibus, lineis tribus elytrorum, una suturali, pedibus
femorumque basi exceptis, abdomine apice, nigris."
(White.)

Long. 5-5¾ lin.

Distinguished from the typical species of *Rhinotragus*
by its more slender form, the elytra narrowing more towards the apex, and by its slender antennæ; but agreeing with them in the glossy elevations of the thorax, the
swollen lateral rim of the elytra, and the distinctly serrate
antennæ.

Hab.—Villa Nova, Amazons ; on flowers.

Genus Agaone.
Pascoe, Trans. Ent. Soc., 2 ser., v. 22.

Lacordaire unites this genus to *Ommata* ; from which,
it appears to me distinct in the short, slender, filiform antennæ, the short cylindrical thorax, and the much shorter

legs, especially the hind pair. It forms a very natural assemblage of small delicate species, all having the same style of colouring. The thorax is free from glossy elevations, and is thickly punctured.

1. *Agaone notabilis.*

Rhinotragus notabilis, White, Cat. Longic. Brit. Mus. p. 199.

"R. luteus; antennis nigris, articulis 8 ultimis basi flavis; thorace macula magna dorsali nigra; elytris nigris, singulis vitta elongata basali et fascia transversali subapicali sulphureis; pedibus nigris, femoribus basi et apice intus flavis." (White.)

Long. 3¾-4¼ lin. ♂ ♀.

The elytra are broad at the apex, and truncate, with each angle briefly spinous; they are distinctly narrowed in the middle, a little widened behind, and narrowed again to the apex, more so in the ♀ than in the ♂. The basal sulphur-yellow vitta of the elytra is very variable in form, and is sometimes only a rounded spot.

Hab.—Tapajos and Ega.

Found sometimes at flowers, and sometimes hovering in numbers over the trunks of felled trees.

2. *Agaone molorchoides.*

Rhinotragus molorchoides, White, Cat. Longic. Brit. Mus. p. 200.

"R. gracilis, luteus; capite, thoracis macula magna irregulari transversa, elytrorum marginibus nigris; antennis nigris, segmentis 8 ultimis basi pallidis; elytris vitreis." (White.)

Long. 3¼ lin. ♂ ♀.

The elytra are much narrowed, and are slightly dehiscent; the apex obliquely truncate, with the angles scarcely produced; the form of the elytra, and their glassy discs, show an approach towards *Odontocera.*

Hab.—River Tapajos; on flowers.

3. *Agaone colon,* n. sp.

Gracilis, testaceo-fulva, antennis, maculis duabus thoracis, et vitta laterali abbreviata elytrorum nigris; antennis articulis basi testaceis; thorace elongato, lateribus paululum rotundato, grosse reticulato-punctato; elytris angustatis, dehiscentibus, apice recte truncatis, supra lateribus dense et disco sparsim punctatis; femoribus supra (posticis apice), tibiisque apice nigris.

Long. 4 lin. ♂.
Hab.—Pará.

4. *Agaone malthinoides,* n. sp.

Tenuis, linearis, testaceo-flava, aureo-tomentosa, vortice maculaque transversa thoracis nigris; hoc sparsim grosse punctato, spatio dorsali laevi; elytris linearibus, haud attenuatis, grosse et dense punctatis, purpureo-fuscis, fascia ante-apicali testacea; antennis corpore multo brevioribus, fuscis, articulis basi testaceis; pedibus annulo femorali, tibiisque apice fuscis; abdomine apice nigro.

Long. 3 lin. ♂ ♀.
Hab.—Ega; on flowers.

5. *Agaone ruficollis,* n. sp.

Tenuis, linearis, nigra, thorace laete rufo, grosse sparsim punctato; elytris medio angustatis, utrinque carina abbreviata laterali, passim grosse punctatis, apice oblique truncatis, basi fulvis; antennis articulis basi, femoribus basi, tibiisque apice flavo-testaceis.

Long. 3 lin.
Hab.—Ega; on flowers.

Genus OMMATA.

White, Cat. Longic. Brit. Mus. p. 194; Lacord. Gen. viii. 502.

Distinguished from the allied genera by the length (longer than the body in the ♂) and clavate form of the antennae, by the long cylindrical thorax, and elongate hind legs.

1. *Ommata aurata*, n. sp.

Viridi-aurata, metallica, dense fortiter (thorace rugoso)
punctata; elytris haud angustatis, basi excepta cæruleis;
abdomine segmentis 2-4 cupreo-aureis splendidis; antennis nigris; pedibus chalybeis.

Long. 5 lin. ♀.

Hab.—Villa Nova.

I found one example only of this beautiful species.

2. *Ommata smaragdina*, n. sp.

Gracilis, læte viridi-cyanea, elytris macula magna humerali aurantiaca et vitta laterali violacea; capite rugoso-
punctato; thorace antice angustato, supra transversim
rugoso-punctato; elytris apice oblique truncatis, angulo
externo spinoso, supra passim dense punctatis; antennis
piceo-violaceis, basi pedibusque chalybeis.

Long. 5 lin. ♂

A still more richly-coloured species than *O. aurata*.
One example.

Hab.—Ega.

Genus ODONTOCERA.

Serville, Ann. Soc. Ent. Fr. 1833, p. 546; Lacord. Gen.
viii. 503.

1. *Odontocera chrysostetha*, n. sp.

Gracilis, melleo-flava, subtus aureo-pubescens; capite
grosse crebre punctato; thorace cylindrico, dorso paulo
longitudinaliter elevato, grosse et dense punctato, nigro
4-maculato vel bi-vittato; elytris longitudine corporis,
paulo attenuatis, apice recte truncatis, disco politissimis,
lateribus dense punctatis, sutura marginibusque (apice
excepto) nigris; antennis piceo-rufis, nigro-maculatis;
femoribus anticis supra nigro-lineatis, posticis medio
nigro-annulatis; abdomine vespiformi.

Long. 6-6½ lin. ♂ ♀.

Hab.—Pará and Ega.

2. *Odontocera pœcilopoda.*

White, Cat. Longic. Brit. Mus. p. 191.

" Nigra, elytris vitreis pallidis nigro-marginatis; abdominis basi subtus et apice pallidis; pedibus nigris, femorum tibiarumque basi albâ; tibiis posticis pilis nigris hirtis." (White.)

Long. 5¼ lin. ♂.

The above description applies only to the ♂. In the ♀ the abdomen is wholly testaceous-red; in both sexes it is much narrower than the metasternum, but in the ♂ it is exceedingly slender and linear in form. The hind tarsi in the ♀ are pale testaceous. In both sexes the head and thorax are black, and the elytra are one-third shorter than the abdomen. The long black hairs of the hind tibiæ form a brush all round the joint from the middle to the apex.

Hab.—River Tapajos.

3. *Odontocera dispar,* n. sp.

O. pœcilopodœ valde affinis; ♂ differt solum vittâ laterali elytrorum ante apicem desinente; ♀ valde diversa, thorace fulvo, maculâ dorsali posticâ nigrâ; abdomine testaceo-fulvo, segmento 2ndo maculâ laterali nigrâ; tarsis omnibus flavis.

Long. 4¼-5¼ lin. ♂ ♀.

Hab.—Ega; abundant.

4. *Odontocera cinctiventris,* n. sp.

Minus elongata, nigra; capite grosse punctato, nitido; thorace breviori, postice angustiori, rotundato, æqualiter convexo, dense reticulato-punctato, marginibus antico et postico lineaque curvatâ laterali aureo-tomentosis; scutello aureo-tomentoso; elytris abdomine vix brevioribus, grosse et dense punctatis, medio utrinque flavo vitreo, vix punctulato; abdomine vespiformi, segmento basali rufo, reliquis argenteo-marginatis; pedibus nigris, femoribus tibiisque basi testaceis, femoribus gradatim incrassatis.

Long. 4 lin. (♂ ?).

Hab.—Ega and Tapajos.

5. *Odontocera parallela.*

White, Cat. Longic. Brit. Mus. p. 189.

"Pallide flava; antennis nigro-annulatis; thoracis dorso lineis duabus parallelis nigris; elytris apice angustatis, singulis linea marginali nigro-fusca; pedibus posticis subhirsutis, femoribus tibiisque apice fusconigris." (White.)

Long. 3¼ lin. ♂ ♀.

The sexes, which I took *in copulâ*, do not differ in colour, and very little in the length of the antennæ, or form of abdomen. The antennæ are scarcely those of the typical *Odontocera*, being slender, with the 7th-11th joints shortened and thickened, and not at all serrate; they are, however, even in the ♂, decidedly shorter than the abdomen, which character separates the species from *Ommata*, while the thickened apices distinguish it from *Agaone*, to which the species bears some resemblance. The elytra, however, are shorter by one-fourth than the abdomen, and have vitreous discs. The hind femora are abruptly clavate.

Hab.—River Tapajos.

6. *Odontocera mellea.*

White, Cat. Longic. Brit. Mus. p. 188.

"Melleo-flava, antennis nigro-annulatis; elytris corpore multo brevioribus, basi punctato, membranaceo, tunc vitreo, apice attenuato nigro; abdomine subtus medio nigro." (White.)

Long. 5¼ lin. ♂.

White's description applies only to the ♂; the ♀ is totally different in coloration, being sooty-black, with the head and antennæ fulvous, spotted with black, the legs dusky, with the middle part of the femora pitchyred, and the apex of the abdomen yellowish. The abdomen, in most examples of the ♂, is black from the base, with the apex yellow, and the hind femora are black at the base.

The antennæ, in this species, are perfectly filiform, being neither thickened nor serrate towards the apex.

The elytra are shorter by one-third than the abdomen, and widely dehiscent at the suture; the vitreous part does not reach to near the base, which is thickly punctured; the femora are gradually but strongly clavate.

Generally distributed throughout the Amazons; flying about branches of newly-felled trees.

7. *Odontocera punctata.*

Stenopterus punctatus, Klug, Nov. Act. Ac. Cæs. L. C. Nat. Cur. xii. 471, pl. xliv. f. 4.

Nigra, thorace supra læte rufo-coccineo, crebre reticulato-punctato, femoribus posticis basi albo-testaceis; antennis incrassatis et subserratis; pedibus posticis valde elongatis, femoribus clavatis; elytris abdomine triento brevioribus, valde dehiscentibus, disco vitreo-flavis.

Long. 4½ lin.
Hab.—River Tapajos.

8. *Odontocera ornaticollis,* n. sp.

Nigra, læte aureo-tomentosa, antennis pedibus abdomineque testaceo-rufis; thorace magno, ovato, postice multo angustato, supra reticulato-punctato, margine antico et vittis quatuor antice abbreviatis læte aureotomentosis; elytris abdomine paulo brevioribus, angustis, mox pone scutellum dehiscentibus, fulvis, vitreis prope basin, fascia obliqua et margine laterali nigris.

Long. 6 lin. ♂.
Hab.—Santarem; at flowers. A superb species; the hind femora are very gradually clavate.

9. *Odontocera furcifera,* n. sp.

Robusta, flava, nigro-setosa, antennis abdomine pedibusque testaceo-rufis; thorace transversim ovato, litura furcata basi annexa maculaque laterali nigris; elytris abdomine paulo brevioribus, a medio abrupte attenuatis, disco flavo-vitreis, marginibus rufis, basin versus nigris, prope scutellum macula obliqua nigra; antennis brevibus, ab articulo 3io incrassatis; pedibus posticis robustis, longe hirsutis, femoribus valde haud abrupte clavatis.

Long. 6 lin. ♀.
Hab.—River Tapajos.

10.   *Odontocera trililurata*, n. sp.

Flava, vertice fascia nigra; antennis apicem versus incrassatis et serratis, testaceo-rufis; thorace breviter cylindrico, crebre grossissime punctato, marginibus antico et postico, vitta dorsali et fascia mediana (vittam dorsalem haud attingente), nigris; olytris abdomine brevioribus, apicem versus dehiscentibus, flavis, dorso vitreis, sutura marginibusque fuscis, macula sub-basali prope humeros nigra; pedibus testaceo-rufis, femoribus basi albo-testaceis, abrupte clavatis.

Long. 5 lin. ♀.
Hab.—Pará.

11.   *Odontocera fasciata.*

Necydalis fasciata, Oliv. Ent. No. 74, p. 10, pl. i. f. 9.

Odontocera chrysozona, White, Cat. Longic. Brit. Mus.
p. 192, pl. v. f. 5.

White gives the reference to Olivier's figure with a mark of doubt; but on comparing the description as well as the figure, there can be no uncertainty about his species being the same.

It is generally distributed throughout the Amazons, and not uncommon in the dry season, at sweet-smelling flowers. Like all the other species of these beautifully varied and interesting little creatures, it flies nimbly from flower to flower, deceiving the eye of the beholder by its strong resemblance to a wasp.

12.   *Odontocera compressipes.*

White, Cat. Longic. Brit. Mus. p. 191.

"Lutea; capite flavo, gula mento et vertice nigris; antennis ferrugineis articulis 1 et 2 intus nigro lineatis; thoracis margine antica et annulo transverso, in dorso crassiores, nigris; elytris vitreis luteis, basi scabris nigro variegatis, apice oblique truncatis; tibiis posticis apice subdilatatis nigris hirtulis." (White.)

Long. 5½ lin. ♂ ♀.

White's description is drawn up from a female specimen. The ♂ differs in having a narrow black margin to the vitreous central part, which is quite continuous, except for a small space at the apex. The elytra are a little shorter than the abdomen, and dehiscent from the middle; the femora are abruptly and strongly clavate; the antennæ are but very slightly thickened towards the apex, almost filiform. The species, however, is very closely allied to *O. fasciata*, in which the antennæ are very strongly thickened. It is still more closely related to *O. triliturata*, in which the hind femora have no brush-like hairs.

Hab.—Ega, flying about dead trees; also Tapajos.

13. *Odontocera simplex.*

White, Cat. Longic. Brit. Mus. p. 189.

"Nigra, punctata; elytris elongatis vitreis pallidis, margine late sutura anguste nigris; femoribus pedum duorum posticorum basi pallidis; tibiis posticis gracilibus nudis; antennis nigris, articulis 5-8 basi pallidis." (White.)

Long. 4 lin. ♂.

The antennæ from the fifth joint are dilated, compressed, and serrate. The elytra are shorter than the abdomen, dehiscent, and sublinear.

Hab.—Pará.

14. *Odontocera cercerina*, n. sp.

Postice, attenuata, nigra, opaca; capite dense punctato, fronte lineis duabus aureo-tomentosis; antennis brevibus, rufo-piceis, versus apicem valde incrassatis; thorace elytris latiori, rotundato-quadrato, haud profunde sed dense punctato, nigro, opaco, marginibus antico et postico aureo-tomentosis; elytris extus mox pone humeros valde attenuatis, sutura dehiscentibus, abdomine paulo brevioribus, apice subacutis, supra vitta valde curvata abbreviata pallido-vitrea, pone hanc macula transversa alba, plaga magna triangulari circum-scutellari nigra opaca grosse punctata, reliquis nigris opacis, margine basali aureo-tomentoso; corpore subtus nigro opaco, pectore utrinque

fasciis duabus et abdomine vittis quatuor albo-tomentosis ;
pedibus rufo-piceis, posticis elongatis, femoribus gradatim
clavatis, tibiis apicem versus incrassatis, pilosis.

Long. 3½ lin. ♀.

Hab.—Pará and Ega.

15. *Odontocera bisulcata*, n. sp.

Minus elongata, nigra, nitida ; capite rufo-piceo, rostro
elongatissimo ; antennis brevibus, apicem versus incrassatis et serratis, rufo-piceis ; thorace oblongo, lateribus rotundatis, supra late bisulcato, grossissime sub-confluenter
passim punctato ; elytris abdomine paululum brevioribus,
gradatim angustatis, apice oblique truncatis, toto disco a
basin usque ad apicem lævissimo, vitreo, vittam abbreviatam testaceo-albam includente, marginibus grosse punctatis ; pedibus omnibus brevibus, robustis, femoribus
fortiter clavatis, nigris, tibiis posticis incrassatis, haud
pilosis, his tarsisque læte fulvis.

Long. 5 lin.

A remarkable species, which, from the inequalities of
the thorax, might be considered to belong to *Acyphoderes*,
but which differs from that genus in the elongate rostrum ;
in the last-named feature it exceeds all other species of
Odontocera.

Hab.—River Tapajos.

Genus ISTHMIADE.

Thomson, Syst. Ceramb. p. 166 ; Lacord. Gen. viii. 504.

The striking resemblance which the species of this genus
bear to species of *Bracon* (Fam. *Ichneumonidæ*) is increased
by the filiform prolongation of the abdomen in the ♀,
which imitates the ovipositor, and by the yellow and
black wings.

1. *Isthmiade ichneumoniformis*, n. sp.

Nigra, nitida, elytris testaceo-flavis, pectore abdomineque (apice excepto) rufis ; pedibus testaceo-rufis, femoribus posticis late fusco-annulatis, tibiis apice tarsisque
fuscis ; alis flavis, fascia apiceque nigris.

Long. 5-7 lin. ♂ ♀.

Hab.—Ega.

The posterior coxæ are black, and the breast spotted on the sides with black in some examples. The species has a most deceptive analogy to species of *Bracon*. It flies nimbly over decaying branches of felled trees.

Stenopterus braconides (Perty) belongs to this genus. It has recently been sent home by Mr. Rogers from Minas Geraes.

Genus PHYGOPODA.

Thomson, Syst. Ceramb. p. 164; Lacord. Gen. viii. 509.

Distinguished by its long and slender form, short muzzle, and very elongate and clavate hind femora, the tibiæ tufted with long black hairs. The following species agree with the definition as given by Lacordaire, but I am unacquainted with the typical species, *Phygopoda fugax* of Thomson.

1. *Phygopoda albitarsis.*

Stenopterus albitarsis, Klug, Nov. Act. Ac. Cæs. L. C. Nat. Cur. xii. 475, pl. xliv. f. 12.

Nigro-chalybea, pectore et annulis abdominis argenteo-tomentosis; thorace nudo, plagis tribus politis elevatis, interstitiis grosse punctatis; elytris apicem segmenti primi abdominis attingentibus, abrupte attenuatis, acuminatis, plaga discoidali albo-testacea vitrea; pedibus posticis longissimis, tibiis nigro-scopariis, tarsis posticis albis.

Long. 4½-7 lin. ♂ ♀.

Hab.—Ega; also Tapajos.

Sometimes in great abundance at sweet-smelling flowers, and looking like a large *Oulea.*

2. *Phygopoda subvestita.*

Odontocera subvestita, White, Cat. Longic. Brit. Mus. p. 190.

Melleo-flava, aureo-tomentosa; capite nigro, dense punctato; antennis fuscis, basi flavo-testaceis; thorace densæ aureo-tomentoso, plagis tribus parvis discoidalibus elevatis politis; elytris apicem segmenti primi abdominis attingentibus, abrupte attenuatis, obtuse acuminatis, melleo-flavis, disco vitreo concolori, marginibus prope basin fuscis; pedibus flavis, tarsis fuscis, femoribus posticis valde elongatis, gradatim clavatis, clava melleo-flava, tibiis posticis apice nigro-scopariis.

Long. 4–6 lin. ♂ ♀.

Resembles *Ph. albitarsis* closely in form, in the small thorax and subulate elytra; but differs in the less abruptly clavate hind femora. It mimics a pale species of bee of the genus *Melipona*, even to the black hairy tufts of the hind tibiæ.

Hab.—River Tapajos.

Genus ACYPHODERES.

Serville, Ann. Soc. Ent. Fr. 1833, p. 549; Lacord. Gen. viii. 505.

1. *Acyphoderes Olivieri.*

Necydalis abdominalis, Oliv. Ent. No. 74, p. 8, pl. i. f. 5 (?).

Niger, sericeo-tomentosus; thorace oblongo-ovato, dorso depresso, lineis elevatis tribus, interstitiis aureo-tomentosis; elytris apicem segmenti ventris secundi attingentibus, subulatis, flavo-testaceis, vitreis, macula suturali pone scutellum marginibusque punctatis nigris, his juxta humerum lineolam testaceam includentibus; pectore argenteo, abdomine rufescente, sparsim griseo-piloso; pedibus anticis et intermediis nigris, femoribus crassis, basi et apice et tibiis lineis testaceis; pedibus posticis fulvo-rufis.

Long. 8¼ lin. ♂ ♀.

Closely allied to the common Brazilian *A. aurulentus;* differs in its slenderer form, black head, thorax, breast, and four anterior legs, and in the red untomentose abdomen. It agrees well with Olivier's description of *N. abdominalis*, but differs from his figure in the much longer and subulate elytra. If we might assume that the figure is incorrect in this respect, the species would be the one described by him.

Hab.—River Tapajos; also Cayenne, Peru, &c.

2. *Acyphoderes odyneroides.*

White, Cat. Longic. Brit. Mus. p. 196, pl. v. f. 3.

Angustatus, niger; thorace oblongo-ovato, costis tribus elevatis grosse punctatis, interstitiis aureo-tomentosis; elytris apicem segmenti ventris tertii attingentibus, attenuatis, valde dehiscentibus, apice acuminatis, pallide testaceo-fulvis, vitreis, margine prope humeros fusco, haud profunde punctato; abdomine vespiformi, cinctubus quatuor flavis; pedibus gracilioribus, femoribus clavatis.

Long. 7 lin. ♂ ♀.

Found on the flowers of a low tree called *Pitomba*, and bears the most deceptive resemblance to a species of wasp seen on the same flowers. I was never sure whether I had captured the beetle or the wasp, until I had closely examined the insect in the bottom of the net.

Hab.—Santarem.

Genus Tomopterus.

Serville, Ann. Soc. Ent. Fr. 1833, p. 544; Lacord. Gen. viii. 509.

Distinguished from *Odontocera* by the short subquadrate elytra, which do not pass the level of the posterior coxæ; and from *Acyphoderes* by the same character, and by the convex closely punctured thorax.

1. *Tomopterus obliquus,* n. sp.

T. staphylino valde affinis, differt elytris prope suturam oblique truncatis. Niger, opacus; antennis rufescentibus, fusco-maculatis; thorace grosse reticulato-punctato, marginibus antico et postico fasciaque mediana medio interrupta aureo-tomentosis; elytris brevibus, extus et apice rotundatis, prope suturam oblique truncatis, basi fascia rufescente, disco linea obliqua flava; scutello nigro, apice aureo-tomentoso; pectore utrinque aureo-bifasciato, abdominis segmento primo rufo, reliquis aureo-marginatis; pedibus rufescentibus, femoribus basi albo-testaceis.

Long. 3¼ lin. ♂.

Hab.—River Tapajos.

2. *Tomopterus bispeculifer*.

Odontocera bispeculifera, White, Cat. Longic. Brit. Mus.
p. 190.

Niger; thorace rotundato, grosse reticulato-punctato, margine postico scutelloque aureo-tomentosis; elytris apice rotundatis nec truncatis, grosse punctatis, disco macula oblonga sub-obliqua flavo-testacea vitrea; corpore subtus nigro, nitido, epimeris aureo-tomentosis, segmento ventris primo interdum (♀) rufo; pedibus nigris, femoribus posticis interdum (♂) albo-testaceis.

Long. 5-6¼ lin. ♂ ♀.

Hab.—River Tapajos.

3. *Tomopterus larroides*.

White, Cat. Longic. Brit. Mus. p. 177.

Brevis, robustus; thorace valde transverso, convexo, lateribus rotundatis, piloso, punctato, opaco; elytris brevibus, apice recte truncatis, extus valde rotundatis, vitta obliqua obscure flava; abdominis segmentorum marginibus flavo-pilosis; femoribus posticis gradatim incrassatis, supra valde arcuatis.

♂. Supra niger, alisque nigris.

♀. Supra thorace medio flavescente, elytris lateribus castaneis, alis fulvis.

Long. 3-4 lin. ♂ ♀.

Abundant once at flowers; closely resembles a small bee of the genus *Megachile*.

Hab.—Santarem.

EPIMELITTA, nov. gen.

Tomoptero affinis, differt corpore toto piloso, elytris abdominis basin attingentibus, apice valde attenuatis, subacuminatis. Rostrum breve, latum. Antennæ paullo incrassatæ, serratæ. Thorax brevis, valde transversus, convexus, hirtus. Tibiæ posticæ longe pilosæ.

The tapering apices of the elytra, widely dehiscent at the suture, and the pilose body and limbs, distinguish this genus from *Tomopterus*. In facies the species bear very little resemblance to any of the other genera, and, in fact, remind one more of bees of the *Melipona* group.

Molorchus scoparius, Klug, Nov. Act. Ac. Cæs. Nat. Cur. xii. 469, pl. xliv. f. 2, belongs to this genus. It is stated to be found at Cametá, on the Tocantins, where I collected for two months, but did not meet with it.

1. *Epimelitta meliponica*, n. sp.

Obscure fulva, densissime fulvo-pilosa, abdomine fusco-testaceo, nudo, nitido, apice segmenti primi aureo-marginato; elytris fulvo-testaceis, nitidis, basi plaga communi nigra punctata, pilis rufis elongatis dense marginata; femoribus et tibiis posticis pilosis, his intus ante medium et extus versus apicem dense nigro-hirsutis; thorace brevi, lato, dense grosse punctato, nigricante, dorso crista pilorum nigricantium.

Long 6 lin. ♀.

Hab.—Ega.

Flying about decaying trees. Not distinguishable from a common species of *Melipona* when on the wing.

2. *Epimelitta rufiventris*, n. sp.

Nigra, abdomine rufo, opaco, griseo-sericeo, segmento primo tomento griseo dense marginato; thorace sub-globuloso, postice constricto, grosse punctato, dorso spatio transverso lævi nitido, nigro-hirsuto, dorso antico cano-pubescenti; elytris vitta sub-obliqua a basi ad apicem albo-testacea vitrea, intus vitta nigro-velutina marginata, plaga scutellari nitida; pedibus piceo-rufis, griseo-pilosis, femoribus posticis infra barbatis, tibiis intus et extus longe griseo-hirsutis.

Long. 6½ lin. ♀.

Hab.—Ega; on the trunk of a dead tree.

Genus Æcmutes.

Bates, Entom. Monthly Mag. iv. 23 (1867); Lacord. Gen. viii. 511.

(Charac. emend.). Facies *Lycorum* (Fam. *Lycidarum*). Corpus oblongum, depressum, postice dilatatum, elytris carinatis, nigris, ochreo-fasciatis. Caput *Odontoceræ* et *Rhinotragi*, rostro elongato. Thorax antice angustatus, æqualis, haud constrictus. Antennæ breves, medio valde dilatatæ, articulis subserratis, tertio brevissimo transverso. Pedes breves; femora gradatim clavata; tarsi breves. Coxæ anticæ subglobosæ, exsertæ, extus haud angulatæ; acetabula intermedia late aperta; prosterno et mesosterno inter coxas arcuatis.

A re-examination of my specimen of this curious insect confirms the supposition I formerly expressed, that it belongs to the *Rhinotraginæ*; the structure of its sternal pieces agreeing with the definition of the group, as given by Lacordaire.

Æchmutes lycoides, Bates, *l. c.*

Depressus, nudus, opacus, testaceo-rufus, antennis nigris; thorace strigis duabus utrinque nigris, usque ad oculos extensis; elytris macula magna communi pentagona prope basin quartaque parte apicali nigris; tarsis tibiisque nigris, femoribus testaceis, in medio nigris; supra totus creberrime punctatus; elytris linea elevata ab humeris fere ad apicem extensa, hoc late undulatim truncato, angulis truncaturæ ambobus spinosis.

Long. 4¼ lin. ♀.

Hab.—Ega.

Genus PANDROSOS.

Bates, Entom. Monthly Mag. iv. 23 (1867); Lacord. Gen. viii. 510.

Corpus gracile, tenue; rostro quam in *Rhinotrago* breviori et latiori. Antennæ (♀) longitudine corporis, apicem versus crassiores, serratæ; articulus 3ius 4to duplo longior. Thorax elongatus, cylindricus. Elytra linearia, haud angustata, sed pygidium haud tegentia, apice oblique truncata, angulo externo dentato; supra plana, grosse punctata, carina laterali obtusa a medio ad apicem. Pedes elongati, tenues; femora longe pedunculata et clavata; tarsi postici articulo primo reliquis conjunctis longiori. Coxæ anticæ conicæ, exsertæ, extus haud angulatæ; acetabula intermedia extus perparum aperta.

1. *Pandrosos exilis.*

Rhinotragus exilis, White, Cat. Longic. Brit. Mus.
p. 201.

Linearis, tenuis, fulvo-testacea; elytris passim punctatis; antennis fuscis; femoribus posticis apice, tibiis tarsisque mediis et posticis, nigro-fuscis.

Long. 3 lin. ♀.

The eyes are widely distant in front and on the crown, and prominent.

Hab.—Villa Nova, Amazons; on flowers.

Sub-fam. CALLICHROMATINÆ.

Genus CALLICHROMA.

Latreille, Règne Anim. (ed. 1) iii. 341; Lacord. Gen.
ix. 16. ·

1. *Callichroma suturale.*

Cerambyx suturalis, Fabricius, Sp. Ins. i. 212; Oliv. Ent.
No. 67, p. 25, pl. vi. f. 40.

Nigrum, subtus paulo virescens, supra velutinum, elytris vitta suturali alteraque discoidali (postice co-euntibus) cupreo-aureis; pedibus nigris, tibiis posticis dilatatis et compressis.

Long. 10-19 lin. ♂ ♀.

Hab.—Pará and Tapajos.

2. *Callichroma porphyrogenitum,* n. sp.

Magnum, robustum, subtus viride, dense argenteo-velutinum; supra capite thorace et scutello violaceis, hoc nigro-bivittato, elytris lute cyaneis, ad latera violaceis, utrinque vitta discoidali prope suturam nigro-velutina; pedibus nigris, femoribus basi grosse punctatis, posticis sericeis, tibiis dilatatis et compressis.

Long. 1 un. 9 lin. ♂.

A magnificent species, with the hind tibiæ dilated in the same manner as *C. suturale.* The legs, however, are more robust, and the tarsi much broader. The head, thorax, and scutellum are of a beautiful violet colour, and the ground-colour of the elytra is of greenish-blue metallic lustre graduating into violet on the sides.

Hab.—Manaos, Rio Negro.

One example only.

8. *Callichroma brachiale*, n. sp.

C. vittato* (Fabr.) affine; subtus viridi-æneum, argenteo-griseo-sericeum; supra capite et thorace cyaneis splendidis, hoc maculis duabus velutinis violaceis; scutello viridi-æneo, splendido; elytris nigro-velutinis, apice violaceis, sutura angusta vittaque discoidali a humero incipiente viridibus; pedibus nigris, femoribus quatuor anticis rufis, basi et apice fuscis, tibiis posticis ♂ ut in *C. vittato* compressis at minus dilatatis.

Long. 18 lin. ♂.

Hab.—River Japurá, near Ega.

One example.

4. *Callichroma rugicolle.*

C. rugicollis, Guérin, Icon. R. A. iii. 220.

C. assimilatum, White, Cat. Longic. Brit. Mus. p. 158.

I can perceive no difference between Pará specimens of White's *C. assimilatum* and Mexican examples of *C. rugicolle.* The species belongs to the group in which the hind tibiæ are only very slightly and very gradually dilated from base to apex. The colour is a brilliant metallic-green, silvery-gray-tomentose on the under surface, marked on the thorax with two short velvetty-black vittæ, and on the elytra by a similar vitta, extending from the middle of the base very nearly to the apex; the sides of the elytra become gradually darker green and velvetty. The thorax is crossed throughout by fine rugæ, and the antenniferous tubers are longitudinally rugose. All the thighs are tawny-red.

---

* *C. vittatum*, described by Fabricius from Banks' collection, is a common South Brazilian insect, having the thighs red, the under-surface clothed with golden tomentum, and the suture generally with a golden tinge: the hind tibiæ are only slightly and gradually dilated from base to apex.

Long. 7-10 lin. ♂ ♀.

Hab.—Pará, Santarem, and Ega, on flowers; also Cayenne, Mexico, and probably widely distributed in Tropical America.

5. *Callichroma aureotinctum*, n. sp.

C. rugicolli valde affine, paulo robustius, magis aureo-viride, præcipue elytris apice læte aureis; capite viridi-aureo, splendido, sparsim punctato, tuberibus antenniforis haud strigosis; thorace multo minus striato.

Long. 7-10 lin. ♂ ♀.

Hab.—Santarem, Amazons; on flowers.

Possibly only a variety of *C. rugicolls*, found on the same trees in company.

6. *Callichroma versatum*, n. sp.

C. rugicolli affine, majus, tarsis omnibus apiceque tibiarum pallide fulvis; robustum, læte viridi-æneum, subtus argenteo-tomentosum; thorace transversim sub-tiliter strigoso, nigro-velutino bivittato; scutello splendido, aureo; elytris utrinque vitta dorsali nigro-volutina; femoribus rufis, tibiis (apice fulvo excepto) nigris, tarsis ochreo-fulvis.

Long. 18 lin. ♂ ♀.

Hab.—Lower Napo, near Pebas, Upper Amazons.

A score examples, all alike.

XIX. *Contributions to an Insect Fauna of the Amazon Valley* (Coleoptera, Cerambycidæ). By H. W. Bates, F.Z.S., late Pres. Ent. Soc.

[Read 7th November, 1870].

I beg now to lay before the Society the conclusion of my descriptions of Longicorn Beetles from the Amazons (continued from p. 335).

Fam. CERAMBYCIDÆ.

Sub-fam. COMPSOCERINÆ.

Genus ORTHOSCHEMA.

Thomson, Classif. des Ceramb. p. 561; Lacord. Gen. ix. 85.

Syn. *Orthostoma*, Serville (nom. præ-occ.).

1. *Orthoschema albicorne.*

Cerambyx albicornis, Fab. Syst. El. ii. 269.

Elongatum, depressum, viridi-æneum, supra opacum, subtus nitidius, griseo tenuiter pubescens; capite ♂ latitudine thoracis, hoc angulis posticis porrectis; antennis articulis tribus terminalibus testaceo-albis; elytris sub-cyaneis, apice anguste emarginatis, angulo suturali producto.

Long. 7¼–9 lin. ♂ ♀.

This species seems to have been overlooked by authors, although the description of Fabricius is tolerably good. He gives the *four* last joints of the antennæ as white; showing that he had the male only before him, in which the eleventh joint is "appendiculate." The species resembles *O. abdominale* of Serville, the type of the genus, but is rather narrower, has green abdomen, emarginate apices to the elytra, &c. The antennal joints 3-8 have a short spine at the apex within, most prominent in the ♂.

Hab.—Pará.

2.   *Orthoschema cyaneum.*

Orthostoma cyanea, Pascoe, Journ. Entom. i. 62.

"Læte cærulea, thorace luteo; antennarum articulis tribus ultimis albis." (Pasc.)

Long. 7-8½ lin. ♂ ♀.

Closely allied to *O. albicorne;* the apex of the elytra is less deeply emarginate, the thorax above and beneath of a bright red colour, and the elytra blue. In all my examples (three), half of the eighth antennal joint is yellowish-white, as well as the remaining three.

Hab.—Ega (not Parú, as stated by Mr. Pascoe).

3.   *Orthoschema Tarnieri,* n. sp.

O. albicorni affine; differt antennis totis nigris, haud spinosis, elytrisque apice integris. Viridi-æneum, infra subnitidum, griseo-tomentosum; supra capite thoraceque subnitidis, elytris opacis.

Long. 7 lin. ♀.

Hab.—Parú.

4.   *Orthoschema tenuicorne,* n. sp.

Parvum, paulo minus elongatum, depressum, læte viridi-æneum, abdomine rufo; antennis tenuibus, valde elongatis, haud ciliatis; thorace (♂ ♀) antice valde angustato, subtiliter confertissime transversim rugoso, nitido; elytris confertissime granulato-rugosis, apice leviter emarginatis, ad suturam dehiscentibus; pedibus nigris.

Long. 4-5 lin. ♂ ♀.

A small slender-limbed species, differing from *O. rufiventre,* Germ. (a common and small species of Rio Janeiro) in being much less linear in form, in the thorax in both sexes narrowing greatly to the front, in the long and very slender antennæ, which in well-developed males are three times the length of the body, and also in its brighter brassy-green hue.

Hab.—Ega. Very common on the branches of dead trees, in company with numerous species of *Chrysoprasis* of similar colour and size, from which it is readily distinguishable by the very short hind tarsi.

5. *Orthoschema Chryseis*, n. sp.

O. tenuicorni simillimum; differt solum thorace igneo-cupreo, elytrisque obscure nigro-æneis, apice viridi-sericeis.

Long. 4-5 lin. ♂ ♀.

Of similar form to *O. tenuicorne*; antennæ long, very slender, and nearly destitute of cilia; the thorax narrowed in front in both sexes, of a glowing purple-coppery hue; the elytra brassy-black, greenish and more shining near the apex.

Hab.—Pará, Cametá, and banks of the Tapajos; on dead trees.

6. *Orthoschema cardinale*, n. sp.

Curtum, depressum, postice paulo dilatatum, saturate cæruleum, nitens, elytris (apice nigro excepto) coccineis, opacis; capite grosse rugoso-punctato; thorace ovato, lateribus grosse rugoso-punctatis, medio lævi; scutello ferrugineo, polito; antennis purpureis, longe ciliatis, basin versus robustis, apice valde attenuatis.

Long. 6 lin. ♂.

A very beautiful species, unlike any other in colours, but undoubtedly belonging to this genus.

Hab.—Ega. One example only.*

---

* The following undescribed species of *Orthoschema* are common in Collections:—

Orthoschema ruficeps.

O. viridipenni (Thoms.) proxime affine. Rufum, antennis articulis 3-11 nigro-piceis, abdomine nigro-æneo, thorace infuscato, elytris violaceis vel obscure cærulcis.

Long. 3 lin. ♂ ♀.
Hab.—Brasilia merid.

Orthoschema nigricorne.

O. viridipenni (Thoms.) proxime affine. Fulvum, antennis articulis 3-11 pedibusque nigris, pectore infuscato, opaco, abdomine nigro, nitido, coxis femoribusque basi fulvis; elytris viridi- vel cyaneo-sericeis, apice nitidis.

Long. 3 lin.
Hab.—Brasilia merid.

Genus CHLORETHE.

Bates, Entom. Monthly Mag. iv. 24 (1867); Lacord.
Gen. viii. 398.

Lacordaire, misled no doubt by the character given of
"eyes coarsely facetted," placed this genus among the
doubtful forms of the first section of *Cerambycidæ*. On
re-examination, I find that the facets of the eyes would
be more correctly described by Lacordaire's term of
"subfinement granulés." They are very similar to
the same organs in *Orthoschema*, near which I stated
the genus should be placed. The genus, in fact, possesses all the essential characters of *Orthoschema* except
the antennæ, which are short (very little longer than
the body even in the ♂) and have the 3-5th joints
thickened, and furnished with long cilia beneath. In
general form the genus differs from *Orthoschema* in
being cylindrical and not depressed; the thorax is
rounded, and without porrect hind angles.

1.   *Chlorethe ingæ.*

Bates, *loc. cit.*

Parva, cylindrica, setosa, viridi-ænea, elytris suturate
sericeo-viridibus, apice rufo-marginatis truncatis; thorace
æqualiter reticulato-punctato; antennis nigris, articulo
basali viridi; pedibus nigro-æneis; abdomine rufo; metasterno sparsim punctato, nitido.

Long. 3½-4½ lin. ♂ ♀.
Hab.—Ega; on felled Inga trees.

Genus COREMIA.

Serville, Ann. Soc. Ent. Fr. 1834, p. 22; Lacord. Gen.
ix. 42.

This name clashes with one of Guénée's genera of *Lepi-
doptera*, over which, however, it has ten years' priority.

1.   *Coremia hirtipes.*

Saperda hirtipes, Oliv. Entom. No. 68, p. 14, pl. i. f. 8.

Linearis, gracilis, nigra; pedibus posticis valde elongatis, femoribus apice clavatis, tibiis apice longe nigro-
hirsutis.

Long. 3½-5 lin. ♂ ♀.

Found throughout the Amazons, flying slowly over dead timber in new clearings. It resembles a large *Culex.*

Sub-fam. CLYTINÆ.

Genus CYLLENE.

Newman, Entom. i. 7; Lacord. Gen. ix. 62.

1. *Cyllene amazonica,* n. sp.

C. cayennensi (Lap. & Gory) proxime affinis; differt solum elytris prope apicem linea transversa alteraque suturali griseis. Elongata, postice attenuata, nigro-velutina; thorace fasciis tribus flavis; elytris fascia prope basin arcuata, alteris duabus ante medium versus scutellum abrupte recurvis, tertia postica arcuata ad suturam interrupta punctiformi, flavis, apice sutura et fascia brevi conjuncta obliqua griseis.

Long. 5-7½ lin. ♂ ♀.

Common throughout the Amazons, on branches of dead trees. It resembles in markings *C. caraccasensis* (Chevr.), but is decidedly broader and more robust in form, in which character it agrees more with *C. cayennensis.*

Genus NEOCLYTUS.

Thomson, Musée Scientifique, p. 67; Lacord. Gen. ix. 75.

1. *Neoclytus tapajonus,* n. sp.

N. guyanensi (Lap. & Gory) proxime affinis, vix postice attenuatus, nigricans vel obscure piceus, partim griseo-tomentosus; thorace oblongo-ovato, lineis tribus elevatis tuberculatis; elytris apice truncatis et utrinque bispinosis, supra prope basin vitta lata obliqua et parte apicali griseo-tomentosis, fasciis tribus flavis, prima pone vittam basalem griseam valde obliqua ad scutellum ascendente, secunda (primæ proxima) recta transversa, tertia longe distante versus apicem; pedibus piceo-rufis vel nigris, femoribus ut in *N. rufo* (Oliv.) gradatim crasse clavatis.

Long. 3½-7 lin. ♂ ♀.

Hab.—Santarem, Tapajos, Ega.

Abundant occasionally on wooden fences of gardens. The yellow belts of the elytra are all of nearly equal width, and form moderately wide fasciæ, and not fine lines as in *N. rufus* and other allied species.

Genus MECOMETOPUS.

Thomson, Classif. des Ceramb. p. 222.

Lacordaire unites this genus with *Neoclytus*, but it seems to me to form a distinct and very natural group, distinguished from *Neoclytus* by the very much shorter and less robust hind legs, which are in due proportion to the anterior and middle pair. The muzzle is in almost all the species longer and narrower than in *Neoclytus*. All the known species are from tropical America.

1. *Mecometopus Batesii.*

Clytus Batesii, White, Cat. Longic. Brit. Mus. p. 257.

Robustus, niger, corpore subtus thoraceque tomento ochreo variegatis, elytris læte croceo-flavis, macula elongata humerali, altera obliqua rhomboidea laterali pone medium, et apice nigris; thorace magno, elytris multo latiore, subgloboso, dorso linea lata elevata transversim rugosa.

Long. 6 lin.

Hab.—Banks of the Iruri, Santarem.   On dead trees.

2. *Mecometopus festivus.*

Clytus festivus, Fab. Syst. El. ii. 348.

Cylindricus, ater; thorace breviter oblongo-rotundato, elytris haud latiore, vage late cinereo fasciato; scutello, elytrorum vitta abbreviata obliqua prope basin, macula

triangulari communi vittam approximante, et fascia angustiori versus apicem, læte flavis; subtus macula magna
metasterni, ventrisque segmentis duobus basalibus, flavo-
tomentosis; antennis subclavatis, nigris.

Long. 4¼ lin.

Hab.—Obydos, Guiana side of Lower Amazons.

3. *Mecometopus Wallacii.*

Clytus Wallacei, White, Cat. Longic. Brit. Mus. p. 259.

Cylindricus, ater; thorace breviter oblongo-rotundato,
elytris paulo angustiori, cinereo vage fasciato; scutello,
elytrorum vitta abbreviata obliqua prope basin, et macula
triangulari communi vittam approximante, læte flavis,
vitta lata ante apicem grisea; subtus macula magna metasterni, ventrisque segmentis duobus basalibus, flavo-
tomentosis; antennis tenuibus, apice subclavatis, piceo-
rufis, clavo pallida.

Long. 4½ lin.

Hab.—Ega.

4. *Mecometopus triangularis.*

Clytus triangularis, Lap. & Gory, Monogr. p. 31, pl.
vii. f. 38.

Gracilis, cylindricus, niger, capite rufo-piceo, antennis
dimidio apicali pallido; thorace oblongo-ovato, obscure
griseo, medio nigro; elytris margine basali, fascia valde
obliqua abbreviata, macula triangulari communi ante
medium, et triente apicali, griseis; metasterno fasciisque
duabus ventris basalibus flavo-cinereis.

Long. 4 lin.

Hab.—Pará.

5. *Mecometopus lætus.*

Clytus lætus, Fabr. Syst. El. ii. 348.

Cylindricus, niger, capite antennis et thorace fulvo-
rufis; hoc oblongo, lateribus vix rotundato, postice

utrinque macula magna flava; scutello flavo; elytris humeris, regione scutellari, macula prope basin elongata obliqua triangulari, alteraque triangulari communi huic adjacente, ot fascia abbreviata angusta versus apicem, læte flavis; pectore flavo-rufo, flavo-tomentoso; abdomine fasciis quatuor flavis; tarsis rufescentibus.

Long. 6 lin.

Hab.—Pará.

6. *Mecometopus rubefactus*, n. sp.

Cylindricus, fulvo-piceus, capite antennis et thorace fulvo-rufis; hoc oblongo-ovato, postice utrinque macula magna flava; elytris fascia basali (scutellum includente), litura sub-humerali, macula prope basin elongata obliqua triangulari, alteraque huic adjacente communi triangulari, et fascia postica ad suturam haud interrupta, læte flavis, parte apicali rufo-tincta; pectore et fasciis quatuor ventralibus flavis; pedibus fulvo-piceis.

Long. 4½-5¼ lin.

Hab.—Ega; on branches of dead trees.

Described from four examples. Allied to *M. amabilis*, Chevrolat, which wants the posterior fascia, and has the anterior oblique spot of a different form.

7. *Mecometopus latecinctus*, n. sp.

Cylindricus; capite, antennis (clava nigra excepta), et thorace fulvo-rufis; hoc oblongo-ovato, postice utrinque griseo-sericeo; scutello flavo; elytris nigris, macula prope basin elongata obliqua triangulari, alteraque adjacente communi triangulari, et fascia lata haud distante, læte flavis, parte apicali flavo-cinerea; pectore et fasciis quatuor ventralibus flavis; pedibus rufis.

Long. 4¼ lin.

Hab.—St. Paulo, Upper Amazons.

The black ground colour of the elytra in this species forms bands much narrower than the yellow belts and spots.

8. *Mecometopus purus*, n. sp.

M. læto valde affinis; differt elytris basi nigerrimis, macula prima antice haud truncata, antennis clava nigra;

cylindricus, niger, capite thorace antennisque basi fulvo-
rufis, his clava piceo-nigra; thorace postice utrinque
macula magna cinereo-flava; acutello læte flavo; elytris
humeris, macula obliqua triangulari basi angulata versus
basin ascendente, macula communi triangulari, et fascia
angusta biarcuata postica, læte flavis; prothorace subtus,
pectore, fasciisque quatuor ventralibus flavis; pedibus
nigris.

Long. 4½ lin.
Hab.—Ega.

9. *Mecometopus Flavius*, n. sp.

Cylindricus, niger, capite, thorace, antennisque basi
fulvo-rufis, his clava piceo-nigra; thorace postice utrinque
macula cinerea; acutello læte flavo; elytris macula magna
rotundata (prope humerum incisa), altera communi rhom-
boides, fasciaque (prope suturam valde dilatata), læte
flavis; pectore segmentisque quatuor ventralibus flavis;
pedibus nigris.

Long. 5 lin.
Hab.—Santarem.*

10. *Mecometopus troglodytes.*

Clytus troglodytes, Lap. & Gory, Monogr. p. 33, pl. vii.
f. 41.

Breviter cylindricus, niger; thorace ovato; elytris
linea angusta abbreviata obliqua prope basin, macula
parva communi triangulari, linea transversa pone medium,
albis; antennis brevibus, clavatis.

Long. 3½ lin.
Hab.—Pará.

* The following is a new species allied to *M. lotus*, but different from
the preceding, and from all those described by MM. Chevrolat and
Thomson.

Mecometopus Jansoni, n. sp.

Cylindricus, niger, capite et thorace ferrugineo-rufis; hoc subgloboso,
postice angustato; scutello flavo; elytris macula magna ovata prope
humerum, altera parva communi obcordata, fasciaque postica lata recta,
læte flavis; pectore segmentisque duobus ventralibus cinereo-flavis;
pedibus nigris.

Long. 5 lin.
Hab.—Chontales, Nicaragua (Janson fil.)

11. *Mecometopus globicollis.*

Clytus globicollis, Lap. & Gory, Monogr. p. 82, pl. vii. f. 39.

Cylindricus, niger; scutello albo; elytris linea prope basin obliqua curvata, macula communi triangulari, lineaque transversa postica, albis; antennis clavatis.

Long. 4¼ lin.
Hab.—Pará.

12. *Mecometopus polygenus.*

Thomson, Classif. des Ceramb. p. 223.

Breviter cylindricus, robustus, niger; antennis valde clavatis; thorace sphærico, elytris latiori; pedibus robustis, tibiis compressis; elytris linea abbreviata obliqua flexuosa maculaque communi triangulari flavis, trieute apicali griseo-sericea.

Long. 3-5 lin.
Hab.—Ega.   Abundant on dead trees.

Sub-fam. TILLOMORPHINÆ.

Genus EPROPETES, nov. gen.

Corpus lineare, longe pilosum. Caput supra planum, tubera antennifera obsoleta. Oculi reniformes. Antennæ lineares, longe pilosæ; ♂ corpore multo longiores, articulo tertio elongato, cæteris subæqualibus; ♀ corpore breviores, articulis 8-11 multo abbreviatis. Thorax longissimus, elytris æqualis, inermis, ante basin valde late constrictus, quasi pedunculatus, parte antica valde convexa. Elytra curta, depressa, apice obtuse rotundata. Pedes curti, longe pilosi; femora clavata; tarsi breves, articulo primo secundo et tertio conjunctis longiori. Acetabula intermedia extus clausa.

The species on which this genus is founded was placed by White in the genus *Ozodes*, with which it has no near affinity whatever, and scarcely any external resemblance. It is evidently a member of the sub-fam. *Tillomorphinæ*, and is allied to the Australian genus *Ipomoria*; differing chiefly in the extreme relative length of the thorax and in the length and proportions of the antennal joints.

1. *Epropetes latifascia.*

Ozodes latifascia, White, Cat. Longic. Brit. Mus. p. 218.

Niger, longe hirsutus; antennis (basi excepta) pedibusque rufo-piceis; capite thoraceque creberrime punctatis, hoc dorso reticulato; elytris argenteo-griseo pubescentibus, medio fascia lata nigro-velutina, antice et postice albo-marginata.

Long. 3-4 lin. ♂ ♀.

Hab.—Dry forests of the Tapajos; on dead branches.

Sub-fam. CLEOMENINÆ.

Genus EUPEMPELUS, nov. gen.

Genus *Listropteræ* affine; differt elytris linearibus, apice obtusis, truncatis. Corpus elongatum, lineare. Caput parvum, thorace angustius, rostro paulo elongato. Antennæ (♂) corpore longiores, tenues, sparsim ciliatæ, articulis subæqualibus. Elytra valde elongata, linearia, apice obtusa, truncata, angulis truncaturæ distinctis, supra leviter recte bicostata. Pedes elongati; femora gradatim incrassata.

Closely allied to *Listroptera,* especially in the form of the head and thorax; but differing, even from the elongate species of that genus (e. g. *L. collaris*) by the linear form, and abruptly rounded and truncate apex of the elytra, which, besides, are destitute of the gray tomentum and curved costæ that distinguish all the *Listropteræ*.

1. *Eupempelus olivaceus,* n. sp.

Elongatus, olivaceo-viridis, sub-opacus; thorace lætu rufo-sericeo, dorso quinque-tuberculato; elytris creberrime rugoso-granulatis, et passim punctatis, costis rectis utrinque duabus vix distinctis, apice transversim truncatis; corpore subtus leviter cinereo-tomentoso.

Long. 6 lin. ♂.

Hab.—Ega.

At fragrant flowers in the forest, in company with species of *Odontocera* and *Agaona.*

Genus LISTROPTERA.

Serville, Ann. Soc. Ent. Fr. 1834, p. 71; Lacord. Gen. ix. 107.

1. *Listroptera tenebrosa.*

Callidium tenebrosum, Fabr. Ent. Syst. I. ii. 322.

Brevis, depressa, nigra, opaca; thorace rufo, medio dorsi margineque antico nigris; elytris postice rotundatis, apice conjunctim acute rotundatis, margine serratis, dorso postice cano-tomentoso; abdomine cinereo-argenteo.

Long. 5 lin.

Hab.—River Tapajos.

2. *Listroptera aterrima.*

Callichroma aterrimum, Germ. Ins. Sp. Nov. p. 497.

L. tenebrosæ valde affinis; differt thorace nigro.

Long. 5 lin. ♂ ♀.

Hab.—Ega. Common on dead branches.

3. *Listroptera angulata.*

White, Cat. Longic. Brit. Mus. p. 208.

"Nigerrima; thorace curtulo, quadrinodoso, angulis posticis prominulis rubro-notatis; elytris basi nigro oblique angulatis, parte cinereo-tomentosa basi solum punctata." (White.)

Long. 4½ lin.

Hab.—Pará. In Coll. Brit. Mus.

4. *Listroptera collaris.*

Cerambyx collaris, Klug, Nov. Ac. Cæs. L. C. Nat. Cur. xii. 459, pl. xliii. f. 8.

A *L. tenebrosa* et *aterrima* differt corpore et antennis longioribus, gracilioribus; antennis ♂ corpore multo longioribus, articulo quarto multo abbreviato; nigra, thorace antice et postice læte rufo; elytris ante apicem rotundatis, apice conjunctim acute rotundatis, prope suturam spina acuta armatis, marginibus haud serratis.

Long. 5 lin. ♂ ♀.

Hab.—Caripi, near Pará. On dead trees.

Genus DIHAMMOPHORA.

Chevrolat, in Thoms. Arc. Nat. p. 50; Lacord. Gen. ix.
108.

1. *Dihammophora nitidicollis*, n. sp.

Nigra, opaca; thorace læte rufo, sericeo-nitente, elongato, inæquali, medio dorsi convexo, postice bituberou-
lato; elytris ante apicem dilatato-rotundatis, supra grosse
lineatim punctatis, bicostatis; antennis corpore multo
brevioribus, articulis 3-11 subæqualibus, leviter serratis;
abdomine argenteo-tomentoso.

Long. 2¾ lin.
Hab.—St. Paulo, Upper Amazons.

2. *Dihammophora pusilla*, n. sp.

Angustissima, linearis, nigro-picea, opaca; thorace
angusto, cylindrico, haud tuberculato, rufo-opaco; elytris
ante apicem gradatim rotundatis, grosse lineatim punctatis, bicostatis; antennis corpore multo brevioribus,
decem-articulatis, articulo 10mo longiori, crassiori; abdomine argenteo-sericeo.

Long. 2½ lin.
Hab.—Villa Nova; on flowers.

Allied to *D. perforata*, Klug, from which it differs, *inter
alia*, in the head being entirely black.

Sub-fam. RHOPALOPHORINÆ.

Genus RHOPALOPHORA.

Serville, Ann. Soc. Ent. Fr. 1834, p. 100; Lacord. Gen.
ix. 110.

1. *Rhopalophora atramentaria.*

Listroptera atramentaria, White, Cat. Longic. Brit. Mus.
p. 208.

Rhopalophora vidua, Chevrolat, in Thoms. Arc. Nat. p. 59.

Elongata, plana, nigro-velutina, antennis pedibusque
nitidis; elytris utrinque vitta latissima suturali griseo-
tomentosa; corpore subtus argenteo-tomentoso.

Long. 6-7 lin. ♂ ♀.
Hab.—Altar do Chaõ, River Tapajos. Abundant.

Genus COSMISOMA.

Serville, Ann. Soc. Ent. Fr. 1834, p. 19; Lacord. Gen.
ix. 112.

1.   *Cosmisoma Diana*, n. sp.

Cerambyx Ammiralis, Lin. Syst. Nat. (ed. xii) ii. 625 (?).

Robustum, lineare, planum, nigrum; thorace antice et
postice constricto, medio lateribus tumido, utrinque vitta
lata læte argentea; elytris macula humerali clare fulvo,
vitta lata pone medium læte argentea; antennis articulis
3io et 4to apice infra nigro-penicillatis, 5to scopa magna
nigra, 6to scopa minore alba; corpore subtus argenteo-
tomentoso.

Long. 6½-8 lin. ♂ ♀.

This superb insect was referred by White to the *C.
Ammiralis* of Linnæus; but the original description in
the Systema Naturæ does not at all agree with the Amazonian specimens.   It is true that Linnæus described it
from a figure only, sent from Surinam by Dr. L'Ammiral,
and this may not have been accurate.   So palpable a
difference, however, as "Thoracis latera rufa" in
L'Ammiral's insect, cannot be assumed to be an inaccuracy,
and in the absence of Surinam specimens, the present
species must be regarded as distinct.   The *C. formo-
sum* (Blanchard, in D'Orbigny's Voyage), from Santa
Cruz, in Bolivia, has also been assumed to belong to the
same species, although both in the description and figure
the sides of the thorax, and the humeral spots and belt
of the elytra, are given as "yellow."

Hab.—Ega.   On flowers of *Myrtaceæ*; a large number
of examples offering no variation.

2.   *Cosmisoma fasciculatum*.

Saperda fasciculata, Oliv. Ent. No. 68, p. 14, pl. i. f. 3.

Cosmisoma Leprieurii, Buquet, Guér. Icon. R. A. p. 231.

Minus robustum, elongatum, depressum, nigrum;
thorace nitido, antice et postice constricto, medio tumido

et dorso trinodoso; scutello argenteo; elytris velutinis,
pone medium utrinque prope suturam macula obliqua
argentea; antennis articulo 3io apice infra nigro-penicillato, 5to scopa magna nigra, 6to scopa parva alba.
Long. 6. lin. ♂.
Hab.—River Tapajos.
Olivier's description is good, but his figure is very
bad.

3. *Cosmisoma argyreum*, n. sp.

Minus robustum, elongatum, depressum, nigrum; thorace subcylindrico, vix constricto, medio haud tumido,
nitido, supra æquali, punctulato; scutello argenteo; elytris utrinque medio linea longitudinali argentea; antennis
articulo 3io apice infra nigro-penicillato, 5to scopa magna
nigra, 6to sparsim argenteo-pubescente; corpore subtus
tenuiter argenteo-pubescente; abdomine subglabro.
Long. 4-5½ lin.
Hab.—Ega. Very abundant, occasionally, at flowers.

4. *Cosmisoma speculiferum.*

Cerambyx speculifer, Gory, in Guér. Icon. R. A. p. 231.

Elongatum, depressum, nigrum; thorace antice et
postice constricto, medio tumido, dorso quadrinodoso;
scutello argenteo; elytris apud medium plaga magna
communi subquadrata argentea; antennis articulis 1-4
sparsim ciliatis, 5to scopa magna nigra, 6to scopa parva
argentea; corpore subtus argenteo-tomentoso.
Long. 6 lin. ♂.
Hab.—Pará.

5. *Cosmisoma lineellum*, n. sp.

Parvum, gracile, nigrum; thorace subcylindrico, elongato, vix constricto, medio haud tumido, supra paulo
inæquali, lateribus inæqualiter grosse punctatis, parte

antica impunctata; scutello argenteo; elytris utrinque medio linea longitudinali argentea; antennis rufo-piceis, articulis 1-4 pilis elongatis paucis, 5to scopa magna nigra; corpore subtus argenteo.

Long. 3¼ lin.

Hab.—Ega.

Bears the closest resemblance to *C. argyreum*, but distinguished by its smaller size, and by the absence of hair-pencil from the tip of the third antennal joint.

6.   *Coemisoma scopulicorne.*

Saperda scopulicornis, Kirby, Trans. Linn. Soc. xii. 442.

Elongatum, postice attenuatum, fulvo-ferrugineum; capite thoraceque densissime punctulatis, opacis; hoc elongato, haud constricto, lateribus paululum rotundato, utrinque vitta argentea; elytris tomentosis, opacis, linea flavo-argentea a basi usque ad apicem; antennis articulis 1-4 sparsissime pilosis, 5to apice scopa parva nigra, 6-11 elongatis; corpore subtus argenteo, prothorace antice abdomineque glabris exceptis; pedibus rufis.

Long. 4½-5 lin.

Hab.—River Tapajos.

A common and well-known Brazilian insect, found as far south as Rio Janeiro. *C. ochraceum* (Perty), confounded with it by some authors, is a very distinct species.

7.   *Coemisoma pulcherrimum,* n. sp.

Elongatum, postice attenuatum, nigrum; thorace antice et postice valde constricto, medio lateribus haud tumidis, supra paulo inæquali, creberrime punctulato, pubescente, linea dorsali lævi; elytris linea alba à basi usque ad apicem; antennis articulis 2-4 infra ciliatis, 5to scopa magna nigra; femoribus clavis læte rufis.

Long. 6½ lin.

Hab.—St. Paulo, Upper Amazons.

8. *Coemisoma cneicolle.*

C. cneicollis, Erichson, in Schomb. Reise Brit. Guy. iii.
572.

C. subvirescens, White, Cat. Longic. Brit. Mus. p. 214.

C. semicupreum, Chevrolat, Rev. et Mag. Zool. 1859,
p. 28.

Parvum, postice attenuatum, viride; capite thoraceque
supra auratis, nitidis; illo sparsim punctato; hoc medio
dorsi crebre grosse rugoso-punctato, antice valde et postice
paulo constricto, parte basali transversim strigosa; elytris creberrime punctulato-rugosis, obscure viridi-sericeis,
sub-opacis; antennis articulo basali viridi-aeneo, scabroso,
reliquis nigris, 2-4 dilatatis, 5to apice infra et lateribus
nigro-penicillato; corpore subtus griseo-sericeo; pedibus
nigris, nitidis.

Long. 4¼ lin. ♂.

Hab.—Pará. *

* The following new species of this beautiful tropical American genus
may be added.

Coemisoma humerale.

Parvum; capite, antennis et pedibus fulvo-ferruginetis; antennis articulo basali scabroso, 2-4 sparsim ciliatis, 5to scopa magna nigra, 7-11
curtis (♀ ?); thorace subcylindrico, antice et postice paulo constricto,
medio paulo rotundato, crebre punctato, opaco, nigro-piceo, linea dorsali
laevi rufescente; elytris nigro-piceis, sericeis, macula humerali fulvo-ferru-
ginea; corpore subtus (capite excepto) nigro, nitido, metasterno opaco.

Long. 3¼ lin.

Hab.—Brasilia. (Rev. Hamlet Clark).

Coemisoma Titania.

Elongatum, gracile; capite nigro, subnitido; thorace elongato, fere
cylindrico, perparum constricto, tomento fulvo-aureo dense vestito; elytris late fulvo-aureo-tomentosis, triente apicali nigro-velutina, parte
nigra linea medio transversa alba; antennis nigris, articulo primo apice
abrupte fortiter elevato, infra nigro-barbato, 2ndo et 4to infra apice
nigro-penicillatis, 3io et 5to scopa magna nigra, 6to testaceo, scopa magna
fulva, 7-11 paulo elongatis, albo-tomentosis; pedibus nigris, nitidis, longe
hirsutis; corpore subtus nigro, subnitido, fusco-piloso.

Long. 6 lin.

Hab.—Chontales, Nicaragua. (Dom. Janson fil.)

C. plumicorni (Drury) coloribus simile.

Genus ARGYRODINES.

Bates, Entom. Monthly Mag. iv. 24 (1867); Lacord. Gen. ix. 118.

(Charac. emend.). Corpus sublineare, elytris depressis ut in *Cosmisomate.* Caput rostro modice elongato, lato; palpis lobis elongatis, exsertis. Antennæ filiformes, simplices, longitudine corporis (♀), setosæ, articulo 3io valde elongato, apice incrassato, 4to quam 5to vel 6to minore. Thorax elongatus, elytris longitudine fere æqualis, antice et postice fortissime constrictus, medio valde rotundatus. Elytra plana, medio angustata, apice obtuse rotundata. Pedes breves, setosi; femora apice clavata; tibiæ compressæ; tarsi curti. Mesosternum latum, planum; acetabula extus clausa.

1. *Argyrodines pulchella,* Bates, *l. c.*

Nigra, nitida; capite crebre punctato; thorace antice lævi, medio reticulato-punctato, plaga utrinque lævi, parte postica transversim strigosa; elytris creberrime punctatis, utrinque fasciis tribus argenteis impunctatis, duabus angustioribus ante et 3ia latiori post medium; mesosterno utrinque argenteo-piloso; tarsis posticis argenteo-pilosis.

Long. 4 lin.

Hab.—Ega.

Genus LISSOZODES, nov. gen.

Genus *Ozodi* (Serv.) affine; differt thorace cylindrico, æquali, et coloribus valde diversis. Caput plus quam in *Ozodis* exsertum, rostro brevissimo, verticali. Palpi articulo ultimo ovato. Thorax elongatus, cylindricus, absque tuberculis. Elytra linearia, plana, apice rotundata. Pedes modice elongati et robusti; femora gradatim et (præcipue ♂) grosse incrassata; tibiæ angustæ; tarsi modice elongati. Antennæ filiformes, robustæ, hirsutæ, corpore ♂ duplo, ♀ sesqui longiores; articulo 1mo brevi, crasso, 3-11 longitudine subæqualibus, 4-6 vix perspicue sulcatis. Coxæ antice globoso-conicæ, exsertæ; pro- et meso-sterna angusta; acetabula extus clausa.

1. *Lissozodes basalis.*

Oycnoderus basalis, White, Cat. Longic. Brit. Mus. p. 213,
pl. vii. f. 5.

Subtus viridi-æneus, griseo-pilosus, supra cyaneus;
capite thoraceque dense transversim strigosis, vix nitidis;
elytris creberrime punctato-rugosis, opacis, macula humerali aurantiaca; pedibus viridi-æneis, femoribus basi
coxisque posticis albo-testaceis.
Long. 4-6 lin. ♂ ♀.
Hab.—Ega; common on dead trees. The legs break
off almost at a touch, as in *Ozodes.*

Genus OZODES.

Serville, Ann. Soc. Ent. Fr. 1834, p. 98; Lacord. Gen.
ix. 116.

1. *Ozodes infuscatus,* n. sp.

O. nodicolli (Serv.) simillimus; differt capite, thorace,
corpore subtus, femorum dimidio basali, et antennarum
articulis apice, nigro-fuscis. Robustus, nigro-fuscus,
cinereo-sericeus; elytris obscure ferrugineis, fascia lata
postica antice obliqua obscuriori; antennis rufo-testaceis,
articulis 3-11 apice fuscis; pedibus rufo-testaceis, femoribus annulo lato nigro-fusco.
Long. 7½ lin. ♀.
Hab.—River Tapajos.*

2. *Ozodes ibidiinus,* n. sp.

Parvus, nigro-obscurus; capite thoraceque creberrime
punctulato-scabrosis, hoc supra inæquali, tuberculis duobus antico-discoidalibus parvis; elytris fulvo-testaceis,

* The following large species is also distinct from *O. nodicollis.*

Ozodes multituberculatus, n. sp.

Elongatus, supra planus, ferrugineo-fuscus, sericeus; thorace quam in
O. nodicolli longiore et angustiore, dorso utrinque tuberculis duobus altis
valde compressis, quinto parvo medio acuto, et angulis tuberculiformibus;
elytris fasciis tribus angustis, valde undulatis, sericeo-griseis.
Long. 6 lin. ♂.
Hab.—Chontales, Nicaragua.

vitta lata pone medium, maculis nonnullis angulatis anterioribus nigris; coxis, femoribus basi, tarsis, et antennarum articulis 2-11, rufo-testaceis.

Long. 4¼ lin. ♂.

Hab.—River Tapajos.

3. *Ozodes malthinoides*, n. sp.

Angustatus, linearis, flavo-testaceus, tomentosus; capite macula elongata frontali nigra; thorace postice gradatim angustato, prope basin constricto, supra haud tuberculato, sericeo, plaga postica nigra flavo-bilineata; elytris sericeis, linea obliqua ante alteraque pone medium et apice late sericeo-albidis; pedibus gracilibus.

Long. 6 lin.

Hab.—Ega.

Sub-fam. HETEROPSINÆ.

Genus MALLOSOMA.

Serville, Ann. Soc. Ent. Fr. 1834, p. 68; Lacord. Gen. ix. 123.

1. *Mallosoma scutellare*.

White, Cat. Longic. Brit. Mus. p. 110.

M. zonato minus, et magis depressum, fulvo-testaceum, macula pone oculos, thoracis vitta lata dorsali, antennis, pedibus, et elytrorum fasciis duabus latissimis, nigris; antennis articulis 3-6 nigro-ciliatis et apice unispinosis; thorace lateribus læte aureo-sericeis; elytris apice late rotundatis et unidentatis, vitta nigra basali, spatio elongato scutellari flavo interrupto.

Long. 5½ lin.

Hab.—Pará.

2. *Mallosoma rubricolle*, n. sp.

Subcylindricum, convexum, griseo-nigrum; thorace rotundato, lateribus medio breviter spinoso, sanguineo, opaco; elytris haud costatis, dense nigro-setosis, apice breviter truncatis, et extus spinosis; antennis articulis 3-6 apice spinosis, 7-11 rufescentibus.

Long. 5 lin.

Hab.—River Tapajos.

Genus CHRYSOPRASIS.

Serville, Ann. Soc. Ent. Fr. 1834, p. 5; Lacord. Gen.
ix. 125.

1. *Chrysoprasis auronitens,* n. sp.

Minus elongata, robusta, læte viridi-ænea, thorace supra
cupreo-aurato, abdomine rufo; capite grosse punctato-
rugoso; thorace lato, supra creberrime punctato; elytris
subdepressis, breviter decumbenti-setosis, apice truncatis;
prothorace subtus crebre punctato-rugoso, griseo-piloso;
metasterno grosse crebre foveolato-punctato, griseo-
piloso; pedibus robustis, dense punctatis, nigris, femoribus viridi-æneis; antennis ♂ corpore paulo longioribus,
robustis, nigris, articulo basali crebre foveolato-punctato,
3-6 apice unispinosis.

Long. 5-5½ lin. ♂ ♀.

Hab.—Pará.

I have seen this species in some collections named
"*C. rufiventris,* Dej. MS.*"*

2. *Chrysoprasis Sthenias,* n. sp.

C. auronitenti valde affinis, differt thorace concolori;
minus elongata, robusta, læte viridi-ænea, abdomine
rufo; capite grosse scabroso-punctato; thorace lato,
supra creberrime reticulato-punctato; elytris subdepressis, breviter decumbenti-setosis, apice truncatis;
prothorace et mesothorace subtus crebre punctato-rugosis, griseo-pilosis; metasterno grosse crebre foveolato-
punctato, griseo-piloso; pedibus robustis, dense punctatis, nigris, femoribus viridi-æneis; antennis ♂ corpore
paulo longioribus, crassis, apice attenuatis, nigris, articulo
1mo grosse punctato, viridi-æneo, 3-6 apice unispinosis.

Long. 4-6½ lin. ♂ ♀.

Hab.—Ega and St. Paulo. Very abundant on branches
of felled trees.

3. *Chrysoprasis rotundicollis,* n. sp.

Minus elongata, depressa, læte viridi-ænea, abdomine
rufo, antennis pedibusque totis nigris, metasterno nigro,

cinereo-tomentoso; capite grosse punctato; thorace
brevi, transverso, lateribus fortiter et regulariter rotundatis, supra crebre reticulato-punctato; elytris incum-
benti-setosis, apice truncatis; prothorace subtus crebre
scabroso, metasterno et coxis posticis nigris, obscuris,
grosse punctatis; antennis (♀) longitudine corporis,
robustis, nigris, articulo 1mo grosse punctato, 3-6 apice
breviter unispinosis.

Long. 4½ lin. ♀.

Hab.—River Tapajos.

4. *Chrysoprasis ruficoxis*, n. sp.

Elongata, minus robusta, læte viridi-ænea, capite
thoraceque aureo-tinctis, abdomine et pedum posticorum
coxis femorumque basi rufis; capite grosse subrugose
punctato; thorace sub-elongato, antice gradatim attenuato, lateribus vix rotundatis, basi perparum angustato,
supra sericeo, haud profunde transversim strigoso; elytris
breviter sub-erecte setosis, apice truncatis, angulo exteriori
dentiformi producto; prothorace subtus punctato-rugoso,
metasterno grosse foveato-punctato, interstitiis nitidis
punctulatis, sparsim cano-decumbenti-piloso; pedibus
elongatis, subgracilibus, nigris, femoribus nigro-æneis;
antennis tenuibus, ♂ corpore plusquam triplo longioribus, ♀ corpus paulo superantibus, nigris, articulo 1mo
punctato-scabroso, viridi-æneo, 3-6 apice unispinosis.

Long. 4-5½ lin. ♂ ♀.

Hab.—Obydos, Villa Nova, Ega, St. Paulo. Abundant
on dead trees.

5. *Chrysoprasis longicornis*, n. sp.

C. ruficoxi proxime affinis, differt coxis posticis viridi-
æneis, femoribus basi nigris. Læte viridi-ænea; thorace
antice attenuato, supra haud profunde transversim strigoso, sericeo; elytris truncatis, angulo externo dentato;
abdomine rufo; metasterno lævi, nitido, punctis magnis
sparsis; antennis ♂ corpore quadruplo longioribus.

Long. 5-5½ lin. ♂ ♀. (anten. maris majoris, 19 lin.).

Hab.—Ega and Villa Nova.

6. *Chrysoprasis auripes*, n. sp.

C. ruficoxi affinis, differt femoribus læte cupro-aureis. Viridi-ænea, sericeo-nitens, elytris æque nitentibus; thorace antice attenuato, lateribus rotundatis, dorso transversim flexuoso-strigato; elytris erecto-setosis, apice truncatis, angulo externo dentato; metasterno nitido, grosse discrete punctato; femoribus infra læte viridi-æneis, supra cupreo-aureis, posticis dimidio basali rufo; abdomine rufo; antennis ♂ corpore plusquam duplo longioribus, nigris, articulis 3-6 apice unispinosis.

Long. 5-5½ lin. ♂ ♀.

Hab.—St. Paulo; rare.*

7. *Chrysoprasis nigriventris*, n. sp.

C. igneæ affinis, differt abdomine nigro; capite et thorace viridi-aureis, sericeis, splendidis, illo crebre grosse punctato, hoc leviter transversim plicato; elytris viridibus, sericeis, certo situ nigrescentibus, setosis, apice truncatis, angulo exteriori dentato; sternis nitidis, metasterno grosse discrete foveato, interstitiis punctulatis; pedibus abdomineque nigris, femoribus supra viridi-tinctis.

Long. 4½ lin. ♀.

Hab.—Pará.

8. *Chrysoprasis punctulata*, n. sp.

Modice elongata, capite thoraceque cupreo-auratis, creberrime reticulato-punctulatis, hoc prope basin subiter dilatato, deinde usque ad apicem attenuato; elytris nigro-

---

* The following belongs to this group:—

Chrysoprasis ignea, n. sp.

C. ruficoxi forma et sculptura simillima; capite et thorace supra cupreo-auratis splendidis, illo grosse crebro punctato, hoc leviter transversim plicato; elytris nigro-viridibus, subsericeis, setosis, apice truncatis, angulo exteriori dentato; metasterno viridi-aurato, nitido, grosse foveato, interstitiis scabrosulis; abdomine rufo; pedibus nigris, femoribus viridi-æneis; antennis ♂ corpore plusquam duplo longioribus, articulo 1mo grosse punctato, æneo, 3-6 apice unispinosis.

Long. 4½ lin. ♂.

Hab.—Cayenne.

viridibus, setosis, apice truncatis, angulo exteriori dentato; sternis nitidis, crebre punctulato-scabrosis; abdomine rufo; pedibus nigris; antennis nigris, ♂ corpore
sesqui longioribus, articulis apice haud spinosis.

Long. 4–4¼ lin.

Hab.—Cameta, Tocantins; abundant on dead trees.

9. *Chrysoprasis mœrens.*

White, Cat. Longic. Brit. Mus. p. 150.

Angustior, subcylindrica, capite et thorace cupreo-auratis, supra crebre grosse reticulato-punctatis, hoc orbiculato;
elytris nigro-viridibus, sericeis, setosis, apice truncatis;
prothorace subtus cupreo-aurato-scabroso; metasterno
viridi-æneo, lævi, grosse haud profunde discrete punctato;
abdomine piceo-nigro; pedibus nigris, femoribus viridi-
tinctis; antennis brevibus, nigris, articulis 3–6 apice
unispinosis.

Long. 3½ lin. ♀.

Hab.—Pará.

10. *Chrysoprasis melanostetha,* n. sp.

Supra planata, viridi-ænea, capite thoraceque sub-
auratis, creberrime grosse reticulato-punctatis, hoc prope
basin dilatato-rotundato; elytris truncatis, angulo exteriori dentato; prosterno minutissime scabroso, meso-
et meta-sternis nigris, hoc cano-tomentoso, grosse punctato; abdomine rufo; pedibus nigris; antennis nigris, ♂
corpore plusquam duplo longioribus, articulis 3–6 apice
unispinosis.

Long. 5–5¼ lin. ♂ ♀.

Hab.—Upper and Lower Amazons.

11. *Chrysoprasis nana,* n. sp.

Parva, tenuis, capite thoraceque auratis, grosse punctato-reticulatis, hoc angusto, postice vix dilatato; elytris
olivaceis, sericeis, longe setosis, apice oblique truncatis;
prosterno viridi-æneo, opaco, scabroso; meso- et meta-
sternis medio nigris, hoc punctulato; abdomine rufo;
pedibus nigris, longe setosis; antennis tenuibus, brevibus,
nigris, articulis simplicibus.

Long. 2¼ lin. (♀ ?).

Hab.—River Tapajos; one example.

12. *Chrysoprasis aureicollis.*

White, Cat. Longic. Brit. Mus. p. 149.

Parva, capite thoraceque cupreo-aureis, creberrime reticulato-punctulatis, hoc prope basin subiter dilatato-rotundato, deinde usque ad apicem attenuato; elytris viridibus, lætе sericeis, erecto-setosis, apice truncatis, angulo exteriori dentato; sternis viridi-æneis, nitidis, omnibus crebre minute punctulatis, fulvo-decumbenti-pilosis; abdomine rufo; pedibus nigris, longe setosis; antennis utroque sexu haud corpore longioribus, nigris, articulis haud spinosis.

Long. 3½ lin. ♂ ♀.

Hab.—Amazons; generally distributed and common.

13. *Chrysoprasis floralis,* n. sp.

Nigra, capite supra et prothorace toto igneo-cupreis, vix nitidis, grosse reticulato-punctatis; hoc postice dilatato-rotundato; elytris sericeis, setosis, truncatis, angulo exteriori spinoso; metasterno æneo-tincto, punctulato, insterstitiis lævibus; antennis nigris, utroque sexu vix corpore longioribus, articulis simplicibus.

Long. 3-4 lin. ♂ ♀.

Hab.—Santarem, and River Tapajos; at flowers, occasionally in great numbers.

14. *Chrysoprasis brevicornis,* n. sp.

Lætе viridi-ænea, capite thoraceque auratis, crebre reticulato-punctatis, hoc juxta basin rotundato-ampliato, deinde usque ad apicem attenuato; elytris sericeo-nitidis, setosis, truncatis, angulo exteriori dentato; sternis nitidis, crebre punctulatis, sparsim fulvo-pilosis; abdomine rufo; pedibus nigris; antennis nigris, utroque sexu vix corpore longioribus, articulis simplicibus.

Long. 3-4 lin. ♂ ♀.

Hab.—River Tapajos, Ega, St. Paulo; common.

15. *Chrysoprasis sobrina,* n. sp.

C. brevicorni proxime affinis; differt solum statura majori, antennis ♂ corpore sesqui longioribus. Lætе

410   Mr. H. W. Bates *on Cerambycidæ*

viridi-ænea, capite thoraceque auratis: prosterno subtiliter rugoso et grosse punotato, metasterno punotulato,
fulvo, sparsim piloso; antennis pedibusque nigris;
abdomine rufo.

Long. 4-5 lin. ♂ ♀. (anten. ♂ 6-7 lin.).

Hab.—Upper Amazons.

16. *Chrysoprasis hispidula*, n. sp.

C. brevicorni affinis; differt thorace medio rotundato-
dilatato. Læte viridi-ænea, elytris cyanescentibus;
thorace grosse reticulato-punotato; elytris truncatis,
angulo exteriori haud producto; sternis nitidis, punotatis;
abdomine rufo; pedibus nigris; antennis nigris, corpore
haud longioribus, articulis simplicibus; elytris, antennis
pedibusque longe setosis.

Long. 3¾ lin.

Hab.—Ega.*

---

* The following species, for the most part common in Collections, have
not previously been described:—

Chrysoprasis valida, n. sp.

Magna, robusta, supra planata, viridi-ænea; capite thoraceque creberrime reticulato-punotatis, hoc lateribus regulariter sed paululum rotundatis; elytris apice truncatis, angulo exteriori dentato; sternis nitidis,
crebre subrugose punotulatis, fulvo-hirsutis; abdomine aureo- vel cupreo-
splendido; antennis (♀ ?) corpore brevioribus, nigris, articulis 3-7 apice
unispinosis; pedibus nigris.

Long. 8 lin.
Hab.—Brasilia merid.

Chrysoprasis chrysogastra, n. sp.

Elongata, gracilis, læte viridi-ænea; capite thoraceque creberrime haud
profunde subrugose reticulato-punotatis, hoc subcylindrico; elytris apice
truncatis, angulo exteriori dentato; sternis nitidis, vix hirsutis, prosterno
rugoso, metasterno sparsim punotulato; abdomine cupreo-aureo, splendido; pedibus valde elongatis (♂), nigris; antennis (♂) corpore duplo
longioribus, nigris, articulis haud spinosis.

Long. 7 lin. ♂.
Hab.—Brasilia merid.

Chrysoprasis æneiventris, n. sp.

Elongata, linearis, viridis, obscurior, interdum sub-olivacea; capite
thoraceque creberrime punotato-reticulatis, hoc subquadrato, lateribus
medio paulo rotundatis; elytris apice truncatis; sternis grosse foveato-
reticulatis, medio griseo-pilosis; abdomine æneo; pedibus nigris, antennis ♂ corpore paululum longioribus, ♀ brevioribus, nigris, articulis
3-7 apice unispinosis.

Long. 5-6 lin. ♂ ♀.
Hab.—Brasilia merid.

Chrysoprasis punctiventris, n. sp.

Elongata, linearis, olivaceo-viridis; capite thoraceque crebre grosse reticulato-punctatis, hoc medio dilatato-rotundato; elytris apice truncatis; corpore subtus nigro, vix æneo-tincto, sternis creberrime grosse punctatis; abdomine crebre passim punctato; pedibus nigris, antennis nigris, haud spinosis.

Long. 4½ lin. ♀.
Hab.—Brasilia merid.

Chrysoprasis nymphula, n. sp.

Elongata, læte viridi-ænea, capite thoraceque reticulato-punctatis, hoc prope basin dilatato, deinde usque ad apicem rotundato-attenuato; elytris apice truncatis, angulo exteriori longe dentato; sternis nitidis, prosterno scabroso, metasterno discrete grosse punctato, interstitiis lævibus; abdomine rufo; pedibus elongatis, gracilibus, femoribus viridi-æneis; antennis ♂ corpore duplo, ♀ paulo longioribus, nigris, articulis 3-6 apice unispinosis.

Long. 5-6 lin. ♂ ♀.
Hab.—Brasilia merid.

Chrysoprasis rugulicollis, n. sp.

Elongata, læte viridi-ænea, capite thoraceque creberrime punctulato-reticulatis, hoc dorso transversim ruguloso, antice angustato; elytris apice truncatis, angulo exteriori dentato; sternis nitidis, crebre punctulatis; abdomine rufo; pedibus gracilibus, femoribus anticis æneis; antennis utroque sexu corpore paulo longioribus, haud spinosis.

Long. 6 lin. ♂ ♀.
Hab.—Brasilia merid.

Chrysoprasis linearis, n. sp.

Linearis, angustissima, læte viridi-ænea; capite thoraceque creberrime reticulato-punctatis, hoc lateribus prope medium rotundatis; elytris apice truncatis, angulo exteriori dentato; metasterno nitido, irregulariter haud confertim punctato; pedibus gracilibus, nigris, femoribus viridi-æneis; abdomine rufo; antennis ♂ corpore sesqui longioribus, ♀ corpori æqualibus, articulis haud spinosis.

Long. 3½ lin. ♂ ♀.
Hab.—Brasilia merid.

Chrysoprasis nigrina, n. sp.

Elongata, linearis, supra planata, nigra, opaca, elytris subcœruleis; capite thoraceque creberrime subtiliter haud profunde punctulato-reticulatis, subrugosis; hoc subquadrato, lateribus rectis, juxta basin subiter angustato; elytris apice truncatis; sternis opacis, subtiliter creberrime punctulatis,-breviter pallido-hirsutis; abdomine rufo; pedibus nigris; antennis ♀ corpore multo brevioribus, articulis haud spinosis.

Long. 4½ lin. ♂.
Hab.—Brasilia merid.

Chrysoprasis ignicollis, n. sp.

Brevior, convexa, nigro-sericea, thorace igneo-aureo, medio dorsi plaga nigro-velutina; capite thoraceque grosse punctato-reticulatis, hoc rotundato, antice paulo angustato; elytris apice truncatis, angulo exteriori dentato; pectore et abdomine nigro-nitidis, sparsim punctulatis; antennis (♀?) corpori æqualibus, robustis, articulis 3-6 apice brevissime unispinosis; pedibus nigris, nitidis.

Long. 4 lin. (♀?).
Hab.—Cayenne.

Genus MICROSPILOMA.

Bates, Entom. Monthly Mag. iv. 24 (1867); Lacord.
Gen. ix. 129.

Genus *Pronuba* (Thoms.) proxime affine; differt antennis brevibus, articulis 3-6 crassioribus, infra longe dense ciliatis; capite rostro paulo elongato; thorace angustiore, subcylindrico, medio paulo dilatato et acute spinoso; pedibus longis, gracilibus, posticis longioribus, femoribus linearibus, apice breviter dentatis; elytris cylindricis, maculis eburneis.

1. *Microspiloma Dorilis.*

Bates, *lib. cit.*, p. 25.

Fulvo-testacea, pubescens; capite rugoso; thorace medio dorsi transversim plicato, tuberculis duobus anterioribus; elytris truncatis, angulis haud productis, supra dense punctatis, utrinque maculis parvis eburneis tribus, una basali, alteris duabus conjunctis discoidalibus.
Long. 5 lin. ♀.
Hab.—Ega; on leaves of trees.

Sub-fam. ANCYLOCERINÆ.

Genus ANCYLOCERA.

Serville, Ann. Soc. Ent. Fr. 1834, p. 107; Lacord.
Gen. ix. 136.

1. *Ancylocera Waterhousei.*

White, Cat. Longic. Brit. Mus. p. 211.

A. cardinalis angustior; thorace antice magis angustato; antennis ♂ dimidium corporis vix attingentibus, articulis brevibus, compressis, serratis. Nigra, grosse punctata, erecte fulvo-hirsuta; elytris (apice excepto), pectore et abdomine, rufis.
Long. 5 lin.
Hab.—Tapajos.

2. *Ancylocera seticornis*, n. sp.

Angustissima, grosse punctata, fulvo-pilosa, rufa, thorace supra, antennis, pectore, elytris triente apicali,

femoribus apice, et tibiis, nigris; antennis (♂ ?) corporis longitudine, setiformibus, articulis à 3io paulo compressis; elytris apice truncatis, angulo exteriori late productis; femoribus gracilibus, vix incrassatis.
Long. 3½ lin. (♂ ?).
Hab.—Santarem.

Genus CALLOPISMA.

Thomson, Syst. Ceramb. p. 212; Lacord. Gen. ix. 137.

1. *Callopisma ruficollis*, n. sp.

Linearis, minus elongata, nigra, thorace late rufo; capite grossissime acabroso-punctato, collo transversim strigoso; thorace curto, sub-ovato, basi constricto et marginato, supra inæquali, grossissime crebre punctato, setoso, opaco; elytris depressis, apice late rotundatis et margine explanato, supra grosse creberrime sub-ordinate punctatis; pectore, abdomine et pedibus nigris, nitidis, femoribus intermediis et posticis haud clavatis, apice intus spinosis; antennis corpore paulo longioribus, nigris, nitidis, articulis 3-10 subæqualibus, apice paulo tumidis, 11mo longiori, apice curvato.
Long. 4½ lin. (♂ ?).
Hab.—Pará.

I place this in the genus *Callopisma* on account of the short thorax, constricted at the base. In the curved apical joint of the antennæ it does not agree with the definition given by Lacordaire.

Sub-fam. PLATYARTHRINÆ.

Genus STENYGRA.

Serville, Ann. Soc. Ent. Fr. 1834, p. 95; Lacord. Gen. ix. 140.

1. *Stenygra angustata.*

Callidium angustatum, Oliv. Ent. No. 70, p. 32, pl. vi. f. 71 (1795).
Clytus coarctatus, Fabr. Syst. El. ii. 49 (1801).

Elongata, elytris medio coarctatis et depressis, nigro-castanea, sparsim longe hirsuta; thorace antice subgloboso, postice contracto, dorso longitudinaliter rugoso-plicato; elytris nitidis, apice arcuatim truncatis, supra lineola obliqua prope basin maculaque subtriangulari pone medium flavo-testaceis; pectore et abdominis segmentis 1-2 argenteo-fasciatis; antennis grossis, compressis, utroque sexu dimidium corporis vix superantibus.

Long. 9 lin. ♂ ♀.

Hab.—Upper Amazons. Found motionless on leaves of trees in the day-time.

2. *Stenygra contracta.*

Pascoe, Journ. of Entom. i. 355.

Elongata, elytris medio coarctatis et depressis; a *S. angustata* differt macula unica elytrorum pone medium elongata obliqua; antennis ♂ filiformibus, longitudine corporis; elytris truncatis, angulo exteriori longe spinoso.

Long. 8 lin. ♂.

Hab.—Ega.

3. *Stenygra cosmocera.*

White, Cat. Longic. Brit. Mus. p. 221.

Elongata, elytris medio haud coarctatis, supra vix depressis, castanea, polita; thorace ovato, basi constricto, dorso lævi, binodoso; elytris apice truncatis, angulo exteriori dentato, supra macula elongata obliqua prope basin, altera triangulari marginali apud medium, flavo-testaceis.

Long. 7½-8 lin.

Hab.—Pará.

Genus PHRMOSIA, nov. gen.

Gen. *Platyarthro* (Guér. = *Cœlarthron*, Thoms., Lacord.) proxime affine; differt articulis antennarum quadrangulatis, dilatato-compressis, subtus dense ciliatis. Caput pone oculos valde elongatum, antice incrassatum; mandibulis utroque sexu fortibus, bidentatis, abrupte curvatis.

Oculi supra longe distantes. Tubera antennifera vix elevata, distantia, lata, supra sulcata. Antennæ articulo 1mo crasso, curvato, 3-10 elongatis, dilatatis, apice angulis productis, 3io cæteris singulis duplo longiori, 3-7 supra et infra late sulcatis et infra longe ciliatis, ♂ magis, ♀ minus elongatis. Thorax prope basin lateraliter profunde constrictus, deinde usque ad apicem angustatus, ibique anguste lateraliter constrictus, supra lævissimus. Elytra elongata, apice obtuse truncata. Pedes breves, validi; femora paulo incrassata; tarsi breves, articulo primo triangulari. Mesosternum in medio tumidum.

The difference in the form and clothing of the antennal joints of itself would scarcely warrant the separation of this genus from *Platyarthron;* but the form of the anterior part of the head, the short and extremely thickened muzzle, the very strong and sharply curved mandibles, and the wide flattened shape of the antenniferous tubercles, form a combination of characters which forbid the association of the form with the genus in question.

1. *Phimosia obscura,* n. sp.

Elongata, antice angustior, nigra, nitida, glabra; antennis, pedibus, et corpore subtus interdum piceis; capite thoraceque vix punctulatis; elytris subtiliter coriaceis et punctulatis, utrinque linea angusta recta a basi usque ultra medium, lineolisque duabus exterioribus, albis.

Long. 6¼-7¼ lin. ♂ ♀.
Hab.—Pará. One pair taken *in copulâ.*

Genus STRÆPTOLADIS.

Bates, Entom. Monthly Mag. iv. 23 (1867); Lacord. Gen. ix. 159.

Oblonga, elytris oblongo-quadratis, postice dilatatis, glabra. Caput parvum, angustum, mandibulis horizontaliter porrectis et recurvis. Oculi haud prominentes, reniformes, supra longe distantes. Antennæ robustæ, glabræ, articulo 3io cæteris paulo longiori, 3-10 subserratis, supra et infra sulcatis. Thorax transversus, antice rotundato-attenuatus, antice et postice sulco profundo

422

Mr. H. W. Bates on *Cerambycidæ*

constrictus, supra lævia. Elytra oblongo-quadrata, postice rotundato-dilatata, apice late obtuse rotundata, prope suturam leviter truncata, supra grosse reticulata. Pedes breves, validi ; femora clavata, prope apicem subtus valide spinosa ; tarsi breves, articulo 1mo breviter cordato. Prosternum apice acute tuberculatum. Mesosternum latum, paulo concavum. Acetabula antica extus angulata ; coxæ anticæ haud exsertæ. Acetabula intermedia extus paululum aperta.

The structure of the prothorax, with its sharp constriction near the fore and hind margins, similar to *Phimosia* and *Trachelia,* joined to other minor characters, seem to indicate the place of this anomalous genus to be in the present sub-family, rather than in the *Trachyderinæ,* where I formerly was inclined to place it, or in the *Tropidosomatinæ,* where Lacordaire has preferred to leave it.

1. *Streptolabis hispoides.*

Bates, Entom. Monthly Mag. iv. 23.

Subdepressa, corallino-rufa, antennis (articulo 1mo excepto) nigris; elytris nigris, subtiliter rugosis, opacis, lineis elevatis lævibus reticulatis, utrinque maculis magnis sex apiceque coccineis; capite et mandibulis rugoso-punctatis; thorace sparsim punctulato.

Long. 7½ lin.

Hab.—Ega ; on the trunk of a dead tree.

Has a great resemblance to the Hispid, *Cephalodonta spinipes,* and also, in colour and form, to *Erythroplatys corallifer* (sub-fam. *Rhinotraginæ*).

Sub-fam. Pœcilopeplinæ.

Genus Pœcilopeplus.

Thomson, Classif. des Ceramb. p. 205 ; Lacord. Gen. ix. 147.

1. *Pœcilopeplus Batesii.*

White, Cat. Longic. Brit. Mus. p. 56, pl. iii. f. 1, ♂.

Niger, elytris læte rufis, fasciis quatuor angustis nigris abrupte flexuosis, prima interrupta, quarta in medio dupla, annulum formante.

♂. Thorace castaneo-rufo, supra fossato, punctulato-opaco, lateribus medio angulatis; abdomine griseo-lanuginoso.

♀. Thorace nigro, nitido, absque foesulis distinctis; abdomine nigro, nudo.

Long. 10-12 lin. ♂ ♀.

Hab.—Santarem; on bushes in the Campo or open districts.

Genus GEORGIA.

Thomson, Archiv. Entom. i. 21; Lacord. Gen. ix. 148.

1. *Georgia xanthomelas.*

Phædinus xanthomelas, White, Proc. Zool. Soc. 1856, p. 408.

Georgia citrina, Thoms. Arch. Ent. i. 21, pl. ix. f. 1, 2.

Nigra, elytris stramineis, tertia parte apicali maculaque utrinque discoidali ante medium nigro-velutinis; abdomine testaceo.

Long. 8 lin.

Hab.—Villa Nova.

Found on one occasion, flying low across a pathway in the forest. Four examples only were taken.

Sub-fam. TROPIDOSOMATINÆ.

Genus TROPIDOSOMA.

Perty, Del. An. Art. Bras. p. 85; Lacord. Gen. ix. 150.

1. *Tropidosoma penniferum*, n. sp.

T. dilaticorni (Gory) simile; elongato-ovatum, valde convexum, fulvo-ochraceum, capitis vertice, thoracis maculis duabus magnis dorsalibus, clytrorum marginibus lateralibus, macula post-humerali et tertia parte posteriori antice obliquata, nigris; antennis brevissimis, articulis a 3io compresso-dilatatis, imprimis latissimis, apicem versus cito angustatis, articulis duobus flavis basalibus exceptis, densissime breviter nigro-hirsutis; thorace transverso, quadrato, lateribus medio et prope angulum posticum profunde emarginatis, margine postico medio lobato, lobo emarginato, supra dorso modice convexo, costis tribus longitudinalibus, duabus lateralibus magis elevatis et antice abbreviatis; elytris subtiliter dense scabrosis, opacis, sutura et costa longitudinali glabris, nitidis; pectore et abdomine nigro-variegatis; pedibus nigris, femorum et tibiarum basibus ochraceis.

Long. 12 lin. ♀.

Hab.—Ega. One example on foliage.

It is possible, notwithstanding the great differences in the antennæ, thorax and elytra, that this species is the ♀ of *Ctenodes isabellina*; if so, *Tropidosoma dilaticorne* is the ♀ of some unknown species of *Ctenodes.*

Genus CTENODES.

Olivier, Entom. No. 95 bis, vol. vi. p. 779.

1. *Ctenodes isabellina*, n. sp.

Paulo convexa, postice valde dilatato-rotundata, fulvo-ochracea, nigro-varia; capite macula frontali alteraque post oculos nigris; antennis omnino nigris, ab articulo 3io regulariter pectinatis, opacis; thorace quadrato, lateribus utrinque antice lobis magnis duobus obtusis, angulo-que postico in lobum acutum producto, margine postico bisinuato, supra dorso valde convexo et quinque-tuberoso, grosse punctato-scabroso, maculis duabus magnis nigris; scutello valde elongato, triangulari, nigro; elytris lateribus explanatis, mox pone basin gradatim dilatatis, ante apicem angustatis, apicibus acute conjunctim rotundatis et paulo sinuatis, supra creberrime subtiliter rugosis, opacis, utrinque costis quatuor et margine laterali elevato glabris, costa prima juxta suturam, quarta minus distincta et ante apicem cum tertia conjuncta, fulvo-ochraceis, triente posteriori nigra, nigredine antice ad suturam profunde sinuata; pectore, lateribus, pedibusque nigris.

Long. 12 lin. ♂.

Hab.—Ega. One example, flying in the forest.

2. *Ctenodes zonata.*

Klug, Nov. Acta Ac. Cæs. L. C. Nat. Cur. xii. 454,
pl. xlii. f. 1.

Elongato-ovata, nigra, thoracis lobis lateralibus et elytrorum fascia mediana obliqua luteis; elytris lateribus vix explanatis, apice oblique subtruncatis.

Long. 10¼ lin. ♂.

Hab.—Pará. On foliage in the dense forest.

3. *Ctenodes miniata.*

Klug, lib. cit., p. 455, pl. xlii. f. 2.

Oblonga, paulo convexa, laete corallina; elytris apice obtusissime rotundatis, supra costatis, interstitiis crebre grosse scabrosis, nigris, utrinque maculis magnis quinque, margine laterali medio interrupto, suturaque ad apicem, corallinis; scutello nigro.

Long. 9 lin. ♂.

Hab.—Villa Nova. One example, on foliage.

Sub-fam. STERNACANTHINÆ.

Genus STERNACANTHUS.

Serville, Ann. Soc. Ent. Fr. 1832, p. 172; Lacord.
Gen. ix. 154.

1. *Sternacanthus Batesii.*

Pascoe, Journ. of Entom. i. 355.

Oblongus, niger, glaber, nitidus; elytris fasciis latis tribus haud dentatis coccineis, tertia interdum ad suturam interrupta, callo humerali nigro; antennis omnino nigris. A *S. undato* (Oliv.) differt fasciis haud fortiter dentatis.

Long. 7-12 lin. ♂ ♀.

Hab.—Pará; on foliage in the forest, at the end of the dry season.

2. *Sternacanthus sexmaculatus*, n. sp.

S. Batesii proxime affinis, differt fasciis coccineis nec
suturam nec marginem lateralem attingentibus, margine
incrassato elytrorum nigro.

Long. 12 lin. ♀.
Hab.—River Tapajos.

3. *Sternacanthus picticornis.*

Pascoe, Trans. Ent. Soc., 2 ser., iv. 95.

S. Batesii forma simillimus, differt antennarum articulis 3-4 vel 3-5 flavis apice nigris, et elytrorum callo
humerali haud nigro. Niger, glaber; elytris coccineis,
fasciis duabus et apice nigris.

Variat. Fasciis nigris angustis, interruptis; fascia
secunda latiori, cum nigredine apicali per suturam
conjuncta; denique fascia prima obliterata, fasciis apicalibus conjunctis.

Long. 9-12 lin. ♂ ♀.
Hab.—Ega and St. Paulo.

In the Andean Valleys of Equador the species was
taken abundantly by Mr. Buckley offering no variation,
the elytra having three belts of red and three of black.

4. *Sternacanthus Allstoni*, n. sp.

Oblongus, niger, glaber; antennarum articulis 3-6,
thorace, tibiis et tarsis fulvis, elytris fasciis duabus latis
curvatis fulvo-testaceis; thorace ut in *S. Batesii*, dorso
valde convexo, 5-tuberculato, sed tuberibus tribus intermediis latis rotundatis, haud (ut in illo) compressis angustis; capite collo fulvo; elytrorum fascia prima intus
ad scutellum extensa.

Long. 10 lin. ♀.
Hab.—Montes Aureos, in the interior East of Pará.
Taken by Dr. Allston.

Genus LOPHONOCERUS.

Serville, Ann. Soc. Ent. Fr. 1834, p. 83; Lacord. Gen.
ix. 156.

1. *Lophonocerus barbicornis.*

Cerambyx barbicornis, Linn. Mus. Lud. Ulr. p. 68.

Fulvus, thoracis vitta lata laterali et elytrorum margine
exteriori, sutura postice, fascia obliqua pone medium valde

flexuosa, maculisque tribus basalibus interdum partim
confluentibus, nigris; antennis articulis 1mo et 3-5 nigris,
dense hirsutis, apice rufis glabris, 6-11 flavis.

Long. 12-16 lin. ♂ ♀.

Hab.—Amazons; general, but not common. Flying
heavily along pathways in the forest.

The description of Linnæus applies to the Amazons
insect, which belongs to the darker Guiana form. The
figure of Olivier (No. 67, pl. vii. f. 48) seems rather to
apply to the distinct South Brazilian form (*L. Latreillei,*
White).

Genus CERAGENIA.

Serville, Ann. Soc. Ent. Fr. 1834, p. 32; Lacord. Gen.
ix. 158.

1. *Ceragenia bicornis.*

Cerambyx bicornis, Fabr. Syst. El. ii. 274.

Cerambyx striatus, Oliv. Ent. No. 67, pl. v. f. 31.

Fulva, tomento sericeo-aureo vestita; antennis articulis
2-6 apice et 7-11 totis fuscis; thorace tuberculis duobus
disci compressis, linea mediana, tuberculisque lateralibus
nigris; elytris apice breviter truncatis, supra vitta marginali, altera discoidali, et maculis duabus interioribus,
nigris.

Long. 8-9 lin. ♂ ♀.

Hab.—Amazons, general; at sweet sap on trunks of
trees, common. Olivier confounded this species with *C.*
(*Trachyderes*) *striatus;* his description, however (No. 67,
p. 27) applies exclusively to the *Trachyderes.*

2. *Ceragenia spinipennis,* n. sp.

C. bicorni simillima, minor, differt solum colore paulo
pallidiori et elytris apice utrinque longe unispinosis.

Long. 7 lin. ♂ ♀.

The form, sculpture, and markings are the same as in
C. bicornis, but the colour is decidedly paler, and the
silky pubescence is paler golden. The apex of the elytra
is briefly truncate, with the exterior angle prolonged
into a rather long acute spine, which does not exist in
any of the numerous specimens I have examined of *C.
bicornis.*

Hab.—Ega.

Genus Athetesis, nov. gen.

Paristemia (sensu Lacord.) proxime affinis; differt
corpore valdo elongato, cylindrico, scutello lato, semi-
ovato, mesosterno lato plano, postice inciso. Antennæ
(♀) dimidium corporis paulo superantes, robustæ, distincto serratæ. Thorax subquadratus, lateribus medio
valide spinosis, prope marginem posticum constrictus.
Elytra valde elongata, convexa, postice paululum rotundato-dilatata, apice obtuse rotundata, marginibus ciliatis.

1. *Athetesis prolixa*, n. sp.

Capite nigro, opaco, punctato, antennis nigris; thorace
ochraceo-fulvo, dorso vittis duabus nigris; scutello fulvo;
elytris nigris, sericeo-opacis, macula suboblongo humerali, lateribus usque ultra medium, et fascia lata post
medium, ochraceo-fulvis, prope suturam linea elevata,
disco costis indistinctis duabus; pedibus nigris; corpore
subtus ochraceo-fulvo, pectore et abdomine fuscis, griseo-
sericeis, ventris segmento ultimo latissimo truncato.

Long. 8½ lin. ♀.
Hab.—St. Paulo. One example.

Genus Pteroplatus.

Buquet, Rev. Zool. 1840, p. 287; Lacord. Gen. ix. 104.

1. *Pteroplatus simulans*, n. sp.

Minus dilatatus, valde depressus, *Lyci*-formis; capite
fulvo-ochraceo, lateribus postice nigris; antennis nigris,
♂ longioribus, versus apicem attenuatis, ♀ corpore
multo brevioribus, ♂ articulis 3-5 et ♀ 3-6 crassioribus,
infra dense ciliatis; thorace lato, lateribus rotundatis,
paulo explanatis, fulvo-ochraceo, dorso utrinque vitta
laterali nigra; elytris deplanatis, postice paululum dilatatis, apice prope suturam breviter truncatis, supra medio
obtuse unicostatis, opacis, nigris, macula triangulari
humerali vittaque lata subdentata pone medium ochraceo-
fulvis; pectore abdomineque cinereo-fuscis; pedibus
fuscis, femoribus basi rufo-testaceis.

Long. 7 lin. ♂ ♀.
Hab.—Ega; on foliage.

Sub-fam. STENASPIDINÆ.

Genus ERIPHUS.

Serville, Ann. Soc. Ent. Fr. 1834, p. 88; Lacord.
Gen. ix. 190.

1. *Eriphus dimidiatus.*

White, Cat. Longic. Brit. Mus. p. 293, pl. vi. f. 7, ♂.

Elongatus, capite nigro, thorace et elytrorum dimidio
basali croceo-fulvis opacis, illo vitta dorsali et scutello
nigris; elytris dimidio apicali nigris, sericeis; pectore
abdomineque nigris, cinereo-pubescentibus; antennis
pedibusque nigris.
Long. 7 lin.
Hab.—Pará.

2. *Eriphus xanthoderus,* n. sp.

Subcylindricus, niger, opacus, pectore et abdomine
dense cinereo-pilosis; thorace croceo-rufo, grosse punctato, opaco, dorso medio leviter infuscato et nitido;
scutello nigro; elytris apice obtuse truncatis, supra crebre
punctatis; pedibus nigris, nitidis, femoribus posticis
apice bispinosis, spina interiori longiori; antennis brevibus (♀), articulis 5-10 subserratis.
Long. 5 lin. ♀.
Hab.—Pará.
Apparently closely allied to *E. collaris,* Erichs.(Schomb.
Reise), which, however, has a yellow scutellum.

3. *Eriphus croceicollis.*

White, Cat. Longic. Brit. Mus. p. 292.

"Niger, crebre et rude punctatus; thorace croceo,
elytris anescenti-nigris."
" Prothorax beneath saffron, mesothorax the same, and
marked with a V-like raised figure; metathorax pitchy,
with a light spot in the middle and behind; underside of
abdomen pitchy-black, shining." (White.)
Long. 5 lin.
Hab.—Pará. In Coll. Brit. Mus.
Although taken by me, I do not find the species
among my own reserved collection of Amazonian Longicorns.

Sub-fam. DORCACERINÆ.

Genus DORCACERUS.

Latr. Règne Anim. (ed. ii.) v. 111; Lacord. Gen. ix. 198.

1. *Dorcacerus barbatus.*

Cerambix barbatus, Oliv. Ent. No. 67, p. 610, pl. xiii. f. 94.

Magnus, purpureo-fuscus, opacus, thoracis marginibus, scutello, suturaque postice aureo-tomentosis; fronte et tuberibus magnis antenniferis longe rufo-hirsutis.

Long. 12-15 lin. ♂ ♀.

Hab.—Santarem; not uncommon on trunks of trees from which sap is exuding.

Sub-fam. TRACHYDERINÆ.

Genus TRACHYDERES.

Dalman, Schön. Syn. Ins. iii. 264; Lacord. Gen. ix. 201.

1. *Trachyderes succinctus.*

Cerambyx succinctus, Linn. Mus. Lud. Ulr. p. 72.

Trachyderes cayennensis, Dupont, Mag. Zool. 1836, p. 34, pl. clvi. f. 1.

Castaneus, glaber; elytris paulo ante medium fascia flavo-testacea, interdum fusco-marginata; antennis nigris, articulis plurimis basi fulvis; pectore abdomineque fulvo-ferrugineis.

♂ articulo basali antennarum clavato, haud dilatato.

Long. 9-14 lin. ♂ ♀.

Hab.—Amazons; generally distributed and common. The larva feeds in the interior wood of trees; the perfect insect is found at sap and on the trunks of felled trees.

Dupont applied the Linnæan name *succinctus* to the South Brazilian species, a local form which has a black abdomen; although Linnæus expressly says "abdomen ferrugineum" and "Habitat Surinami." The true *succinctus* Dupont named *cayennensis*.

I have specimens from Panamá, which do not differ from those of the Amazons.

2. Trachyderes Reichei.

Dupont, Mag. Zool. 1836, p. 31, pl. clv. ♂.

Castaneus, glaber; elytris fascia paulo ante medium, et macula apicali triangulari ad angulum suturalem, flavo-testaceis; thorace angulis posticis testaceis.

♂ articulo basali antennarum maxime dilatato, difformi; ♀ crasso, rotundato.

Long. 10-16 lin. ♂ ♀.

Also generally distributed throughout the Amazon region. The thorax is of the same form as in *T. succinctus*, and the colours of body and limbs offer no constant difference.

3. Trachyderes cingulatus.

Klug, Nov. Act. Ac. Cæs. L. C. Nat. Cur. xii. 456.

T. Reichei proxime affinis, differt colore purpurascenti-nigro, elytrorum macula flava apicali elongata suturali per marginem apicalem haud extensa; corpore subtus omnino nigro-nitido.

Long. 9-14 lin. ♀.

Hab.—Pará. Four examples, all females; quite distinct from *T. Reichei.*

4. Trachyderes rhodopus, n. sp.

T. succincto affinis, thorace elongatiori et angustiori, dorso postice plano, lateribus antice (angulo antico excepto) haud tuberculatis; fascia elytrorum longe ante medium et antice in medio ad scutellum extensa; corpore subtus pedibusque rufis; elytris sparsim punctulatis, apice breviter sinuato-truncatis; antennis ♀ articulis 7-11 totis nigris, 7-10 valde abbreviatis, serratis.

Long. 5½ lin. ♀.

Hab.—Santarem.

Apparently allied to *T. rubripes* (Dupont), but differing from the description in several essential points. By the form of the thorax it belongs to Dupont's fifth division, and not to the first, in which *T. rubripes* is placed.

5. *Trachyderes melas*, n. sp.

T. succincto statura formaque thoracis simillimus, differt colore toto nigro, antennarum articulis 4-6 basi, 10-11 totis rufis exceptis; antennis ♀ multo longioribus.

Long. 10 lin. ♀.

Hab.—Obydos.

The antennæ in the female are half as long again as the body; in the same sex of *T. succinctus* they are very little longer than the body. In the black colour of its legs, it resembles *T. nigripes* (Dupont), but it belongs to a different division of the genus from that species; the form of its thorax is precisely that of *T. succinctus*. *

6. *Trachyderes impunctipennis*, n. sp.

T. succincto similis, differt corpore (præcipue elytris) multo longiori; thorace lateribus antice rotundato et prope angulum anticum haud tuberculato, dorso postico tri-tuberoso (haud plano et in medio depresso, ut in *T. succincto*); elytris omnino impunctatis, apice sinuato-truncatis. Castaneus, elytris fascia ante medium flavo-testacea; antennis ♀ articulis 3-5 basi rufis, 8-11 totis flavis; corpore subtus rufo-castaneo.

Long. 10 lin. ♀.

Hab.—Santarem.

Closely allied to Dupont's *T. Lacordairei*, differing only in the colour of the antennæ and under-surface of the body. It may perhaps be only a variety of that species.

From the nearly allied *T. Latreillei* it differs in many essential respects, being a broader and more robust insect, and very distinct in its colours.

---

* The following very distinct species of the *succinctus* group has not yet been described, although common in collections:—

Trachyderes politus, (Chevr., MS.).

Latior, lævis, planatus, nigro-castaneus, capite, thorace supra, scutello, humeris, et mesosterni medio, rufo-castaneis; thoracis forma ut in *T. succincto*; elytris apice latis, prope suturam obtuse truncatis; antennis nigris, opacis, ♂ longissimis, articulis 10-11 basi fulvis, ♀ articulis 10-11 rufis.

Long. 10-13 lin. ♂ ♀.

Hab.—Venezuela.

7. *Trachyderes globicollis*, n. sp.

T. succincto coloribus simillimus; differt thorace magno, antice valde rotundato, convexo, confertim punctulato, dorso linea transversali lævi medio incrassata apud extremitates tuberosa, antice et postice linea impressa marginato. Castaneus, thorace magis rufo; elytris postice valde attenuatis, apice rotundatis, supra lævissimis, fascia ante medium flavo-testacea; corpore subtus pedibusque rufis, femoribus apice nigro-piceis; antennis ♂ articulis 1-2 nigris, 3 nigro medio rufo, 4-11 rufis apice nigris; prosterno lobis parvis obtusis.

Long. 12 lin. ♂.

Hab.—Ega. One example.

8. *Trachyderes bilineatus.*

Cerambix bilineatus, Oliv. Ent. No. 67, p. 17, pl. xxi. f. 161, ♂.

Trachyderes scabricollis, Dalman, Anal. Ent. p. 64, ♂.

T. Dejeanii, Dupont, Mag. Zool. 1838, p. 15, pl. cxciv. f. 1, ♀.

T. Solieri, Dupont, *lib. cit.*, p. 16, pl. cxciv. f. 2, ♀.

T. Duponti, Dupont, *lib. cit.*, p. 17, pl. cxcv. f. 1, ♂.

Species variabilis, forma thoracis secundum sexum valde diversa. Angustior; capite, thorace, et scutello, castaneo-rufis; elytris nigris, basi plus minusve et lineis 2 vel 3 (interdum obsoletis) rufis.

♂. Thorace crebre scabroso, opaco, nigro-maculato, lateribus bituberculato, dorso plaga pentagona depressa grossius scabrosa tumorem lævem includente; elytris alutaceis, basi anguste rufis; antennis corpore sesqui longioribus, rufis, articulis 1-2 nigris, 3-8 apice nigris, 9-11 piceis.

♂ (minor). Thorace minore, area scabrosa multo minus extensa, plaga dorsali irregulari, lineis et plagulis lævibus fracta, maculis nigris paucis, lateribus tri-tuberculatis; elytris basi late rufis; antennis fulvo-rufis, articulis apice leviter infuscatis.

♀ . Thorace toto rufo, lateribus tri-tuberculatis, dorso
lævi, polito, medio foveis tribus scabrosis, tuberibus
lævibus marginatis et separatis; elytris lævissimis, basi
late rufis; antennis dimidium corporis paulo superantibus,
fulvo-rufis, articulis apice leviter infuscatis.

Long. 5-12 lin. ♂ ♀ .

Generally distributed throughout the Amazons region.
In newly-burnt clearings in the forest, on dead trees,
sometimes abundant.

It is so variable that two individuals can scarcely be
found nearly alike. Dupont failed to notice the sexual
differences in form and colours, and hence described
them as distinct species, besides giving each variety as
distinct. Five or six other of his species are probably
only varieties of this.

9. *Trachyderes conformis.*

Dupont, Mag. Zool. 1838, p. 40, pl. clxiii. f. 2.

Angustus, flavo-testaceus; vertice thoracequo nigro-
maculatis; elytris apice truncatis, angulo exteriori leviter
dentato, supra tertia parte posteriori nigra, nigredine
ramos per marginem fere ad humeros et per suturam usque
ad scutellum emittente; antennis et pedibus fulvo-testaceis, illis apice infuscatis, his femoribus apice nigris.

Long. 7-8½ lin. ♂ ♀ .

Hab.—Santarem.

Closely allied to *T. dimidiatus*, Fabr., the chief difference (which is constant) being that the black colour of
the apical portion of the elytra in *dimidiatus* does not
emit a branch along the suture towards the scutellum.
This speciality is mentioned in Fabricius' description,
and applies to the form from South Brazil. *T. conformis*
occurs also in Venezuela.

Genus OXYMERUS.

Serville, Ann. Soc. Ent. Fr. 1834, p. 50 ; Lacord.
Gen. ix. 204.

1. *Oxymerus basalis.*

Trachyderes basalis, Dalman, Anal. Ent. p. 65.

Oxymerus basalis, Dupont, Mag. Zool. 1838, p. 35, pl. ccviii. f. 1.

Rufo-castaneus, abdomine, elytris (basi excepta), et pedibus posticis, nigris; thorace immaculato; antennis medio fulvis, apice infuscatis.

Long. 7 lin. ♂ ♀.

Hab.—Santarem.

The Amazons specimens differ from the Brazilian typical form in the basal red of the elytra being much larger, extending beyond the scutellum, and in the fore and middle femora, and the basal half of the hind femora, being red.

2. *Oxymerus rivulosus.*

Trachyderes rivulosus, Germar, Ins. Sp. Nov. p. 512,

Oxymerus lineatus, Dupont, Mag. Zool. 1838, p. 41, pl. ccxi. f. 1.

Oxymerus rivulosus, Dup. lib. cit., p. 42, pl. ccxi. f. 2.

Castaneo-fulvus, interdum pallidior, thorace punctis 11 nigris, elytris lineis quatuor et margine pallidis, lineis 1ma prope scutellum et 3ia abbreviatis; antennis pedibusque immaculatis; abdomine interdum basi infuscato.

Long. 5-9 lin. ♂ ♀.

Hab.—Pará. Sometimes abundant in new clearings.

According to Dupont's own description, there is no real difference between his *O. lineatus* and *O. rivulosus.* The size is of no importance in a group where it varies very greatly in almost every species.

Sub-fam. METOPOCŒLINÆ.

Genus METOPOCŒLUS.

Serville, Ann. Soc. Ent. Fr. 1832, p. 170; Lacord. Gen. viii. 244.

The position of this genus is one of the few points in which I venture to depart from the arrangement of

Lacordaire. It is clearly allied in all essential points to the *Trachyderinæ*, and forms an unnecessary exception, in the fine granulation of the eyes, to the section in which the author of the "Genera" has placed it.

1. *Metopocœlus Rojasi.*

Sallé, Ann. Soc. Ent. Fr. 1853, p. 650, pl. xx. f. 1, 2, ♂ ♀.

Magnus, valde elongatus, testaceo-fulvus, nudus, nitidus, supra rugoso-punctatus; thorace vittis duabus nigris; elytris ♂ lineis posticis et margine apicali, ♀ dimidio apicali, nigris; antennis nigris, brevibus, ♂ subserratis, ♀ fortiter serratis.

Long. 15-18 lin. ♂ ♀.

Hab.—Santarem.

On flowers in open grassy districts, at the beginning of the wet season in December. Originally found near Caraccas.

Sub-fam. LISSONOTINÆ.

Genus LISSONOTUS.

Dalman, in Schönh. Syn. Ins. App. p. 364; Lacord. Gen. ix. 209.

1. *Lissonotus Shepherdi.*

Pascoe, Trans. Ent. Soc., 2 ser., v. 16.

Nigerrimus, politus; elytris late recte truncatis, angulo exteriori spinoso, ante medium fascia lata coccinea, prope suturam angustata et abbreviata.

Long. 6½-7 lin. ♂ ♀.

Hab.—Altar do Chaõ, River Tapajos.

2. *Lissonotus fallax*, n. sp.

Nigerrimus, politus, scutello et macula ovali obliqua adjacente, metasterno, abdomine, femorumque basi, coccineis; elytris apice acute conjunctim rotundatis.

Long. 5 lin. ♀.

Hab.—Ega.

3. *Lissonotus rubidus.*

White, Cat. Longic. Brit. Mus. p. 68.

Rufus, politus, immaculatus, antennis et tarsis nigris, tibiis femorumque basi infuscatis; elytris obtuse breviter truncatis.

Long. 8 lin. ♀.

Hab.—Pará.

4. *Lissonotus unifasciatus.*

Gory, in Guér. Icon. Règne Anim. p. 217, pl. xliii. f. 1.

L. abdominalis, Dupont, Mag. Zool. 1836, p. 12, pl. cxlv. f. 1.

Latior, nigerrimus, politus, elytrorum macula obliqua ovata juxta scutellum, metasterno, abdomine, femoribusque intermediis et posticis, coccineis; elytris breviter truncatis, angulo exteriori spinoso.

Long. 8 lin.

Hab.—River Tapajos.

5. *Lissonotus ephippiatus,* n. sp.

L. unifasciato valde affinis, differt corpore angustiori, antennis ♀ multo minus dilatatis; elytrorum macula coccinea minus obliqua, ovali, postice longe ultra apicem scutelli extensa; femoribus intermediis prope basin subtus piceis; elytris apice late recte truncatis, angulo exteriori longe spinoso.

Long. 6½-7 lin. ♀.

Hab.—Ega and St. Paulo, Upper Amazons.

6. *Lissonotus biguttatus.*

Dalman, in Schönh. Syn. Ins. App. p. 150, pl. vi. f. 4.

Rufo-ferrugineus, politus, antennis (articulo basali excepto) tibiis et tarsis nigris; elytris disco vel totis nigris, utrinque ante medium macula ferruginea, apice late truncatis, angulo exteriori spinoso.

Long. 5½-6 lin. ♂ ♀.

Hab.—Pará.

7.   *Lissonotus simplex*, n. sp.

L. biguttato forma similis; niger, femoribus, processu
mesosterni, metasterno toto, abdomineque rufo-ferrugineis; elytris immaculatis, apice late truncatis, angulo
exteriori spinoso.

Long. 5 lin. ♀.
Hab.—Villa Nova.*

Sub-fam. MEGADERINÆ.

Genus MEGADERUS.

Latreille, Règ. An. (ed. ii.) v. 111; Lacord. Gen. ix. 216.

1.   *Megaderus stigma.*

Cerambyx stigma, Linn. Syst. Nat. ii. 635.
Megaderus stigma, Dupont, Mag. Zool. 1838, pl. cxli. f. 1.

Fusco-niger, supra rugoso-punctatus, breviter griseo-
setosus, subtus griseo-tomentosus; thorace magno, rotundato; elytris ad medium fascia angusta obliqua
interdum interrupta flavo-testacea; tibiis tarsisque fulvo-
testaceis.

Long. 6½-12 lin. ♂ ♀.

Common on trunks of newly-felled trees throughout
the Amazon region.†

* The following is a fine new species of this genus:—

Lissonotus princeps.

Thorax antice ut in *L. spadiceo* angustatus. Nigerrimus, nitidissimus;
elytris fascia sub-basali lata, antice et postice recta, prope humeros sinuata,
nec basin nec margines laterales attingente, lætissime coccinea; metasterno et abdomine clare sanguineis; pedibus totis, coxis inclusis, nigerrimis; scutello breviori, æquilatero-triangulari; elytris truncatis, angulo
exteriori valde spinoso; antennis ♂ corpore brevioribus.

Long. 9 lin. ♂.
Hab.—Bolivia. A Dom. Pearce lectus.

† The following is to be added to this genus.

Megaderus latifasciatus.

M. stigmati forma et sculptura simillimus, sed antennis præcipue ♀
robustioribus, multo brevioribus; elytris fascia duplo latiori et leviter
sinuata.

Long. 11 lin. ♂ ♀.
Hab.—Chontales, Nicaragua. A Dom. Ed. Janson, fil., nuper lectus.
Specimina plurima omnino conformia.

Sub-fam. DISTENIINÆ.

Genus DISTENIA.

Serville, Encycl. Méth. x. 485 ; Lacord. Gen. ix. 227.

1. *Distenia agroides*, n. sp.

Elongata, supra violacea, nitida, erecte setosa, subtus
chalybeo-nigra, nitida ; antennis testaceo-rufis, articulis
5-10 (♂) infra longe penicillatis ; pedibus rufo-testaceis,
geniculis infuscatis, femoribus subclavatis, apice haud
spinosis ; thorace supra grosse tuberculato ; elytris apice
valde attenuatis, unispinosis, inter spinam et angulum
suturalem breviter oblique truncatis, supra grosso aspere
striato-punctatis, apicem versus fere lævibus.

Long. 7 lin. ♂.

Hab.—Tapajos.

Concealed within a folded leaf of a tree, like the
species of *Agra (Carabidæ)*, which the metallic *Disteniæ*
somewhat resemble. In repose the antennæ are porrect.

2. *Distenia splendens*, n. sp.

Supra ænea, elytris splendide viridi-æneis, breviter
erecto-setosis, subtus nigro-ænea, nitida ; antennis articulis 1-3 nigro-æneis, cæteris rufo-testaceis, infra (♂)
longe penicillatis ; pedibus flavo-testaceis, femoribus
medio et apice nigris incrassatis, apice haud spinosis ;
antennis articulo 1mo sub-abrupte clavato ; thorace grosse
tuberculato ; elytris apice unispinosis, inter spinam et
angulum suturalem breviter oblique truncatis, supra grosse
striato-punctatis, interstitiis nonnullis costatis, apice
sublævibus.

Long. 8¼ lin. ♂.

Hab.—Ega.

3. *Distenia denticornis*, n. sp.

Robusta, nigro-ænea, nitida, elytris viridi-tinctis ; antennis articulis 1-3 nigro-æneis, cæteris piceo-rufis, infra
(♂) longe sparsim penicillatis, articulo primo gradatim
incrassato, grosse scabroso, infra denticulis validis circiter
6 armato ; thorace grosse tuberculato ; elytris longe
erecto fulvo-setosis, apice unispinosis, angulo suturali

etiam producto acuto, supra grosse striato-punctatis, interstitiis nonnullis costatis, apice sublævibus; pedibus omnino nigro-æneis, trochanteribus pallido-testaceis exceptis; femoribus subclavato-incrassatis, intermediis et posticis apice bispinosis.

Long. 10-12 lin. ♂.

Hab.—Ega. Three examples, one of which is now in the collection of Mr. Alexander Fry.

4. *Distenia suturalis*, n. sp.

Angustata, gracilis, cyanea, nitida, subtus pectore in medio fulvo-testaceo, elytris vitta communi lata saturali purpureo-rufa, pedibus flavo-testaceis, antennis nigris, infra (♂) longe penicillatis; thorace tuberculo elongato mediano distincto, cæteris partibus irregulariter grosse punctatis; elytris sparsissime setosis, apice unispinosis, angulo suturali producto, supra crebre punctatis, vitta suturali postice ante apicem terminata; femoribus omnibus apice spina unica elongata armatis.

Long. 7 lin. ♂.

Hab.—Ega.*

Genus COMETES.

Serville, Encycl. Méth. x. 485; Lacord. Gen. ix. 229.

Syn. *Heteropalpus*, Buquet, Mag. Zool. 1843, pl. cxviii.

The sole constant character which distinguishes this genus from *Distenia* is the relative shortness and thick-

---

* The following are also undescribed species of this genus:—

Distenia rufipes.

Viridi-ænea, pedibus testaceo-rufis, antennis nigris, apicem versus piceis; thorace grosse tuberculato et punctato; elytris apice unispinosis, angulo suturali nullo, supra brevissime setosis, passim subtilissime punctulatis et grosse striato-punctatis, interstitiis nonnullis costatis; femoribus haud spinosis; antennis (♂) sparse penicillatis.

Long. 6½ lin. ♂.

Hab.—Santa Marta, Nova Granada (Bouchard).

Distenia angustata.

Angustata, linearis, capite et thorace præcipue parvis; viridi-ænea, corpore subtus et elytris fundo testaceis, pedibus sordide flavo-testaceis, antennis fusco-æneis, infra longe penicillatis; thorace tuberculato, grosse sparsim punctato; elytris apice unispinosis, angulo suturali producto, acuto, supra sparsim longe setosis, ut in D. *suturali* crebre punctulatis; femoribus omnibus apice spina unica elongata armatis.

Long. 6½ lin. ♂.

Hab.—Cayenna interiore (D. Bar).

ness of the antennæ, which in the males are not much longer than the body, and in both sexes are furnished with the peculiar long soft hairs on the underside of many of the joints. The character derived from the apex of the elytra, spineless in *Cometes*, and spined in *Distenia*, is rendered inapplicable by the discovery of species of *Distenia* (e. g., *D. viridi-cyanea*, Thoms.) which have the elytra obtusely truncate, precisely as in certain species of *Cometes*. The great and abrupt variations in the form of the terminal joint of the maxillary palpi in the *Disteniinæ* are mentioned by Lacordaire as affording no generic distinction; he admitted, however, the genus *Heteropalpus*, which is founded on an extraordinary development of these organs in the males of certain *Cometes*, in which they are excessively elongate, and exhibit, proceeding from the base of the terminal joint, almost at right angles to it, an elongate hairy filament, as long as the joint of the palpus itself. This curious structure might be taken to be a monstrosity, did it not appear, in different form as to points of detail, in three distinct species. It cannot, however, be a generic distinction, for it occurs in the males of *Cometes acutipennis* (Buquet) a species having the closest possible affinity with others (e. g., *C. lætificus*) in which the palpi are of normal form.

1. *Cometes lætificus*, n. sp.

C. acutipenni proxime affinis, differt elytris apice magis obtusis, macula fulva humerali postice rotundata, suturam haud attingente, capite angustiori, etc. Cyaneus, nitidus, elytris læte purpureis, vitta lata discoidali cærulea, maculaque rotundata fulva humerali; antennis nigris, (♂) usque ad apicem longe penicillatis; capite angusto; oculis haud prominentibus ; thorace spina laterali obtusa; elytris crebre grosse punctatis, disco unicostatis, apice breviter obtuse truncatis.

Long. 5 lin. ♂.

Hab.—Ega.

2. *Cometes scapularis*, n. sp.

Robustior, viridi-cyaneus, nitidus, elytris macula humerali sanguinea, femoribus dimidio basali flavo-testaceis;

antennis (♀ ?) grossis, subtus articulis 3-8 penicillatis ; thorace sparsim grosse foveato-punctato, spina laterali obtusa ; elytris apice breviter truncatis, angulo suturali producto acuto, supra regulariter grosse striato-punctatis, macula humerali a scutello et sutura longe distante ; pedibus brevibus, validis, femoribus medio incrassatis.

Long. 6¼ lin. (♀ ?).

Hab.—Ega. Allied to *C. argutulus* (Buq.), in which the red at the base of the elytra extends as a fascia from side to side.

8. *Cometes cæruleus*, n. sp.

Angustus, læte cæruleus, femoribus basi flavo-testaceis, abdomine piceo, antennis nigris, usque ad apicem (♂) infra penicillatis; thorace medio grosse foveato-punctato, spina laterali obtusa ; elytris apice obtuse truncatis, supra grosse lineatim punctatis.

♂ palpis maxillaribus valde elongatis, articulo ultimo apice clavato, basi ramum rectum hirsutum emittente.

Long. 4¼ lin. ♂.

Hab.—Ega.

Addendum.

The following was accidentally omitted (*ante*, p. 285).

Sub-fam. IBIDIINÆ.

8. *Hexoplon prætermissum*, n. sp.

Angustum, lineare, nigro-castaneum, nitidum; elytris macula triangulari laterali ante medium, antice rufotincta, mox pone hanc fascia obliqua angusta, et apice flavo-testaceis, apice truncatis et extus unispinosis, supra punctis sparsis lineatim ordinatis ; pedibus antennisque testaceo-piceis, his basi obscurioribus.

Long. 4 lin.

Hab.—Tapajos. Almost identical in colours and sculpture with *Gnomidolon humerale* (*ante*, p. 287).

The following Tables shew the numbers of Genera and Species of Amazonian Longicorns. (The *Prionidæ* will be found described in Trans. Ent. Soc. 1869, p. 37 ; the *Lamiidæ* in Ann. & Mag. Nat. Hist., 1861-66).

PRIONIDÆ.

Group.	Number of Genera.	Number of Species.
Prionides aberrantes	1	1
Prionides vari, subterranei...	1	1
„ sylvani	10	14
„ pœcilosomi ..	4	10
Total	16	26

CERAMBYCIDÆ.

Sub-family.	Number of Genera.	Number of Species.
Œminæ	6	6
Achrysinæ	1	4
Tornsatinæ	1	1
Cerambycinæ	5	14
Hesperophaninæ	4	7
Bburiinæ	3	9
Sphæriinæ...................	9	12
Piezocerinæ	3	8
Ibidiinæ	9	54
Obriinæ	3	3
Lepturinæ	2	6
Necydalinæ	1	1
Molorchinæ	1	1
Necydalopsiæ	1	1
Rhinotragiæ	13	38
Callichromatinæ	1	6
Compsocerinæ	3	8
Clytiæ	3	14
Tillomorphinæ	1	1
Cleomeninæ.................	3	5
Rhopalophorinæ	5	14
Heteropsinæ	3	19
Ancylocerinæ	2	3
Platyarthrinæ	3	5
Pœcilopeplinæ	2	2
Tropidosomatinæ	2	4
Stornacanthinæ	5	9
Stenaspidinæ	1	8
Dorcacerinæ	1	1
Trachyderinæ	2	11
Metopocœlinæ	1	1
Lissonotinæ................	1	7
Megaderinæ	1	1
Distenlinæ	2	7
Total.........	104	288

LAMIIDÆ.

Sub-family.	Number of Genera.	Number of Species.
Acanthoderinæ	18	59
Anisocerinæ	9	14
Lagocheirinæ	2	6
Leiopodinæ	25	117
Colobotheinæ	3	44
Tæniotinæ	1	4
Onoiderinæ	15	46
Hippopsinæ	2	6
Exocentrinæ	6	8
Tapeininæ	1	2
Compsosomatinæ	3	5
Desmiphorinæ	1	6
Pogonocherinæ	4	10
Apomecyninæ	1	3
Calliinæ	6	11
Astatheinæ	1	1
Amphionychinæ	6	20
Phytœciinæ	1	1
Saperdinæ	1	1
Total	**101**	**365**

SUMMARY.

		Genera.	Species.
COLEOPTERA LONGICORNIA }	Prionidæ	16	26
	Cerambycidæ	104	288
	Lamiidæ	101	365
	Grand Total	221	679

www.ingramcontent.com/pod-product-compliance
Lightning Source LLC
Chambersburg PA
CBHW021346210326
41599CB00011B/766